T0344883

LINEAR STATE-SPACE
CONTROL SYSTEMS

Robert L. Williams II

Douglas A. Lawrence
Ohio University

JOHN WILEY & SONS, INC.

Published by John Wiley & Sons, Inc., Hoboken, New Jersey
Published simultaneously in Canada.

For general information on our other products and services or for technical support, please contact our Customer Care Department within the United States at (800) 762-2974, outside the United States at (317) 572-3993 or fax (317) 572-4002.

Wiley also publishes its books in a variety of electronic formats. Some content that appears in print may not be available in electronic formats. For more information about Wiley products, visit our web site at www.wiley.com.

Library of Congress Cataloging-in-Publication Data:

Williams, Robert L., 1962-
   Linear state-space control systems / Robert L. Williams II and Douglas A. Lawrence.
      p. cm.
   Includes bibliographical references.
   ISBN 0-471-73555-8 (cloth)
   1. Linear systems.   2.   State-space methods.   3.   Control theory.   I. Lawrence, Douglas A.   II.   Title.
   QA402.W547 2007
   629.8′32—dc22

                                                                2006016111

Printed in the United States of America

SKY10079418_071124

To Lisa, Zack, and especially Sam, an aspiring author. — R.L.W.

To Traci, Jessica, and Abby. — D.A.L.

CONTENTS

Preface   **ix**

1 Introduction   **1**

    1.1   Historical Perspective and Scope / 1
    1.2   State Equations / 3
    1.3   Examples / 5
    1.4   Linearization of Nonlinear Systems / 17
    1.5   Control System Analysis and Design using
         MATLAB / 24
    1.6   Continuing Examples / 32
    1.7   Homework Exercises / 39

2 State-Space Fundamentals   **48**

    2.1   State Equation Solution / 49
    2.2   Impulse Response / 63
    2.3   Laplace Domain Representation / 63
    2.4   State-Space Realizations Revisited / 70
    2.5   Coordinate Transformations / 72
    2.6   MATLAB for Simulation and Coordinate
         Transformations / 78
    2.7   Continuing Examples for Simulation
         and Coordinate Transformations / 84
    2.8   Homework Exercises / 92

3 Controllability 108

3.1 Fundamental Results / 109
3.2 Controllability Examples / 115
3.3 Coordinate Transformations
and Controllability / 119
3.4 Popov-Belevitch-Hautus Tests for
Controllability / 133
3.5 MATLAB for Controllability and Controller Canonical
Form / 138
3.6 Continuing Examples for Controllability
and Controller Canonical Form / 141
3.7 Homework Exercises / 144

4 Observability 149

4.1 Fundamental Results / 150
4.2 Observability Examples / 158
4.3 Duality / 163
4.4 Coordinate Transformations and Observability / 165
4.5 Popov-Belevitch-Hautus Tests for Observability / 173
4.6 MATLAB for Observability and Observer Canonical
Form / 174
4.7 Continuing Examples for Observability and Observer
Canonical Form / 177
4.8 Homework Exercises / 180

5 Minimal Realizations 185

5.1 Minimality of Single-Input, Single Output
Realizations / 186
5.2 Minimality of Multiple-Input, Multiple Output
Realizations / 192
5.3 MATLAB for Minimal Realizations / 194
5.4 Homework Exercises / 196

6 Stability 198

6.1 Internal Stability / 199
6.2 Bounded-Input, Bounded-Output Stability / 218
6.3 Bounded-Input, Bounded-Output Stability Versus
Asymptotic Stability / 220
6.4 MATLAB for Stability Analysis / 225

6.5   Continuing Examples: Stability Analysis / 227
6.6   Homework Exercises / 230

7   Design of Linear State Feedback Control Laws   234

7.1   State Feedback Control Law / 235
7.2   Shaping the Dynamic Response / 236
7.3   Closed-Loop Eigenvalue Placement via State
       Feedback / 250
7.4   Stabilizability / 263
7.5   Steady-State Tracking / 268
7.6   MATLAB for State Feedback Control Law Design / 278
7.7   Continuing Examples: Shaping Dynamic Response
       and Control Law Design / 283
7.8   Homework Exercises / 293

8   Observers and Observer-Based Compensators   300

8.1   Observers / 301
8.2   Detectability / 312
8.3   Reduced-Order Observers / 316
8.4   Observer-Based Compensators and the Separation
       Property / 323
8.5   Steady-State Tracking with Observer-Based
       Compensators / 337
8.6   MATLAB for Observer Design / 343
8.7   Continuing Examples: Design of State
       Observers / 348
8.8   Homework Exercises / 351

9   Introduction to Optimal Control   357

9.1   Optimal Control Problems / 358
9.2   An Overview of Variational Calculus / 360
9.3   Minimum Energy Control / 371
9.4   The Linear Quadratic Regulator / 377
9.5   MATLAB for Optimal Control / 397
9.6   Continuing Example 1: Linear Quadratic
       Regulator / 399
9.7   Homework Exercises / 403

Appendix A Matrix Introduction   407

A.1   Basics / 407
A.2   Matrix Arithmetic / 409
A.3   Determinants / 412
A.4   Matrix Inversion / 414

Appendix B Linear Algebra   417

B.1   Vector Spaces / 417
B.2   Subspaces / 419
B.3   Standard Basis / 421
B.4   Change of Basis / 422
B.5   Orthogonality and Orthogonal Complements / 424
B.6   Linear Transformations / 426
B.7   Range and Null Space / 430
B.8   Eigenvalues, Eigenvectors, and Related Topics / 435
B.9   Norms for Vectors and Matrices / 444

Appendix C Continuing MATLAB Example m-file   447

References   456

Index   459

PREFACE

This textbook is intended for use in an advanced undergraduate or first-year graduate-level course that introduces state-space methods for the analysis and design of linear control systems. It is also intended to serve practicing engineers and researchers seeking either an introduction to or a reference source for this material. This book grew out of separate lecture notes for courses in mechanical and electrical engineering at Ohio University. The only assumed prerequisites are undergraduate courses in linear signals and systems and control systems. Beyond the traditional undergraduate mathematics preparation, including calculus, differential equations, and basic matrix computations, a prior or concurrent course in linear algebra is beneficial but not essential.

This book strives to provide both a rigorously established foundation to prepare students for advanced study in systems and control theory and a comprehensive overview, with an emphasis on practical aspects, for graduate students specializing in other areas. The reader will find rigorous mathematical treatment of the fundamental concepts and theoretical results that are illustrated through an ample supply of academic examples. In addition, to reflect the complexity of real-world applications, a major theme of this book is the inclusion of continuing examples and exercises. Here, practical problems are introduced in the first chapter and revisited in subsequent chapters. The hope is that the student will find it easier to apply new concepts to familiar systems. To support the nontrivial computations associated with these problems, the book provides a chapter-by-chapter

tutorial on the use of the popular software package MATLAB and the associated Control Systems Toolbox for computer-aided control system analysis
and design. The salient features of MATLAB are illustrated in each chapter
through a continuing MATLAB example and a pair of continuing examples.

This textbook consists of nine chapters and three appendices organized
as follows. Chapter 1 introduces the state-space representation for linear time-invariant systems. Chapter 2 is concerned primarily with the
state equation solution and connections with fundamental linear systems
concepts along with several other basic results to be used in subsequent
chapters. Chapters 3 and 4 present thorough introductions to the important topics of controllability and observability, which reveal the power of
state-space methods: The complex behavior of dynamic systems can be
characterized by algebraic relationships derived from the state-space system description. Chapter 5 addresses the concept of minimality associated
with state-space realizations of linear time-invariant systems. Chapter 6
deals with system stability from both internal and external (input-output)
viewpoints and relationships between them. Chapter 7 presents strate-
gies for dynamic response shaping and introduces state feedback control
laws. Chapter 8 presents asymptotic observers and dynamic observer-
based compensators. Chapter 9 gives an introduction to optimal control,
focusing on the linear quadratic regulator. Appendix A provides a sum-
mary of basic matrix computations. Appendix B provides an overview of
basic concepts from linear algebra used throughout the book. Appendix
C provides the complete MATLAB program for the Continuing MATLAB
Example.

Each chapter concludes with a set of exercises intended to aid
the student in his or her quest for mastery of the subject matter.
Exercises will be grouped into four categories: Numerical Exercises,
Analytical Exercises, Continuing MATLAB Exercises, and Continuing
Exercises. Numerical Exercises are intended to be straightforward
problems involving numeric data that reinforce important computations.
Solutions should be based on hand calculations, although students are
strongly encouraged to use MATLAB to check their results. Analytical
Exercises are intended to require nontrivial derivations or proofs of facts
either asserted without proof in the chapter or extensions thereof. These
exercises are by nature more challenging than the Numerical Exercises.
Continuing MATLAB Exercises will revisit the state equations introduced
in Chapter 1. Students will be called on to develop MATLAB m-files
incrementally for each exercise that implement computations associated
with topics in each chapter. Continuing Exercises are also cumulative
and are patterned after the Continuing Examples introduced in Chapter
1. These exercises are based on physical systems, so the initial task will

be to derive linear state equation representations from the given physical descriptions. The use of MATLAB also will be required over the course of working these exercises, and the experience gained from the Continuing MATLAB Exercises will come in handy .

1

INTRODUCTION

This chapter introduces the state-space representation for linear time-invariant systems. We begin with a brief overview of the origins of state-space methods to provide a context for the focus of this book. Following that, we define the state equation format and provide examples to show how state equations can be derived from physical system descriptions and from transfer-function representations. In addition, we show how linear state equations arise from the linearization of a nonlinear state equation about a nominal trajectory or equilibrium condition.

This chapter also initiates our use of the MATLAB software package for computer-aided analysis and design of linear state-space control systems. Beginning here and continuing throughout the book, features of MATLAB and the accompanying Control Systems Toolbox that support each chapter's subject matter will be presented and illustrated using a Continuing MATLAB Example. In addition, we introduce two Continuing Examples that we also will revisit in subsequent chapters.

1.1 HISTORICAL PERSPECTIVE AND SCOPE

Any scholarly account of the history of control engineering would have to span several millennia because there are many examples throughout

---

ancient history, the industrial revolution, and into the early twentieth century of ingeniously designed systems that employed feedback mechanisms in various forms. Ancient water clocks, south-pointing chariots, Watt's flyball governor for steam engine speed regulation, and mechanisms for ship steering, gun pointing, and vacuum tube amplifier stabilization are but a few. Here we are content to survey important developments in the theory and practice of control engineering since the mid-1900s in order to provide some perspective for the material that is the focus of this book in relation to topics covered in most undergraduate controls courses and in more advanced graduate-level courses.

In the so-called classical control era of the 1940s and 1950s, systems were represented in the frequency domain by transfer functions. In addition, performance and robustness specifications were either cast directly in or translated into the frequency domain. For example, transient response specifications were converted into desired closed-loop pole locations or desired open-loop and/or closed-loop frequency-response characteristics. Analysis techniques involving Evans root locus plots, Bode plots, Nyquist plots, and Nichol's charts were limited primarily to single-input, single-output systems, and compensation schemes were fairly simple, e.g., a single feedback loop with cascade compensation. Moreover, the design process was iterative, involving an initial design based on various simplifying assumptions followed by parameter tuning on a trial-and-error basis. Ultimately, the final design was not guaranteed to be optimal in any sense.

The 1960s and 1970s witnessed a fundamental paradigm shift from the frequency domain to the time domain. Systems were represented in the time domain by a type of differential equation called a *state equation*. Performance and robustness specifications also were specified in the time domain, often in the form of a quadratic performance index. Key advantages of the state-space approach were that a time-domain formulation exploited the advances in digital computer technology and the analysis and design methods were well-suited to multiple-input, multiple-output systems. Moreover, feedback control laws were calculated using analytical formulas, often directly optimizing a particular performance index.

The 1980's and 1990's were characterized by a merging of frequency-domain and time-domain viewpoints. Specifically, frequency-domain performance and robustness specifications once again were favored, coupled with important theoretical breakthroughs that yielded tools for handling multiple-input, multiple-output systems in the frequency domain. Further advances yielded state-space time-domain techniques for controller synthesis. In the end, the best features of the preceding decades were merged into a powerful, unified framework.

   The chronological development summarized in the preceding paragraphs correlates with traditional controls textbooks and academic curricula as follows. Classical control typically is the focus at the undergraduate level, perhaps along with an introduction to state-space methods. An in-depth exposure to the state-space approach then follows at the advanced undergraduate/first-year graduate level and is the focus of this book. This, in turn, serves as the foundation for more advanced treatments reflecting recent developments in control theory, including those alluded to in the preceding paragraph, as well as extensions to time-varying and nonlinear systems.

   We assume that the reader is familiar with the traditional undergraduate treatment of linear systems that introduces basic system properties such as system dimension, causality, linearity, and time invariance. This book is concerned with the analysis, simulation, and control of finite-dimensional, causal, linear, time-invariant, continuous-time dynamic systems using state-space techniques. From now on, we will refer to members of this system class as *linear time-invariant systems.*

   The techniques developed in this book are applicable to various types of engineering (even nonengineering) systems, such as aerospace, mechanical, electrical, electromechanical, fluid, thermal, biological, and economic systems. This is so because such systems can be modeled mathematically by the same types of governing equations. We do not formally address the modeling issue in this book, and the point of departure is a linear time-invariant state-equation model of the physical system under study. With mathematics as the unifying language, the fundamental results and methods presented here are amenable to translation into the application domain of interest.

1.2   STATE EQUATIONS

A state-space representation for a linear time-invariant system has the general form

$$\dot{x}(t) = Ax(t) + Bu(t) \qquad x(t_0) = x_0$$
$$y(t) = Cx(t) + Du(t) \qquad\qquad\qquad \text{(1.1)}$$

in which $x(t)$ is the n-dimensional *state vector*

$$x(t) = \begin{bmatrix} x_1(t) \\ x_2(t) \\ \vdots \\ x_n(t) \end{bmatrix}$$

whose n scalar components are called *state variables*. Similarly, the m-dimensional *input vector* and p-dimensional *output vector* are given, respectively, as

$$u(t) = \begin{bmatrix} u_1(t) \\ u_2(t) \\ \vdots \\ u_m(t) \end{bmatrix} \qquad y(t) = \begin{bmatrix} y_1(t) \\ y_2(t) \\ \vdots \\ y_p(t) \end{bmatrix}$$

Since differentiation with respect to time of a time-varying vector quantity is performed component-wise, the time-derivative on the left-hand side of Equation (1.1) represents

$$\dot{x}(t) = \begin{bmatrix} \dot{x}_1(t) \\ \dot{x}_2(t) \\ \vdots \\ \dot{x}_n(t) \end{bmatrix}$$

Finally, for a specified initial time t_0, the *initial state* $x(t_0) = x_0$ is a specified, constant n-dimensional vector.

The state vector $x(t)$ is composed of a minimum set of system variables that uniquely describes the future response of the system given the current state, the input, and the dynamic equations. The input vector $u(t)$ contains variables used to actuate the system, the output vector $y(t)$ contains the measurable quantities, and the state vector $x(t)$ contains internal system variables.

Using the notational convention $M = [m_{ij}]$ to represent the matrix whose element in the ith row and jth column is m_{ij}, the coefficient matrices in Equation (1.1) can be specified via

$$A = [a_{ij}] \quad B = [b_{ij}] \quad C = [c_{ij}]$$
$$D = [d_{ij}]$$

having dimensions $n \times n$, $n \times m$, $p \times n$, and $p \times m$, respectively. With these definitions in place, we see that the state equation (1.1) is a compact representation of n scalar first-order ordinary differential equations, that is,

$$\dot{x}_i(t) = a_{i1}x_1(t) + a_{i2}x_2(t) + \cdots + a_{in}x_n(t)$$
$$+ b_{i1}u_1(t) + b_{i2}u_2(t) + \cdots + b_{im}u_m(t)$$

for $i = 1, 2, \ldots, n$, together with p scalar linear algebraic equations

$$y_j(t) = c_{j1}x_1(t) + c_{j2}x_2(t) + \cdots + c_{jn}x_n(t)$$
$$+ d_{j1}u_1(t) + d_{j2}u_2(t) + \cdots + d_{jm}u_m(t)$$

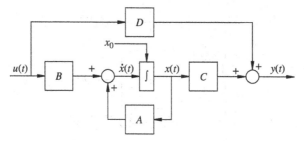

FIGURE 1.1 State-equation block diagram.

for $j = 1, 2, \ldots, p$. From this point on the vector notation (1.1) will be preferred over these scalar decompositions. The state-space description consists of the state differential equation $\dot{x}(t) = Ax(t) + Bu(t)$ and the algebraic output equation $y(t) = Cx(t) + Du(t)$ from Equation (1.1). Figure 1.1 shows the block diagram for the state-space representation of general multiple-input, multiple-output linear time-invariant systems.

One motivation for the state-space formulation is to convert a coupled system of higher-order ordinary differential equations, for example, those representing the dynamics of a mechanical system, to a coupled set of first-order differential equations. In the single-input, single-output case, the state-space representation converts a single nth-order differential equation into a system of n coupled first-order differential equations. In the multiple-input, multiple-output case, in which all equations are of the same order n, one can convert the system of k nth-order differential equations into a system of kn coupled first-order differential equations.

1.3 EXAMPLES

In this section we present a series of examples that illustrate the construction of linear state equations. The first four examples begin with first-principles modeling of physical systems. In each case we adopt the strategy of associating state variables with the energy storage elements in the system. This facilitates derivation of the required differential and algebraic equations in the state-equation format. The last two examples begin with transfer-function descriptions, hence establishing a link between transfer functions and state equations that will be pursued in greater detail in later chapters.

Example 1.1 Given the linear single-input, single-output, mass-spring-damper translational mechanical system of Figure 1.2, we now derive the

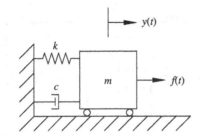

FIGURE 1.2   Translational mechanical system.

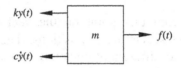

FIGURE 1.3   Free-body diagram.

system model and then convert it to a state-space description. For this system, the input is force $f(t)$ and the output is displacement $y(t)$.

Using Newton's second law, the dynamic force balance for the free-body diagram of Figure 1.3 yields the following second-order ordinary differential equation

$$m\ddot{y}(t) + c\dot{y}(t) + ky(t) = f(t)$$

that models the system behavior. Because this is a single second-order differential equation, we need to select a 2×1 state vector. In general, energy storage is a good criterion for choosing the state variables. The total system energy at any time is composed of potential spring energy $ky(t)^2/2$ plus kinetic energy $m\dot{y}(t)^2/2$ associated with the mass displacement and velocity. We then choose to define the state variables as the mass displacement and velocity:

$$x(t) = \begin{bmatrix} x_1(t) \\ x_2(t) \end{bmatrix} \qquad \begin{aligned} x_1(t) &= y(t) \\ x_2(t) &= \dot{y}(t) = \dot{x}_1(t) \end{aligned}$$

Therefore,

$$\dot{y}(t) = x_2(t)$$
$$\ddot{y}(t) = \dot{x}_2(t)$$

Substituting these two state definitions into the original system equation gives

$$m\dot{x}_2(t) + cx_2(t) + kx_1(t) = f(t)$$

The original single second-order differential equation can be written as a coupled system of two first-order differential equations, that is,

$$\dot{x}_1(t) = x_2(t)$$

$$\dot{x}_2(t) = -\frac{c}{m}x_2(t) - \frac{k}{m}x_1(t) + \frac{1}{m}f(t)$$

The output is the mass displacement

$$y(t) = x_1(t)$$

The generic variable name for input vectors is $u(t)$, so we define:

$$u(t) = f(t)$$

We now write the preceding equations in matrix-vector form to get a valid state-space description. The general state-space description consists of the state differential equation and the algebraic output equation. For Example 1.1, these are

State Differential Equation

$$\dot{x}(t) = Ax(t) + Bu(t)$$

$$\begin{bmatrix} \dot{x}_1(t) \\ \dot{x}_2(t) \end{bmatrix} = \begin{bmatrix} 0 & 1 \\ -\dfrac{k}{m} & -\dfrac{c}{m} \end{bmatrix} \begin{bmatrix} x_1(t) \\ x_2(t) \end{bmatrix} + \begin{bmatrix} 0 \\ \dfrac{1}{m} \end{bmatrix} u(t)$$

Algebraic Output Equation

$$y(t) = Cx(t) + Du(t)$$

$$y(t) = \begin{bmatrix} 1 & 0 \end{bmatrix} \begin{bmatrix} x_1(t) \\ x_2(t) \end{bmatrix} + [0]u(t)$$

The two-dimensional single-input, single-output system matrices in this example are (with $m = p = 1$ and $n = 2$):

$$A = \begin{bmatrix} 0 & 1 \\ -\dfrac{k}{m} & -\dfrac{c}{m} \end{bmatrix} \qquad B = \begin{bmatrix} 0 \\ \dfrac{1}{m} \end{bmatrix} \qquad C = \begin{bmatrix} 1 & 0 \end{bmatrix}$$

$$D = 0$$

In this example, the state vector is composed of the position and velocity of the mass m. Two states are required because we started with one second-order differential equation. Note that $D = 0$ in this example because no part of the input force is directly coupled to the output. $\square$

Example 1.2 Consider the parallel electrical circuit shown in Figure 1.4. We take the input to be the current produced by the independent current source $u(t) = i(t)$ and the output to be the capacitor voltage $y(t) = v(t)$.

It is often convenient to associate state variables with the energy storage elements in the network, namely, the capacitors and inductors. Specifically, capacitor voltages and inductor currents, while not only directly characterizing the energy stored in the associated circuit element, also facilitate the derivation of the required differential equations. In this example, the capacitor voltage coincides with the voltage across each circuit element as a result of the parallel configuration.

This leads to the choice of state variables, that is,

$$x_1(t) = i_L(t)$$
$$x_2(t) = v(t)$$

In terms of these state variables, the inductor's voltage-current relationship is given by

$$x_2(t) = L\dot{x}_1(t)$$

Next, Kirchhoff's current law applied to the top node produces

$$\frac{1}{R}x_2(t) + x_1(t) + C\dot{x}_2(t) = u(t)$$

These relationships can be rearranged so as to isolate state-variable time derivatives as follows:

$$\dot{x}_1(t) = \frac{1}{L}x_2(t)$$

$$\dot{x}_2(t) = -\frac{1}{C}x_1(t) - \frac{1}{RC}x_2(t) + \frac{1}{C}u(t)$$

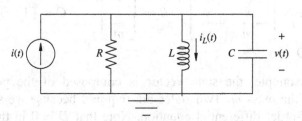

FIGURE 1.4 Parallel electrical circuit.

This pair of coupled first-order differential equations, along with the output definition $y(t) = x_2(t)$, yields the following state-space description for this electrical circuit:

State Differential Equation

$$\begin{bmatrix} \dot{x}_1(t) \\ \dot{x}_2(t) \end{bmatrix} = \begin{bmatrix} 0 & \dfrac{1}{L} \\ -\dfrac{1}{C} & -\dfrac{1}{RC} \end{bmatrix} \begin{bmatrix} x_1(t) \\ x_2(t) \end{bmatrix} + \begin{bmatrix} 0 \\ \dfrac{1}{C} \end{bmatrix} u(t)$$

Algebraic Output Equation

$$y(t) = [0 \quad 1] \begin{bmatrix} x_1(t) \\ x_2(t) \end{bmatrix} + [0]u(t)$$

from which the coefficient matrices A, B, C, and D are found by inspection, that is,

$$A = \begin{bmatrix} 0 & \dfrac{1}{L} \\ -\dfrac{1}{C} & -\dfrac{1}{RC} \end{bmatrix} \qquad B = \begin{bmatrix} 0 \\ \dfrac{1}{C} \end{bmatrix} \qquad C = [0 \quad 1]$$

$$D = 0$$

Note that $D = 0$ in this example because there is no direct coupling between the current source and the capacitor voltage. $\qquad\qquad\square$

Example 1.3   Consider the translational mechanical system shown in Figure 1.5, in which $y_1(t)$ and $y_2(t)$ denote the displacement of the associated mass from its static equilibrium position, and $f(t)$ represents a force applied to the first mass m_1. The parameters are masses m_1 and m_2, viscous damping coefficient c, and spring stiffnesses k_1 and k_2. The input is the applied force $u(t) = f(t)$, and the outputs are taken as the mass displacements. We now derive a mathematical system model and then determine a valid state-space representation.

Newton's second law applied to each mass yields the coupled second-order differential equations, that is,

$$m_1 \ddot{y}_1(t) + k_1 y_1(t) - k_2[y_2(t) - y_1(t)] = f(t)$$
$$m_2 \ddot{y}_2(t) + c \dot{y}_2(t) + k_2[y_2(t) - y_1(t)] = 0$$

Here, the energy-storage elements are the two springs and the two masses. Defining state variables in terms of mass displacements and velocities

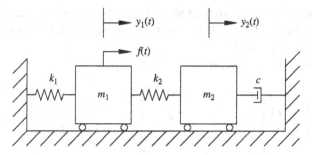

FIGURE 1.5   Translational mechanical system.

yields

$$x_1(t) = y_1(t)$$
$$x_2(t) = y_2(t) - y_1(t)$$
$$x_3(t) = \dot{y}_1(t)$$
$$x_4(t) = \dot{y}_2(t)$$

Straightforward algebra yields the following state equation representation:

State Differential Equation

$$
\begin{bmatrix} \dot{x}_1(t) \\ \dot{x}_2(t) \\ \dot{x}_3(t) \\ \dot{x}_4(t) \end{bmatrix}
=
\begin{bmatrix}
0 & 0 & 1 & 0 \\
0 & 0 & -1 & 1 \\
-\dfrac{k_1}{m_1} & \dfrac{k_2}{m_1} & 0 & 0 \\
0 & \dfrac{-k_2}{m_2} & 0 & -\dfrac{c}{m_2}
\end{bmatrix}
\begin{bmatrix} x_1(t) \\ x_2(t) \\ x_3(t) \\ x_4(t) \end{bmatrix}
+
\begin{bmatrix} 0 \\ 0 \\ \dfrac{1}{m_1} \\ 0 \end{bmatrix}
u(t)
$$

Algebraic Output Equation

$$
\begin{bmatrix} y_1(t) \\ y_2(t) \end{bmatrix}
=
\begin{bmatrix} 1 & 0 & 0 & 0 \\ 1 & 1 & 0 & 0 \end{bmatrix}
\begin{bmatrix} x_1(t) \\ x_2(t) \\ x_3(t) \\ x_4(t) \end{bmatrix}
+
\begin{bmatrix} 0 \\ 0 \end{bmatrix}
u(t)
$$

from which the coefficient matrices $A, B, C,$ and D can be identified. Note that $D = [0 \ \ 0]^T$ because there is no direct feedthrough from the input to the output.

Now, it was convenient earlier to define the second state variable as the difference in mass displacements, $x_2(t) = y_2(t) - y_1(t)$, because this relative displacement is the amount the second spring is stretched. Instead we could have defined the second state variable based on the absolute mass displacement, that is $x_2(t) = y_2(t)$, and derived an equally valid state-space representation. Making this one change in our state variable definitions, that is,

$$x_1(t) = y_1(t)$$

$$x_2(t) = y_2(t)$$

$$x_3(t) = \dot{y}_1(t)$$

$$x_4(t) = \dot{y}_2(t)$$

yields the new A and C matrices

$$A = \begin{bmatrix} 0 & 0 & 1 & 0 \\ 0 & 0 & 0 & 1 \\ -\dfrac{(k_1 + k_2)}{m_1} & \dfrac{k_2}{m_1} & 0 & 0 \\ \dfrac{k_2}{m_2} & \dfrac{-k_2}{m_2} & 0 & -\dfrac{c}{m_2} \end{bmatrix}$$

$$C = \begin{bmatrix} 1 & 0 & 0 & 0 \\ 0 & 1 & 0 & 0 \end{bmatrix}$$

The B and D matrices are unchanged. □

Example 1.4 Consider the electrical network shown in Figure 1.6. We now derive the mathematical model and then determine a valid state-space representation. The two inputs are the independent voltage and current sources $v_{in}(t)$ and $i_{in}(t)$, and the single output is the inductor voltage $v_L(t)$.

In terms of clockwise circulating mesh currents $i_1(t), i_2(t)$, and $i_3(t)$, Kirchhoff's voltage law applied around the leftmost two meshes yields

$$R_1 i_1(t) + v_{C_1}(t) + L\frac{d}{dt}[i_1(t) - i_2(t)] = v_{in}(t)$$

$$L\frac{d}{dt}[i_2(t) - i_1(t)] + v_{C_2}(t) + R_2[i_2(t) - i_3(t)] = 0$$

and Kirchhoff's current law applied to the rightmost mesh yields

$$i_3(t) = -i_{in}(t)$$

In addition, Kirchoff's current law applied at the top node of the inductor gives

$$i_L(t) = i_1(t) - i_2(t)$$

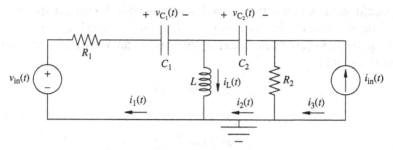

FIGURE 1.6   Electrical circuit.

As in Example 1.2, it is again convenient to associate state variables with the capacitor and inductor energy-storage elements in the network. Here, we select

$$x_1(t) = v_{C_1}(t)$$
$$x_2(t) = v_{C_2}(t)$$
$$x_3(t) = i_L(t)$$

We also associate inputs with the independent sources via

$$u_1(t) = v_{in}(t)$$
$$u_2(t) = i_{in}(t)$$

and designate the inductor voltage $v_L(t)$ as the output so that

$$y(t) = v_L(t) = L\dot{x}_3(t)$$

Using the relationships

$$C_1\dot{x}_1(t) = i_1(t)$$
$$C_2\dot{x}_2(t) = i_2(t)$$
$$x_3(t) = C_1\dot{x}_1(t) - C_2\dot{x}_2(t)$$

the preceding circuit analysis now can be recast as

$$R_1C_1\dot{x}_1(t) + L\dot{x}_3(t) = -x_1(t) + u_1(t)$$
$$R_2C_2\dot{x}_2(t) - L\dot{x}_3(t) = -x_2(t) - R_2u_2(t)$$
$$C_1\dot{x}_1(t) - C_2\dot{x}_2(t) = x_3(t)$$

Packaging these equations in matrix form and isolating the state-variable time derivatives gives

$$
\begin{bmatrix} \dot{x}_1(t) \\ \dot{x}_2(t) \\ \dot{x}_3(t) \end{bmatrix} = \begin{bmatrix} R_1 C_1 & 0 & L \\ 0 & R_2 C_2 & -L \\ C_1 & -C_2 & 0 \end{bmatrix}^{-1}
$$

$$
\left(\begin{bmatrix} -1 & 0 & 0 \\ 0 & -1 & 0 \\ 0 & 0 & 0 \end{bmatrix} \begin{bmatrix} x_1(t) \\ x_2(t) \\ x_3(t) \end{bmatrix} + \begin{bmatrix} 1 & 0 \\ 0 & -R_2 \\ 0 & 0 \end{bmatrix} \begin{bmatrix} u_1(t) \\ u_2(t) \end{bmatrix} \right)
$$

Calculating and multiplying through by the inverse and yields the state differential equation, that is,

$$
\begin{bmatrix} \dot{x}_1(t) \\ \dot{x}_2(t) \\ \dot{x}_3(t) \end{bmatrix} = \begin{bmatrix} \dfrac{1}{(R_1+R_2)C_1} & \dfrac{-1}{(R_1+R_2)C_1} & \dfrac{R_2}{(R_1+R_2)C_1} \\ \dfrac{-1}{(R_1+R_2)C_2} & \dfrac{-1}{(R_1+R_2)C_2} & \dfrac{-R_1}{(R_1+R_2)C_2} \\ \dfrac{-R_2}{(R_1+R_2)L} & \dfrac{R_1}{(R_1+R_2)L} & \dfrac{-R_1 R_2}{(R_1+R_2)L} \end{bmatrix} \begin{bmatrix} x_1(t) \\ x_2(t) \\ x_3(t) \end{bmatrix}
$$

$$
+ \begin{bmatrix} \dfrac{1}{(R_1+R_2)C_1} & \dfrac{-R_2}{(R_1+R_2)C_1} \\ \dfrac{1}{(R_1+R_2)C_2} & \dfrac{-R_2}{(R_1+R_2)C_2} \\ \dfrac{R_2}{(R_1+R_2)L} & \dfrac{R_1 R_2}{(R_1+R_2)L} \end{bmatrix} \begin{bmatrix} u_1(t) \\ u_2(t) \end{bmatrix}
$$

which is in the required format from which coefficient matrices A and B can be identified. In addition, the associated output equation $y(t) = L\dot{x}_3(t)$ can be expanded to the algebraic output equation as follows

$$
y(t) = \begin{bmatrix} \dfrac{-R_2}{(R_1+R_2)} & \dfrac{R_1}{(R_1+R_2)} & \dfrac{-R_1 R_2}{(R_1+R_2)} \end{bmatrix} \begin{bmatrix} x_1(t) \\ x_2(t) \\ x_3(t) \end{bmatrix}
$$

$$
+ \begin{bmatrix} \dfrac{R_2}{(R_1+R_2)} & \dfrac{R_1 R_2}{(R_1+R_2)} \end{bmatrix} \begin{bmatrix} u_1(t) \\ u_2(t) \end{bmatrix}
$$

from which coefficient matrices C and D can be identified.

Note that in this example, there is direct coupling between the independent voltage and current source inputs $v_{in}(t)$ and $i_{in}(t)$ and the inductor voltage output $v_L(t)$, and hence the coefficient matrix D is nonzero.   □

Example 1.5 This example derives a valid state-space description for a general third-order differential equation of the form

$$\dddot{y}(t) + a_2\ddot{y}(t) + a_1\dot{y}(t) + a_0 y(t) = b_0 u(t)$$

The associated transfer function definition is

$$H(s) = \frac{b_0}{s^3 + a_2 s^2 + a_1 s + a_0}$$

Define the following state variables:

$$x(t) = \begin{bmatrix} x_1(t) \\ x_2(t) \\ x_3(t) \end{bmatrix} \quad \begin{array}{l} x_1(t) = y(t) \\ x_2(t) = \dot{y}(t) = \dot{x}_1(t) \\ x_3(t) = \ddot{y}(t) = \ddot{x}_1(t) = \dot{x}_2(t) \end{array}$$

Substituting these state-variable definitions into the original differential equation yields the following:

$$\dot{x}_3(t) = -a_0 x_1(t) - a_1 x_2(t) - a_2 x_3(t) + b_0 u(t)$$

The state differential and algebraic output equations are then
State Differential Equation

$$\begin{bmatrix} \dot{x}_1(t) \\ \dot{x}_2(t) \\ \dot{x}_3(t) \end{bmatrix} = \begin{bmatrix} 0 & 1 & 0 \\ 0 & 0 & 1 \\ -a_0 & -a_1 & -a_2 \end{bmatrix} \begin{bmatrix} x_1(t) \\ x_2(t) \\ x_3(t) \end{bmatrix} + \begin{bmatrix} 0 \\ 0 \\ b_0 \end{bmatrix} u(t)$$

Algebraic Output Equation

$$y(t) = \begin{bmatrix} 1 & 0 & 0 \end{bmatrix} \begin{bmatrix} x_1(t) \\ x_2(t) \\ x_3(t) \end{bmatrix} + [0]u(t)$$

from which the coefficient matrices $A, B, C,$ and D can be identified. $D = 0$ in this example because there is no direct coupling between the input and output.

This example may be generalized easily to the nth-order ordinary differential equation

$$\frac{d^n y(t)}{dt^n} + a_{n-1}\frac{d^{n-1}y(t)}{dt^{n-1}} + \cdots + a_2\frac{d^2 y(t)}{dt^2} + a_1\frac{dy(t)}{dt} + a_0 y(t) = b_0 u(t)$$

$$(1.2)$$

For this case, the coefficient matrices $A, B, C,$ and D are

$$A = \begin{bmatrix} 0 & 1 & 0 & \cdots & 0 \\ 0 & 0 & 1 & \cdots & 0 \\ \vdots & \vdots & \vdots & \ddots & \vdots \\ 0 & 0 & 0 & \cdots & 1 \\ -a_0 & -a_1 & -a_2 & \cdots & -a_{n-1} \end{bmatrix} \qquad B = \begin{bmatrix} 0 \\ 0 \\ \vdots \\ 0 \\ b_0 \end{bmatrix}$$

$$C = [1 \quad 0 \quad 0 \quad \cdots \quad 0] \qquad D = [0] \qquad\qquad (1.3) \;\square$$

Example 1.6   Consider a single-input, single-output system represented by the third-order transfer function with second-order numerator polynomial

$$H(s) = \frac{b_2 s^2 + b_1 s + b_0}{s^3 + a_2 s^2 + a_1 s + a_0}$$

If we attempted to proceed as in the preceding example in defining state variables in terms of the output $y(t)$ and its derivatives, we eventually would arrive at the relationship

$$\dot{x}_3(t) = -a_0 x_1(t) - a_1 x_2(t) - a_2 x_3(t) + b_2 \ddot{u}(t) + b_1 \dot{u}(t) + b_0 u(t)$$

This is not consistent with the state-equation format because of the presence of time derivatives of the input, so we are forced to pursue an alternate state-variable definition. We begin by factoring the transfer function according to $H(s) = H_2(s)H_1(s)$ with

$$H_1(s) = \frac{1}{s^3 + a_2 s^2 + a_1 s + a_0} \qquad H_2(s) = b_2 s^2 + b_1 s + b_0$$

and introducing an intermediate signal $w(t)$ with Laplace transform $W(s)$ so that

$$W(s) = H_1(s)U(s)$$

$$= \frac{1}{s^3 + a_2 s^2 + a_1 s + a_0} U(s)$$

$$Y(s) = H_2(s)W(s)$$

$$= (b_2 s^2 + b_1 s + b_0)W(s)$$

A block-diagram interpretation of this step is shown in Figure 1.7. In the time domain, this corresponds to

$$\dddot{w}(t) + a_2 \ddot{w}(t) + a_1 \dot{w}(t) + a_0 w(t) = u(t)$$

$$y(t) = b_2 \ddot{w}(t) + b_1 \dot{w}(t) + b_0 w(t)$$

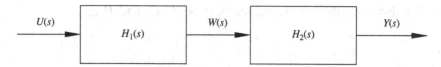

FIGURE 1.7 Cascade block diagram.

Now, the key observation is that a state equation describing the relationship between input $u(t)$ and output $w(t)$ can be written down using the approach of the preceding example. That is, in terms of state variables

$$x_1(t) = w(t)$$
$$x_2(t) = \dot{w}(t) = \dot{x}_1(t)$$
$$x_3(t) = \ddot{w}(t) = \ddot{x}_1(t) = \dot{x}_2(t)$$

we have

$$\begin{bmatrix} \dot{x}_1(t) \\ \dot{x}_2(t) \\ \dot{x}_3(t) \end{bmatrix} = \begin{bmatrix} 0 & 1 & 0 \\ 0 & 0 & 1 \\ -a_0 & -a_1 & -a_2 \end{bmatrix} \begin{bmatrix} x_1(t) \\ x_2(t) \\ x_3(t) \end{bmatrix} + \begin{bmatrix} 0 \\ 0 \\ 1 \end{bmatrix} u(t)$$

$$w(t) = [1 \ \ 0 \ \ 0] \begin{bmatrix} x_1(t) \\ x_2(t) \\ x_3(t) \end{bmatrix} + [0]u(t)$$

As the final step, we recognize that an equation relating the true system output $y(t)$ and our chosen state variables follows from

$$y(t) = b_0 w(t) + b_1 \dot{w}(t) + b_2 \ddot{w}(t)$$
$$= b_0 x_1(t) + b_1 x_2(t) + b_2 x_3(t)$$

which gives the desired state equations:
State Differential Equation

$$\begin{bmatrix} \dot{x}_1(t) \\ \dot{x}_2(t) \\ \dot{x}_3(t) \end{bmatrix} = \begin{bmatrix} 0 & 1 & 0 \\ 0 & 0 & 1 \\ -a_0 & -a_1 & -a_2 \end{bmatrix} \begin{bmatrix} x_1(t) \\ x_2(t) \\ x_3(t) \end{bmatrix} + \begin{bmatrix} 0 \\ 0 \\ 1 \end{bmatrix} u(t)$$

Algebraic Output Equation

$$y(t) = [b_0 \ \ b_1 \ \ b_2] \begin{bmatrix} x_1(t) \\ x_2(t) \\ x_3(t) \end{bmatrix} + [0]u(t)$$

At this point, it should be clear how to extend this approach to systems of arbitrary dimension n beginning with a transfer function of the form

$$H(s) = \frac{b_{n-1}s^{n-1} + \cdots + b_1 s + b_0}{s^n + a_{n-1}s^{n-1} + \cdots + a_1 s + a_0}$$

Notice that the numerator polynomial in $H(s)$ has degree strictly less than the denominator polynomial degree, so $H(s)$ is referred to as a strictly proper rational function (ratio of polynomials in the complex variable s). The preceding state-equation construction can be extended further to handle proper transfer functions

$$H(s) = \frac{b_n s^n + b_{n-1}s^{n-1} + \cdots + b_1 s + b_0}{s^n + a_{n-1}s^{n-1} + \cdots + a_1 s + a_0}$$

in which the numerator and denominator polynomial degrees are equal. The procedure involves first using polynomial division to write $H(s)$ as a strictly proper part plus a constant

$$H(s) = \frac{\hat{b}_{n-1}s^{n-1} + \cdots + \hat{b}_1 s + \hat{b}_0}{s^n + a_{n-1}s^{n-1} + \cdots + a_1 s + a_0} + b_n$$

in which the reader may verify that $\hat{b}_i = b_i - b_n a_i$, for $i = 0, 1, \ldots, n - 1$. Next, the coefficient matrices A, B, and C are found from the numerator and denominator polynomial coefficients of the strictly proper component and, in addition, $D = b_n$.   □

In general, we say that a state equation is a *state-space realization* of a given system's input-output behavior if it corresponds to the relationship $Y(s) = H(s)U(s)$ in the Laplace domain or to the associated differential equation relating $y(t)$ and $u(t)$ in the time domain (for zero initial conditions). The exact meaning of *corresponds to* will be made precise in the next chapter. The preceding example serves to illustrate that a state-space realization of a single-input, single-output system can be written down by inspection simply by plugging the numerator and denominator coefficients into the correct locations in the coefficient matrices C and A, respectively. Owing to its special structure, this state equation is referred to as the *phase-variable canonical form* realization as well as the *controller canonical form* realization.

1.4 LINEARIZATION OF NONLINEAR SYSTEMS

Linear state equations also arise in the course of linearizing nonlinear state equations about nominal trajectories. We begin with a more general nonlinear, time-varying state equation

$$\begin{aligned} \dot{x}(t) &= f[x(t), u(t), t] \\ y(t) &= h[x(t), u(t), t] \end{aligned} \qquad x(t_0) = x_0 \qquad (1.4)$$

where $x(t), u(t)$, and $y(t)$ retain their default vector dimensions and $f(\cdot, \cdot, \cdot)$ and $h(\cdot, \cdot, \cdot)$ are continuously differentiable functions of their $(n + m + 1)$-dimensional arguments. Linearization is performed about a nominal trajectory defined as follows.

Definition 1.1 *For a nominal input signal, $\tilde{u}(t)$, the nominal state trajectory $\tilde{x}(t)$ satisfies*

$$\dot{\tilde{x}}(t) = f[\tilde{x}(t), \tilde{u}(t), t]$$

and the nominal output trajectory $\tilde{y}(t)$ satisfies

$$\tilde{y}(t) = h[\tilde{x}(t), \tilde{u}(t), t]$$

If $\tilde{u}(t) = \tilde{u}$, a constant vector, a special case is an equilibrium state $\tilde{x}$ that satisfies

$$0 = f(\tilde{x}, \tilde{u}, t)$$

for all t. $\qquad\qquad\qquad\qquad\qquad\qquad\qquad\qquad\qquad\qquad\qquad$ □

Deviations of the state, input, and output from their nominal trajectories are denoted by δ subscripts via

$$x_\delta(t) = x(t) - \tilde{x}(t)$$

$$u_\delta(t) = u(t) - \tilde{u}(t)$$

$$y_\delta(t) = y(t) - \tilde{y}(t)$$

Using the compact notation

$$\frac{\partial f}{\partial x}(x, u, t) = \left[\frac{\partial f_i}{\partial x_j}(x, u, t) \right] (n \times n)$$

$$\frac{\partial f}{\partial u}(x, u, t) = \left[\frac{\partial f_i}{\partial u_j}(x, u, t) \right] (n \times m)$$

$$\frac{\partial h}{\partial x}(x, u, t) = \left[\frac{\partial h_i}{\partial x_j}(x, u, t) \right] (p \times n)$$

$$\frac{\partial h}{\partial u}(x, u, t) = \left[\frac{\partial h_i}{\partial u_j}(x, u, t) \right] (p \times m)$$

and expanding the nonlinear maps in Equation (1.4) in a multivariate Taylor series about $[\tilde{x}(t), \tilde{u}(t), t]$ we obtain

$$\dot{x}(t) = f[x(t), u(t), t]$$

$$= f[\tilde{x}(t), \tilde{u}(t), t] + \frac{\partial f}{\partial x}[\tilde{x}(t), \tilde{u}(t), t][x(t) - \tilde{x}(t)]$$

$$+ \frac{\partial f}{\partial u}[\tilde{x}(t), \tilde{u}(t), t][u(t) - \tilde{u}(t)] + \text{higher-order terms}$$

$$y(t) = h[x(t), u(t), t]$$

$$= h[\tilde{x}(t), \tilde{u}(t), t] + \frac{\partial h}{\partial x}[\tilde{x}(t), \tilde{u}(t), t][x(t) - \tilde{x}(t)]$$

$$+ \frac{\partial h}{\partial u}[\tilde{x}(t), \tilde{u}(t), t][u(t) - \tilde{u}(t)] + \text{higher-order terms}$$

On defining coefficient matrices

$$A(t) = \frac{\partial f}{\partial x}(\tilde{x}(t), \tilde{u}(t), t)$$

$$B(t) = \frac{\partial f}{\partial u}(\tilde{x}(t), \tilde{u}(t), t)$$

$$C(t) = \frac{\partial h}{\partial x}(\tilde{x}(t), \tilde{u}(t), t)$$

$$D(t) = \frac{\partial h}{\partial u}(\tilde{x}(t), \tilde{u}(t), t)$$

rearranging slightly, and substituting deviation variables [recognizing that $\dot{x}_\delta(t) = \dot{x}(t) - \dot{\tilde{x}}(t)$] we have

$$\dot{x}_\delta(t) = A(t)x_\delta(t) + B(t)u_\delta(t) + \text{higher-order terms}$$

$$y_\delta(t) = C(t)x_\delta(t) + D(t)u_\delta(t) + \text{higher-order terms}$$

Under the assumption that the state, input, and output remain close to their respective nominal trajectories, the high-order terms can be neglected, yielding the linear state equation

$$\dot{x}_\delta(t) = A(t)x_\delta(t) + B(t)u_\delta(t)$$

$$y_\delta(t) = C(t)x_\delta(t) + D(t)u_\delta(t) \tag{1.5}$$

which constitutes the *linearization* of the nonlinear state equation (1.4) about the specified nominal trajectory. The linearized state equation approximates the behavior of the nonlinear state equation provided that the deviation variables remain small in norm so that omitting the higher-order terms is justified.

If the nonlinear maps in Equation (1.4) do not explicitly depend on t, and the nominal trajectory is an equilibrium condition for a constant nominal input, then the coefficient matrices in the linearized state equation are constant; i.e., the linearization yields a time-invariant linear state equation.

Example 1.7   A ball rolling along a slotted rotating beam, depicted in Figure 1.8, is governed by the equations of motion given below. In this example we will linearize this nonlinear model about a given desired trajectory for this system.

$$\left[\frac{J_b}{r^2} + m\right] \ddot{p}(t) + mg \sin \theta(t) - mp(t)\dot{\theta}(t)^2 = 0$$

$$[mp(t)^2 + J + J_b]\ddot{\theta}(t) + 2\, mp(t)\dot{p}(t)\dot{\theta}(t) + mgp(t)\cos\theta(t) = \tau(t)$$

in which $p(t)$ is the ball position, $\theta(t)$ is the beam angle, and $\tau(t)$ is the applied torque. In addition, g is the gravitational acceleration constant, J is the mass moment of inertia of the beam, and m, r, and J_b are the mass, radius, and mass moment of inertia of the ball, respectively. We define state variables according to

$$x_1(t) = p(t)$$
$$x_2(t) = \dot{p}(t)$$
$$x_3(t) = \theta(t)$$
$$x_4(t) = \dot{\theta}(t)$$

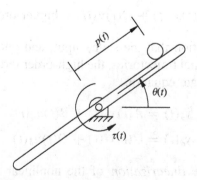

FIGURE 1.8   Ball and beam apparatus.

In addition, we take the input to be the applied torque $\tau(t)$ and the output to be the ball position $p(t)$, so

$$u(t) = \tau(t)$$
$$y(t) = p(t)$$

The resulting nonlinear state equation plus the output equation then are

$$\dot{x}_1(t) = x_2(t)$$
$$\dot{x}_2(t) = b[x_1(t)x_4(t)^2 - g \sin x_3(t)]$$
$$\dot{x}_3(t) = x_4(t)$$
$$\dot{x}_4(t) = \frac{-2mx_1(t)x_2(t)x_4(t) - mgx_1(t)\cos x_3(t) + u(t)}{mx_1(t)^2 + J + J_b}$$
$$y(t) = x_1(t)$$

in which $b = m/[(J_b/r^2) + m]$.

We consider nominal trajectories corresponding to a steady and level beam and constant-velocity ball position responses. In terms of an initial ball position p_0 at the initial time t_0 and a constant ball velocity v_0, we take

$$\tilde{x}_1(t) = \tilde{p}(t) = v_0(t - t_0) + p_0$$
$$\tilde{x}_2(t) = \dot{\tilde{p}}(t) = v_0$$
$$\tilde{x}_3(t) = \tilde{\theta}(t) = 0$$
$$\tilde{x}_4(t) = \dot{\tilde{\theta}}(t) = 0$$
$$\tilde{u}(t) = \tilde{\tau}(t) = mg\tilde{x}_1(t)$$

for which it remains to verify that Definition 1.1 is satisfied. Comparing

$$\dot{\tilde{x}}_1(t) = v_0$$
$$\dot{\tilde{x}}_2(t) = 0$$
$$\dot{\tilde{x}}_3(t) = 0$$
$$\dot{\tilde{x}}_4(t) = 0$$

with

$$\tilde{x}_2(t) = v_0$$
$$b(\tilde{x}_1(t)\tilde{x}_4(t)^2 - g \sin \tilde{x}_3(t)) = b(0 - g \sin(0)) = 0$$
$$\tilde{x}_4(t) = 0$$

$$\frac{-2m\tilde{x}_1(t)\tilde{x}_2(t)\tilde{x}_4(t) - mg\tilde{x}_1(t)\cos\tilde{x}_3(t) + \tilde{u}(t)}{m\tilde{x}_1(t)^2 + J + J_b} = \frac{0 - mg\tilde{x}_1(t)\cos(0) + mg\tilde{x}_1(t)}{m\tilde{x}_1(t)^2 + J + J_b} = 0$$

we see that $\tilde{x}(t)$ is a valid nominal state trajectory for the nominal input $\tilde{u}(t)$. As an immediate consequence, the nominal output is $\tilde{y}(t) = \tilde{x}_1(t) = \tilde{p}(t)$. It follows directly that deviation variables are specified by

$$x_\delta(t) = \begin{bmatrix} p(t) - \tilde{p}(t) \\ \dot{p}(t) - \dot{\tilde{p}}(t) \\ \theta(t) - 0 \\ \dot{\theta}(t) - 0 \end{bmatrix}$$

$$u_\delta(t) = \tau(t) - mg\tilde{p}(t)$$

$$y_\delta(t) = p(t) - \tilde{p}(t)$$

With

$$f(x, u) = \begin{bmatrix} f_1(x_1, x_2, x_3, x_4, u) \\ f_2(x_1, x_2, x_3, x_4, u) \\ f_3(x_1, x_2, x_3, x_4, u) \\ f_4(x_1, x_2, x_3, x_4, u) \end{bmatrix} = \begin{bmatrix} x_2 \\ b(x_1 x_4^2 - g\sin x_3) \\ x_4 \\ \dfrac{-2mx_1 x_2 x_4 - mg x_1 \cos x_3 + u}{mx_1^2 + J + J_b} \end{bmatrix}$$

partial differentiation yields

$$\frac{\partial f}{\partial x}(x, u) = \begin{bmatrix} 0 & 1 & 0 & 0 \\ bx_4^2 & 0 & -bg\cos x_3 & 2bx_1 x_4 \\ 0 & 0 & 0 & 1 \\ \dfrac{\partial f_4}{\partial x_1} & \dfrac{-2mx_1 x_4}{mx_1^2 + J + J_b} & \dfrac{mg x_1 \sin x_3}{mx_1^2 + J + J_b} & \dfrac{-2mx_1 x_2}{mx_1^2 + J + J_b} \end{bmatrix}$$

where

$$\frac{\partial f_4}{\partial x_1} = \frac{[(-2mx_2 x_4 - mg\cos x_3)(mx_1^2 + J + J_b)] - [(-2mx_1 x_2 x_4 - mg x_1 \cos x_3 + u)(2mx_1)]}{(mx_1^2 + J + J_b)^2}$$

$$\frac{\partial f}{\partial u}(x, u) = \begin{bmatrix} 0 \\ 0 \\ 0 \\ 1 \\ \overline{mx_1^2 + J + J_b} \end{bmatrix}$$

$$\frac{\partial h}{\partial x}(x, u) = [1 \quad 0 \quad 0 \quad 0]$$

$$\frac{\partial h}{\partial u}(x, u) = 0$$

Evaluating at the nominal trajectory gives

$$A(t) = \frac{\partial f}{\partial x}[\tilde{x}(t), \tilde{u}(t)]$$

$$= \begin{bmatrix} 0 & 1 & 0 & 0 \\ 0 & 0 & -bg & 0 \\ 0 & 0 & 0 & 1 \\ \dfrac{-mg}{m\tilde{p}(t)^2 + J + J_b} & 0 & 0 & \dfrac{-2m\tilde{p}(t)v_0}{m\tilde{p}(t)^2 + J + J_b} \end{bmatrix}$$

$$B(t) = \frac{\partial f}{\partial u}[\tilde{x}(t), \tilde{u}(t)] = \begin{bmatrix} 0 \\ 0 \\ 0 \\ 1 \\ \dfrac{1}{m\tilde{p}(t)^2 + J + J_b} \end{bmatrix}$$

$$C(t) = \frac{\partial h}{\partial x}[\tilde{x}(t), \tilde{u}(t)] = [1 \quad 0 \quad 0 \quad 0]$$

$$D(t) = \frac{\partial h}{\partial u}[\tilde{x}(t), \tilde{u}(t)] = 0 \qquad (1.6)$$

which, together with the deviation variables defined previously, specifies the linearized time-varying state equation for the ball and beam system.

A special case of the nominal trajectory considered thus far in this example corresponds to zero ball velocity $v_0 = 0$ and, consequently, constant ball position $\tilde{p}(t) = p_0$. The beam must remain steady and level, so the nominal state trajectory and input reduce to

$$\tilde{x}_1(t) = \tilde{p}(t) = p_0$$
$$\tilde{x}_1(t) = \dot{\tilde{p}}(t) = 0$$
$$\tilde{x}_1(t) = \tilde{\theta}(t) = 0$$
$$\tilde{x}_1(t) = \dot{\tilde{\theta}}(t) = 0$$
$$\tilde{u}(t) = \tilde{\tau}(t) = mg\, p_0$$

with an accompanying impact on the deviation variables. Given that the nonlinear ball and beam dynamics are time invariant and that now the nominal state trajectory and input are constant and characterize an equilibrium condition for these dynamics, the linearization process yields a time-invariant linear state equation. The associated coefficient matrices are obtained by making the appropriate substitutions in Equation (1.6) to obtain

$$
A = \frac{\partial f}{\partial x}(\tilde{x}, \tilde{u}) =
\begin{bmatrix}
0 & 1 & 0 & 0 \\
0 & 0 & -bg & 0 \\
0 & 0 & 0 & 1 \\
\dfrac{-mg}{m\,p_0^2 + J + J_b} & 0 & 0 & 0
\end{bmatrix}
$$

$$
B = \frac{\partial f}{\partial u}(\tilde{x}, \tilde{u}) =
\begin{bmatrix}
0 \\
0 \\
0 \\
1 \\
\dfrac{}{m\,p_0^2 + J + J_b}
\end{bmatrix}
$$

$$
C = \frac{\partial h}{\partial x}(\tilde{x}, \tilde{u}) = [\,1 \quad 0 \quad 0 \quad 0\,]
$$

$$
D = \frac{\partial h}{\partial u}(\tilde{x}, \tilde{u}) = 0 \qquad\qquad\qquad \square
$$

1.5   CONTROL SYSTEM ANALYSIS AND DESIGN USING MATLAB

In each chapter we include a section to identify and explain the use of MATLAB software and MATLAB functions for state-space analysis and design methods. We use a continuing example to demonstrate the use of MATLAB throughout; this is a single-input, single-output two–dimensional rotational mechanical system that will allow the student to perform all operations by hand to compare with MATLAB results. We assume that the Control Systems Toolbox is installed with MATLAB.

MATLAB: General, Data Entry, and Display

In this section we present general MATLAB commands, and we start the Continuing MATLAB Example. We highly recommend the use of MATLAB m-files, which are scripts and functions containing MATLAB commands that can be created and modified in the MATLAB Editor and then executed. Throughout the MATLAB examples, **bold Courier New** font indicates MATLAB function names, user inputs, and variable names; this is given for emphasis only. Some useful MATLAB statements are listed below to help the novice get started.

General MATLAB Commands:

`help`	Provides a list of topics for which you can get online help.
`help fname`	Provides online help for MATLAB function **fname** .
`%`	The **%** symbol at any point in the code indicates a comment; text beyond the **%** is ignored by MATLAB and is highlighted in **green** .
`;`	The semicolon used at the end of a line suppresses display of the line's result to the MATLAB workspace.
`clear`	This command clears the MATLAB workspace, i.e., erases any previous user-defined variables.
`clc`	Clears the cursor.
`figure(n)`	Creates an empty figure window (numbered **n**) for graphics.
`who`	Displays a list of all user-created variable names.
`whos`	Same as **who** but additionally gives the dimension of each variable.
`size(name)`	Responds with the dimension of the matrix **name** .
`length(name)`	Responds with the length of the vector **name** .
`eye(n)`	Creates an $n \times n$ identity matrix I_n.
`zeros(m,n)`	Creates a $m \times n$ array of zeros.
`ones(m,n)`	Creates a $m \times n$ array of ones.
`t = t0:dt:tf`	Creates an evenly spaced time array starting from initial time **t0** and ending at final time **tf** , with steps of **dt** .
`disp('string')`	Print the text **string** to the screen.
`name = input('string')`	The **input** command displays a text **string** to the user, prompting for input; the entered data then are written to the variable **name** .

In the MATLAB Editor (not in this book), comments appear in **green**, text strings appear in **red**, and logical operators and other reserved programming words appear in **blue**.

MATLAB for State-Space Description

MATLAB uses a data-structure format to describe linear time-invariant systems. There are three primary ways to describe a linear time-invariant system in MATLAB: (1) state-space realizations specified by coefficient matrices **A**, **B**, **C**, and **D** (**ss**); (2) transfer functions with (**num**, **den**), where **num** is the array of polynomial coefficients for the transfer-function numerator and **den** is the array of polynomial coefficients for the transfer-function denominator (**tf**); and (3) transfer functions with (**z**, **p**, **k**), where **z** is the array of numerator polynomial roots (the zeros), **p** is the array of denominator polynomial roots (the poles), and **k** is the system gain. There is a fourth method, frequency response data (**frd**), which will not be considered in this book. The three methods to define a continuous-time linear time-invariant system in MATLAB are summarized below:

```
SysName = ss(A,B,C,D);
SysName = tf(num,den);
SysName = zpk(z,p,k);
```

In the first statement (**ss**), a scalar 0 in the **D** argument position will be interpreted as a zero matrix **D** of appropriate dimensions. Each of these three statements (**ss, tf, zpk**) may be used to define a system as above or to convert between state-space, transfer-function, and zero-pole-gain descriptions of an existing system. Alternatively, once the linear time-invariant system **SysName** is defined, the parameters for each system description may be extracted using the following statements:

```
[num,den]     = tfdata(SysName);
[z,p,k]       = zpkdata(SysName);
[A,B,C,D]     = ssdata(SysName);
```

In the first two statements above, if we have a single-input, single-output system, we can use the switch **'v'**: **tfdata(SysName,'v')** and **zpkdata(SysName,'v')**. There are three methods to access data from the defined linear time-invariant **SysName**: **set/get** commands, direct structure referencing, and data-retrieval commands. The latter approach is given above; the first two are:

```
set(SysName,PropName,PropValue);
PropValue = get(SysName,PropName);
SysName.PropName = PropValue;
% equivalent to 'set' command
```

```
PropValue   = SysName.PropName;
% equivalent to 'get' command
```

In the preceding, **SysName** is set by the user as the desired name for the defined linear time-invariant system. **PropName** (property name) represents the valid properties the user can modify, which include **A, B, C, D** for **ss, num, den, variable** (the default is 's ' for a continuous system Laplace variable) for **tf**, and **z, p, k, variable** (again, the default is 's ') for **zpk**. The command **set(SysName)** displays the list of properties for each data type. The command **get(SysName)** displays the value currently stored for each property. **PropValue** (property value) indicates the value that the user assigns to the property at hand. In previous MATLAB versions, many functions required the linear time-invariant system input data (**A, B, C, D** for state space, **num, den** for transfer function, and **z, p, k** for zero-pole-gain notation); although these still should work, MATLAB's preferred mode of operation is to pass functions the **SysName** linear time-invariant data structure. For more information, type **help ltimodels** and **help ltiprops** at the MATLAB command prompt.

Continuing MATLAB Example

Modeling   A single-input, single-output rotational mechanical system is shown in Figure 1.9. The single input is an externally applied torque $\tau(t)$, and the output is the angular displacement $\theta(t)$. The constant parameters are motor shaft polar inertia J, rotational viscous damping coefficient b, and torsional spring constant k_R (provided by the flexible shaft). This example will be used in every chapter to demonstrate the current topics via MATLAB for a model that will become familiar. To derive the system model, MATLAB does not help (unless the Symbolic Math Toolbox capabilities of MATLAB are used).

In the free-body diagram of Figure 1.10, the torque resulting from the rotational viscous damping opposes the instantaneous direction of the angular velocity and the torque produced by the restoring spring

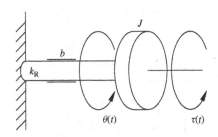

FIGURE 1.9   Continuing MATLAB Example system.

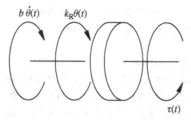

FIGURE 1.10   Continuing MATLAB Example free-body diagram.

opposes the instantaneous direction of the angular displacement. We apply Euler's rotational law (the rotational equivalent of Newton's Second Law), to derive the system model. Euler's rotational law may be stated as $\sum M = J\alpha$, where $\sum M$ is the sum of moments, J is the polar moment of inertia, and α is the shaft angular acceleration.

$$\sum M = J\ddot{\theta}(t) = \tau(t) - b\dot{\theta}(t) - k_{R}\theta(t)$$

This system can be represented by the single second-order linear time-invariant ordinary differential equation

$$J\ddot{\theta}(t) + b\dot{\theta}(t) + k_{R}\theta(t) = \tau(t)$$

This equation is the rotational equivalent of a translational mechanical mass-spring-damper system with torque $\tau(t)$ as the input and angular displacement $\theta(t)$ as the output.

State-Space Description   Now we derive a valid state-space description for the Continuing MATLAB Example. That is, we specify the state variables and derive the coefficient matrices A, B, C, and D. We start with the second-order differential equation above for which we must define two state variables $x_{i}(t), i = 1, 2$. Again, energy-storage elements guide our choice of states:

$$x_1(t) = \theta(t)$$

$$x_2(t) = \dot{\theta}(t) = \dot{x}_1(t)$$

We will have two first-order differential equations, derived from the original second-order differential equation, and $\dot{x}_1(t) = x_2(t)$ from above. The state differential equation is

$$\begin{bmatrix} \dot{x}_1(t) \\ \dot{x}_2(t) \end{bmatrix} = \begin{bmatrix} 0 & 1 \\ \dfrac{-k_R}{J} & \dfrac{-b}{J} \end{bmatrix} \begin{bmatrix} x_1(t) \\ x_2(t) \end{bmatrix} + \begin{bmatrix} 0 \\ \dfrac{1}{J} \end{bmatrix} \tau(t)$$

TABLE 1.1   Numerical Parameters for the Continuing MATLAB Example

Parameter	Value	Units	Name
J	1	kg-m^2	motor shaft polar inertia
b	4	N-m-s	motor shaft damping constant
k_R	40	N-m/rad	torsional spring constant

The algebraic output equation is:

$$y(t) = [1 \quad 0] \begin{bmatrix} x_1(t) \\ x_2(t) \end{bmatrix} + [0]\tau(t)$$

The coefficient matrices A, B, C, D for this Continuing MATLAB Example are thus:

$$A = \begin{bmatrix} 0 & 1 \\ -\dfrac{k_R}{J} & -\dfrac{b}{J} \end{bmatrix} \quad B = \begin{bmatrix} 0 \\ \dfrac{1}{J} \end{bmatrix} \quad C = [1 \quad 0]$$

$$D = 0$$

This specifies a two–dimensional single-input, single-output system with $m = 1$ input, $p = 1$ output, and $n = 2$ states. Let us assume the constant parameters listed in Table 1.1 for the continuing MATLAB example. Then the numerical coefficient matrices are

$$A = \begin{bmatrix} 0 & 1 \\ -40 & -4 \end{bmatrix} \quad B = \begin{bmatrix} 0 \\ 1 \end{bmatrix} \quad C = [1 \quad 0] \quad D = 0$$

Chapter by chapter we will present MATLAB code and results dealing with the topics at hand for the Continuing MATLAB Example. These code segments will be complete only if taken together over all chapters (i.e., ensuing code portions may require previously defined variables in earlier chapters to execute properly). Appendix C presents this complete program for all chapters. To get started, we need to define the coefficient matrices A, B, C, and D in MATLAB. Then we can find the system transfer function and zero-pole-gain descriptions.

```
%-----------------------------------------------------
%    Chapter 1. State-Space Description
%-----------------------------------------------------

J   = 1;
b   = 4;
kR  = 40;
```

```
A = [0 1;-kR/J -b/J];                % Define the
                                     % state-space
                                     % realization

B = [0;1/J];
C = [1 0];
D = [0];

JbkR = ss(A,B,C,D);                  % Define model from
                                     % state-space

JbkRtf  = tf(JbkR);                  % Convert to
                                     % transfer function
JbkRzpk = zpk(JbkR);                 % Convert to
                                     % zero-pole
                                     % description

[num,den]   = tfdata(JbkR,'v');      % Extract transfer
                                     % function
                                     % description
[z,p,k]     = zpkdata(JbkR,'v');     % Extract zero-pole
                                     % description

JbkRss  = ss(JbkRtf)                 % Convert to
                                     % state-space
                                     % description
```

The **ss** command yields

```
a =
                    x1            x2
          x1         0            1
          x2        -40          -4
b =
                    u1
          x1         0
          x2         1
c =
                    x1            x2
          y1         1            0
```

```
d =
                              u1
            y1              0
Continuous-time model.
```

The **tf** and **zpk** commands yield

```
Transfer function:
     1
- - - - - - - - - - - - - -
s^2 + 4 s + 40
```

```
Zero/pole/gain:
     1
- - - - - - - - - - - - - -
(s^2 + 4s + 40)
```

The **tfdata** and **zpkdata** commands yield

```
num =
     0     0     1

den =
   1.0000    4.0000    40.0000

z =
   Empty matrix: 0-by-1

p =
   -2.0000 + 6.0000i
   -2.0000 - 6.0000i

k =
     1
```

Finally, the second **ss** command yields

```
a =
                    x1          x2
          x1        -4          -5
          x2         8           0
```

b =

	u1
x1	0.25
x2	0

c =

	x1	x2
y1	0	0.5

d =

	u1
y1	0

Note that when MATLAB converted from the **tf** to the **ss** description above, it returned a different state-space realization than the one originally defined. The validity of this outcome will be explained in Chapter 2.

1.6 CONTINUING EXAMPLES

Continuing Example 1: Two-Mass Translational Mechanical System

This multiple-input, multiple-output example will continue throughout each chapter of this book, building chapter by chapter to demonstrate the important topics at hand.

Modeling A mechanical system is represented by the two degree-of-freedom linear time-invariant system shown in Figure 1.11. There are two force inputs $u_i(t)$ and two displacement outputs $y_i(t), i = 1, 2$. The constant parameters are masses m_i, damping coefficients c_i, and spring coefficients $k_i, i = 1, 2$. We now derive the mathematical model for this system; i.e., we draw the free-body diagrams and then write the correct number of independent ordinary differential equations. All motion is constrained to be horizontal, as shown in Figure 1.11. Outputs $y_i(t)$ are each measured from the neutral spring equilibrium location of each mass m_i. Figure 1.12 shows the two free-body diagrams.

Now we apply Newton's second law twice, once for each mass, to derive the two second-order dynamic equations of motion:

$$\sum F_1 = m_1 \ddot{y}_1(t) = k_2[y_2(t) - y_1(t)] + c_2[\dot{y}_2(t) - \dot{y}_1(t)]$$
$$- k_1 y_1(t) - c_1 \dot{y}_1(t) + u_1(t)$$
$$\sum F_2 = m_2 \ddot{y}_2(t) = -k_2[y_2(t) - y_1(t)] - c_2[\dot{y}_2(t) - \dot{y}_1(t)] + u_2(t)$$

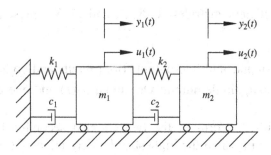

FIGURE 1.11 Continuing Example 1 system.

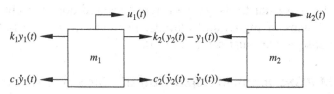

FIGURE 1.12 Continuing Example 1 free-body diagrams.

We rewrite these equations so that the output-related terms $y_i(t)$ appear on the left side along with their derivatives and the input forces $u_i(t)$ appear on the right. Also, $y_i(t)$ terms are combined.

$$m_1\ddot{y}_1(t) + (c_1 + c_2)\dot{y}_1(t) + (k_1 + k_2)y_1(t) - c_2\dot{y}_2(t) - k_2 y_2(t) = u_1(t)$$

$$m_2\ddot{y}_2(t) + c_2\dot{y}_2(t) + k_2 y_2(t) - c_2\dot{y}_1(t) - k_2 y_1(t) = u_2(t)$$

These equations are two linear, coupled, second-order ordinary differential equations. In this type of vibrational system, it is always possible to structure the equations such that the coefficients of $\ddot{y}_i(t)$, $\dot{y}_i(t)$, and $y_i(t)$ are positive in the ith equation, and the coefficients of any $\dot{y}_j(t)$ and $y_j(t)$ terms that appear in the ith equation are negative for $j \neq i$.

Example 1 is a multiple-input, multiple output system with two inputs $u_i(t)$ and two outputs $y_i(t)$. We can express the two preceding second-order differential equations in standard second-order matrix-vector form, $M\ddot{y}(t) + C\dot{y}(t) + Ky(t) = u(t)$, that is,

$$\begin{bmatrix} m_1 & 0 \\ 0 & m_2 \end{bmatrix}\begin{bmatrix} \ddot{y}_1(t) \\ \ddot{y}_2(t) \end{bmatrix} + \begin{bmatrix} c_1 + c_2 & -c_2 \\ -c_2 & c_2 \end{bmatrix}\begin{bmatrix} \dot{y}_1(t) \\ \dot{y}_2(t) \end{bmatrix}$$

$$+ \begin{bmatrix} k_1 + k_2 & -k_2 \\ -k_2 & k_2 \end{bmatrix}\begin{bmatrix} y_1(t) \\ y_2(t) \end{bmatrix} = \begin{bmatrix} u_1(t) \\ u_2(t) \end{bmatrix}$$

State-Space Description Next, we must derive a valid state-space description for this system. That is, we specify the state variables and then derive the coefficient matrices A, B, C, and D. We present two distinct cases:

a. Multiple-input, multiple-output: Both inputs and both outputs
b. Single-input, single-output: One input $u_2(t)$ and one output $y_1(t)$

We start with the form of the two coupled second-order differential equations above in which the highest-order derivatives $\ddot{y}_i(t), i = 1, 2$, are isolated. For both cases, the choice of state variables and the resulting system dynamics matrix A will be identical. This always will be true, i.e., A is fundamental to the system dynamics and does not change with different choices of inputs and outputs. For case a, we use both inputs $u_i(t)$; for case b, we must set $u_1(t) = 0$.

Case a: Multiple-Input, Multiple-Output   Since we have two second-order differential equations, the state-space dimension is $n = 4$, and thus we need to define four state variables $x_i(t), i = 1, 2, 3, 4$. Again, energy-storage elements guide our choice of states:

$$x_1(t) = y_1(t)$$
$$x_2(t) = \dot{y}_1(t) = \dot{x}_1(t)$$
$$x_3(t) = y_2(t)$$
$$x_4(t) = \dot{y}_2(t) = \dot{x}_3(t)$$

We will have four first-order ordinary differential equations derived from the original two second-order differential equations. Two are $\dot{x}_i(t) = x_{i+1}(t)$ from the state variable definitions above, for $i = 1, 3$. The remaining two come from the original second-order differential equations, rewritten by isolating accelerations and substituting the state variable definitions in place of the outputs and their derivatives. Also, we must divide by m_i to normalize each equation.

$$\dot{x}_1(t) = x_2(t)$$
$$\dot{x}_2(t) = \frac{-(k_1 + k_2)x_1(t) - (c_1 + c_2)x_2(t) + k_2 x_3 + c_2 x_4(t) + u_1(t)}{m_1}$$
$$\dot{x}_3(t) = x_4(t)$$
$$\dot{x}_4(t) = \frac{k_2 x_1(t) + c_2 x_2(t) - k_2 x_3(t) - c_2 x_4(t) + u_2(t)}{m_2}$$

The state differential equation is

$$
\begin{bmatrix} \dot{x}_1(t) \\ \dot{x}_2(t) \\ \dot{x}_3(t) \\ \dot{x}_4(t) \end{bmatrix} =
\begin{bmatrix}
0 & 1 & 0 & 0 \\
\dfrac{-(k_1+k_2)}{m_1} & \dfrac{-(c_1+c_2)}{m_1} & \dfrac{k_2}{m_1} & \dfrac{c_2}{m_1} \\
0 & 0 & 0 & 1 \\
\dfrac{k_2}{m_2} & \dfrac{c_2}{m_2} & \dfrac{-k_2}{m_2} & \dfrac{-c_2}{m_2}
\end{bmatrix}
\begin{bmatrix} x_1(t) \\ x_2(t) \\ x_3(t) \\ x_4(t) \end{bmatrix}
$$

$$
+ \begin{bmatrix}
0 & 0 \\
\dfrac{1}{m_1} & 0 \\
0 & 0 \\
0 & \dfrac{1}{m_2}
\end{bmatrix}
\begin{bmatrix} u_1(t) \\ u_2(t) \end{bmatrix}
$$

from which we identify coefficient matrices A and B:

$$
A = \begin{bmatrix}
0 & 1 & 0 & 0 \\
\dfrac{-(k_1+k_2)}{m_1} & \dfrac{-(c_1+c_2)}{m_1} & \dfrac{k_2}{m_1} & \dfrac{c_2}{m_1} \\
0 & 0 & 0 & 1 \\
\dfrac{k_2}{m_2} & \dfrac{c_2}{m_2} & \dfrac{-k_2}{m_2} & \dfrac{-c_2}{m_2}
\end{bmatrix}
\qquad
B = \begin{bmatrix}
0 & 0 \\
\dfrac{1}{m_1} & 0 \\
0 & 0 \\
0 & \dfrac{1}{m_2}
\end{bmatrix}
$$

The algebraic output equation is

$$
\begin{bmatrix} y_1(t) \\ y_2(t) \end{bmatrix} =
\begin{bmatrix} 1 & 0 & 0 & 0 \\ 0 & 0 & 1 & 0 \end{bmatrix}
\begin{bmatrix} x_1(t) \\ x_2(t) \\ x_3(t) \\ x_4(t) \end{bmatrix}
+ \begin{bmatrix} 0 & 0 \\ 0 & 0 \end{bmatrix}
\begin{bmatrix} u_1(t) \\ u_2(t) \end{bmatrix}
$$

from which we identify coefficient matrices C and D:

$$
C = \begin{bmatrix} 1 & 0 & 0 & 0 \\ 0 & 0 & 1 & 0 \end{bmatrix}
\qquad
D = \begin{bmatrix} 0 & 0 \\ 0 & 0 \end{bmatrix}
$$

This is a four-dimensional multiple-input, multiple-output system with $m = 2$ inputs, $p = 2$ outputs, and $n = 4$ states.

Case b: Single-Input, Single-Output: One Input u_2, One Output y_1.
Remember, system dynamics matrix A does not change when considering different system inputs and outputs. For the single-input, single-output

case b, only coefficient matrices B, C, and D change. The state differential equation now is:

$$
\begin{bmatrix} \dot{x}_1(t) \\ \dot{x}_2(t) \\ \dot{x}_3(t) \\ \dot{x}_4(t) \end{bmatrix} = \begin{bmatrix} 0 & 1 & 0 & 0 \\ \dfrac{-(k_1+k_2)}{m_1} & \dfrac{-(c_1+c_2)}{m_1} & \dfrac{k_2}{m_1} & \dfrac{c_2}{m_1} \\ 0 & 0 & 0 & 1 \\ \dfrac{k_2}{m_2} & \dfrac{c_2}{m_2} & \dfrac{-k_2}{m_2} & \dfrac{-c_2}{m_2} \end{bmatrix} \begin{bmatrix} x_1(t) \\ x_2(t) \\ x_3(t) \\ x_4(t) \end{bmatrix}
$$

$$
+ \begin{bmatrix} 0 \\ 0 \\ 0 \\ \dfrac{1}{m_2} \end{bmatrix} u_2(t)
$$

A is the same as that given previously, and the new input matrix is

$$
B = \begin{bmatrix} 0 \\ 0 \\ 0 \\ \dfrac{1}{m_2} \end{bmatrix}
$$

The algebraic output equation now is:

$$
y_1(t) = \begin{bmatrix} 1 & 0 & 0 & 0 \end{bmatrix} \begin{bmatrix} x_1(t) \\ x_2(t) \\ x_3(t) \\ x_4(t) \end{bmatrix} + [0]u_2(t)
$$

so that

$$
C = \begin{bmatrix} 1 & 0 & 0 & 0 \end{bmatrix} \qquad D = 0
$$

This is still a four-dimensional system, now with $m = 1$ input and $p = 1$ output.

Continuing Example 2: Rotational Electromechanical System

This example also will continue throughout each chapter of this book, building chapter by chapter to demonstrate the important topics.

Modeling   A simplified dc servomotor model is shown in Figure 1.13. The input is the armature voltage $v(t)$ and the output is the motor shaft

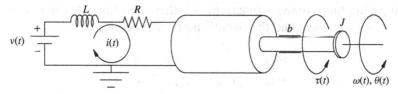

FIGURE 1.13 Continuing Example 2 system.

angular displacement $\theta(t)$. The constant parameters are armature circuit inductance and resistance L and R, respectively, and motor shaft polar inertia and rotational viscous damping coefficient J and b, respectively. The intermediate variables are armature current $i(t)$, motor torque $\tau(t)$, and motor shaft angular velocity $\omega(t) = \dot{\theta}(t)$. In this Continuing Example 2, we have simplified the model; we ignore back emf voltage, and there is no gear ratio or load inertia included. For improvements on each of these issues, see Continuing Exercise 3.

We can derive the dynamic model of this system in three steps: circuit model, electromechanical coupling, and rotational mechanical model. For the circuit model, Kirchhoff's voltage law yields a first-order differential equation relating the armature current to the armature voltage, that is,

$$L\frac{di(t)}{dt} + Ri(t) = v(t)$$

Motor torque is modeled as being proportional to the armature current, so the electromechanical coupling equation is

$$\tau(t) = k_T i(t)$$

where k_T is the motor torque constant. For the rotational mechanical model, Euler's rotational law results in the following second-order differential equation relating the motor shaft angle $\theta(t)$ to the input torque $\tau(t)$:

$$J\ddot{\theta}(t) + b\dot{\theta}(t) = \tau(t)$$

To derive the overall system model, we need to relate the designated system output $\theta(t)$ to the designated system input $v(t)$. The intermediate variables $i(t)$ and $\tau(t)$ must be eliminated. It is convenient to use Laplace transforms and transfer functions for this purpose rather than manipulating the differential equations. Here, we are applying a method similar to Examples 1.5 and 1.6, wherein we use a transfer-function description to derive the state equations. We have

$$\frac{I(s)}{V(s)} = \frac{1}{Ls + R} \qquad \frac{T(s)}{I(s)} = k_T \qquad \frac{\Theta(s)}{T(s)} = \frac{1}{Js^2 + bs}$$

Multiplying these transfer functions together, we eliminate the intermediate variables to generate the overall transfer function:

$$\frac{\Theta(s)}{V(s)} = \frac{k_T}{(Ls + R)(Js^2 + bs)}$$

Simplifying, cross-multiplying, and taking the inverse Laplace transform yields the following third-order linear time-invariant ordinary differential equation:

$$LJ\dddot{\theta}(t) + (Lb + RJ)\ddot{\theta}(t) + Rb\dot{\theta}(t) = k_T v(t)$$

This equation is the mathematical model for the system of Figure 1.13. Note that there is no rotational mechanical spring term in this equation, i.e., the coefficient of the $\theta(t)$ term is zero.

State-Space Description   Now we derive a valid state-space description for Continuing Example 2. That is, we specify the state variables and derive the coefficient matrices A, B, C, and D. The results then are written in matrix-vector form. Since we have a third-order differential equation, the state-space dimension is $n = 3$, and thus we need to define three state variables $x_i(t), i = 1, 2, 3$. We choose

$$x_1(t) = \theta(t)$$
$$x_2(t) = \dot{\theta}(t) = \dot{x}_1(t)$$
$$x_3(t) = \ddot{\theta}(t) = \dot{x}_2(t)$$

We will have three first-order differential equations, derived from the original third-order differential equation. Two are $\dot{x}_i(t) = x_{i+1}(t)$ from the state variable definitions above, for $i = 1, 2$. The remaining first-order differential equation comes from the original third-order differential equation, rewritten by isolating the highest derivative $\dddot{\theta}(t)$ and substituting the state-variable definitions in place of output $\theta(t)$ and its derivatives. Also, we divide the third equation by LJ:

$$\dot{x}_1(t) = x_2(t)$$
$$\dot{x}_2(t) = x_3(t)$$
$$\dot{x}_3(t) = \frac{-(Lb + RJ)}{LJ}x_3(t) - \frac{Rb}{LJ}x_2(t) + \frac{k_T}{LJ}v(t)$$

The state differential equation is

$$\begin{bmatrix} \dot{x}_1(t) \\ \dot{x}_2(t) \\ \dot{x}_3(t) \end{bmatrix} = \begin{bmatrix} 0 & 1 & 0 \\ 0 & 0 & 1 \\ 0 & \dfrac{-Rb}{LJ} & \dfrac{-(Lb + RJ)}{LJ} \end{bmatrix} \begin{bmatrix} x_1(t) \\ x_2(t) \\ x_3(t) \end{bmatrix} + \begin{bmatrix} 0 \\ 0 \\ \dfrac{k_T}{LJ} \end{bmatrix} v(t)$$

from which we identify coefficient matrices A and B:

$$A = \begin{bmatrix} 0 & 1 & 0 \\ 0 & 0 & 1 \\ 0 & \dfrac{-Rb}{LJ} & \dfrac{-(Lb + RJ)}{LJ} \end{bmatrix} \qquad B = \begin{bmatrix} 0 \\ 0 \\ \dfrac{k_T}{LJ} \end{bmatrix}$$

The algebraic output equation is

$$y(t) = \begin{bmatrix} 1 & 0 & 0 \end{bmatrix} \begin{bmatrix} x_1(t) \\ x_2(t) \\ x_3(t) \end{bmatrix} + [0]v(t)$$

from which we identify coefficient matrices C and D:

$$C = \begin{bmatrix} 1 & 0 & 0 \end{bmatrix} \qquad D = 0$$

This is a three-dimensional single-input, single-output system with $m = 1$ input, $p = 1$ output, and $n = 3$ states.

1.7 HOMEWORK EXERCISES

We refer the reader to the Preface for a description of the four classes of exercises that will conclude each chapter: Numerical Exercises, Analytical Exercises, Continuing MATLAB Exercises, and Continuing Exercises.

Numerical Exercises

NE1.1 For the following systems described by the given transfer functions, derive valid state-space realizations (define the state variables and derive the coefficient matrices A, B, C, and D).

a. $G(s) = \dfrac{Y(s)}{U(s)} = \dfrac{1}{s^2 + 2s + 6}$

b. $G(s) = \dfrac{Y(s)}{U(s)} = \dfrac{s + 3}{s^2 + 2s + 6}$

c. $G(s) = \dfrac{Y(s)}{U(s)} = \dfrac{10}{s^3 + 4s^2 + 8s + 6}$

d. $G(s) = \dfrac{Y(s)}{U(s)} = \dfrac{s^2 + 4s + 6}{s^4 + 10s^3 + 11s^2 + 44s + 66}$

NE1.2 Given the following differential equations (or systems of differential equations), derive valid state-space realizations (define the state variables and derive the coefficient matrices A, B, C, and D).

a. $\dot{y}(t) + 2y(t) = u(t)$

b. $\ddot{y}(t) + 3\dot{y}(t) + 10y(t) = u(t)$

c. $\dddot{y}(t) + 2\ddot{y}(t) + 3\dot{y}(t) + 5y(t) = u(t)$

d. $\ddot{y}_1(t) + 5y_1(t) - 10[y_2(t) - y_1(t)] = u_1(t)$

   $2\ddot{y}_2(t) + \dot{y}_2(t) + 10[y_2(t) - y_1(t)] = u_2(t)$

Analytical Exercises

AE1.1 Suppose that A is $n \times m$ and H is $p \times q$. Specify dimensions for the remaining matrices so that the following expression is valid.

$$\begin{bmatrix} A & B \\ C & D \end{bmatrix}\begin{bmatrix} E & F \\ G & H \end{bmatrix} = \begin{bmatrix} AE + BG & AF + BH \\ CE + DG & CF + DH \end{bmatrix}$$

AE1.2 Suppose that A and B are square matrices, not necessarily of the same dimension. Show that

$$\begin{vmatrix} A & 0 \\ 0 & B \end{vmatrix} = |A| \cdot |B|$$

AE1.3 Continuing AE1.2, show that

$$\begin{vmatrix} A & 0 \\ C & B \end{vmatrix} = |A| \cdot |B|$$

AE1.4 Continuing AE1.3, show that if A is nonsingular,

$$\begin{vmatrix} A & D \\ C & B \end{vmatrix} = |A| \cdot |B - CA^{-1}D|$$

AE1.5 Suppose that X is $n \times m$ and Y is $m \times n$. With I_k denoting the $k \times k$ identity matrix for any integer $k > 0$, show that

$$|I_n - XY| = |I_m - YX|$$

Explain the significance of this result when $m = 1$. Hint: Apply AE1.4 to

$$\begin{bmatrix} I_m & Y \\ X & I_n \end{bmatrix} \text{ and } \begin{bmatrix} I_n & X \\ Y & I_m \end{bmatrix}$$

AE1.6 Show that the determinant of a square upper triangular matrix (zeros everywhere below the main diagonal) equals the product if its diagonal entries.

AE1.7 Suppose that A and C are nonsingular $n \times n$ and $m \times m$ matrices, respectively. Verify that

$$[A + BCD]^{-1} = A^{-1} - A^{-1}B[C^{-1} + DA^{-1}B]^{-1}DA^{-1}$$

What does this formula reduce to when $m = 1$ and $C = 1$?

AE1.8 Suppose that X is $n \times m$ and Y is $m \times n$. With I_k denoting the $k \times k$ identity matrix for any integer $k > 0$, show that

$$(I_n - XY)^{-1}X = X(I_m - YX)^{-1}$$

when the indicated inverses exist.

AE1.9 Suppose that A and B are nonsingular matrices, not necessarily of the same dimension. Show that

$$\begin{bmatrix} A & 0 \\ 0 & B \end{bmatrix}^{-1} = \begin{bmatrix} A^{-1} & 0 \\ 0 & B^{-1} \end{bmatrix}$$

AE1.10 Continuing AE1.8, derive expressions for

$$\begin{bmatrix} A & 0 \\ C & B \end{bmatrix}^{-1} \text{ and } \begin{bmatrix} A & D \\ 0 & B \end{bmatrix}^{-1}$$

AE1.11 Suppose that A is nonsingular and show that

$$\begin{bmatrix} A & D \\ C & B \end{bmatrix}^{-1} = \begin{bmatrix} A^{-1} + E\Delta^{-1}F & -E\Delta^{-1} \\ -\Delta^{-1}F & \Delta^{-1} \end{bmatrix}$$

in which $\Delta = B - CA^{-1}D$, $E = A^{-1}D$, and $F = CA^{-1}$.

AE1.12 Compute the inverse of the $k \times k$ Jordan block matrix

$$J_k(\lambda) = \begin{bmatrix} \lambda & 1 & 0 & \cdots & 0 \\ 0 & \lambda & 1 & \cdots & 0 \\ 0 & 0 & \lambda & \ddots & 0 \\ \vdots & \vdots & \vdots & \ddots & 1 \\ 0 & 0 & 0 & \cdots & \lambda \end{bmatrix}$$

AE1.13 Suppose that $A : \mathbb{R}^n \rightarrow \mathbb{R}^m$ is a linear transformation and $\mathbb{S}$ is a subspace of $\mathbb{R}^m$. Verify that the set

$$A^{-1}\mathbb{S} = \{x \in \mathbb{R}^n | Ax \in \mathbb{S}\}$$

is a subspace of $\mathbb{R}^n$. This subspace is referred to as the *inverse image* of the subspace $\mathbb{S}$ under the linear transformation A.

AE1.14 Show that for conformably dimensioned matrices A and B, any induced matrix norm satisfies

$$\|AB\| \leq \|A\|\|B\|$$

AE1.15 Show that for A nonsingular, any induced matrix norm satisfies

$$\|A^{-1}\| \geq \frac{1}{\|A\|}$$

AE1.16 Show that for any square matrix A, any induced matrix norm satisfies

$$\|A\| \geq \rho(A)$$

where $\rho(A) \triangleq \max_{\lambda_i \in \sigma(A)} |\lambda_i|$ is the spectral radius of A.

Continuing MATLAB Exercises

CME1.1 Given the following open-loop single-input, single-output two–dimensional linear time-invariant state equations, namely,

$$\begin{bmatrix} \dot{x}_1(t) \\ \dot{x}_2(t) \end{bmatrix} = \begin{bmatrix} -1 & 0 \\ 0 & -2 \end{bmatrix} \begin{bmatrix} x_1(t) \\ x_2(t) \end{bmatrix} + \begin{bmatrix} 1 \\ \sqrt{2} \end{bmatrix} u(t)$$

$$y(t) = \begin{bmatrix} 1 & -\sqrt{2}/2 \end{bmatrix} \begin{bmatrix} x_1(t) \\ x_2(t) \end{bmatrix} + [0]u(t)$$

find the associated open-loop transfer function $H(s)$.

CME1.2 Given the following open-loop single-input, single-output three–dimensional linear time-invariant state equations, namely

$$\begin{bmatrix} \dot{x}_1(t) \\ \dot{x}_2(t) \\ \dot{x}_3(t) \end{bmatrix} = \begin{bmatrix} 0 & 1 & 0 \\ 0 & 0 & 1 \\ -52 & -30 & -4 \end{bmatrix} \begin{bmatrix} x_1(t) \\ x_2(t) \\ x_3(t) \end{bmatrix} + \begin{bmatrix} 0 \\ 0 \\ 1 \end{bmatrix} u(t)$$

$$y(t) = \begin{bmatrix} 20 & 1 & 0 \end{bmatrix} \begin{bmatrix} x_1(t) \\ x_2(t) \\ x_3(t) \end{bmatrix} + [0]u(t)$$

find the associated open-loop transfer function $H(s)$.

CME1.3 Given the following open-loop single-input, single-output fourth–order linear time-invariant state equations, namely,

$$
\begin{bmatrix} \dot{x}_1(t) \\ \dot{x}_2(t) \\ \dot{x}_3(t) \\ \dot{x}_4(t) \end{bmatrix} = \begin{bmatrix} 0 & 1 & 0 & 0 \\ 0 & 0 & 1 & 0 \\ 0 & 0 & 0 & 1 \\ -962 & -126 & -67 & -4 \end{bmatrix} \begin{bmatrix} x_1(t) \\ x_2(t) \\ x_3(t) \\ x_4(t) \end{bmatrix} + \begin{bmatrix} 0 \\ 0 \\ 0 \\ 1 \end{bmatrix} u(t)
$$

$$
y(t) = [300 \quad 0 \quad 0 \quad 0] \begin{bmatrix} x_1(t) \\ x_2(t) \\ x_3(t) \\ x_4(t) \end{bmatrix} + [0]u(t)
$$

find the associated open-loop transfer function $H(s)$.

CME1.4 Given the following open-loop single-input, single-output four–dimensional linear time-invariant state equations, namely,

$$
\begin{bmatrix} \dot{x}_1(t) \\ \dot{x}_2(t) \\ \dot{x}_3(t) \\ \dot{x}_4(t) \end{bmatrix} = \begin{bmatrix} 0 & 1 & 0 & 0 \\ 0 & 0 & 1 & 0 \\ 0 & 0 & 0 & 1 \\ -680 & -176 & -86 & -6 \end{bmatrix} \begin{bmatrix} x_1(t) \\ x_2(t) \\ x_3(t) \\ x_4(t) \end{bmatrix} + \begin{bmatrix} 0 \\ 0 \\ 0 \\ 1 \end{bmatrix} u(t)
$$

$$
y(t) = [100 \quad 20 \quad 10 \quad 0] \begin{bmatrix} x_1(t) \\ x_2(t) \\ x_3(t) \\ x_4(t) \end{bmatrix} + [0]u(t)
$$

find the associated open-loop transfer function $H(s)$.

Continuing Exercises

CE1.1a A mechanical system is represented by the three degree-of-freedom linear time-invariant system shown in Figure 1.14. There are three input forces $u_i(t)$ and three output displacements $y_i(t)$, $i = 1, 2, 3$. The constant parameters are the masses m_i, $i = 1, 2, 3$, the spring coefficients k_j, and the damping coefficients c_j, $j = 1, 2, 3, 4$. Derive the mathematical model for this system, i.e., draw the free-body diagrams and write the correct number of independent ordinary differential equations. All motion is constrained to be horizontal. Outputs $y_i(t)$ are each measured from the neutral spring equilibrium location of each mass m_i.

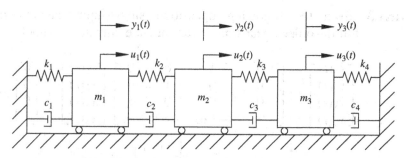

FIGURE 1.14 Diagram for Continuing Exercise 1.

Also express the results in matrix-vector form $M\ddot{y}(t) + C\dot{y}(t) + Ky(t) = u(t)$.

CE1.1b Derive a valid state-space realization for the CE1.1a system. That is, specify the state variables and derive the coefficient matrices A, B, C, and D. Write out your results in matrix-vector form. Give the system order and matrix/vector dimensions of your result. Consider three distinct cases:

    i. Multiple-input, multiple-output: three inputs, three displacement outputs.

    ii. Multiple-input, multiple-output: two inputs [$u_1(t)$ and $u_3(t)$ only], all three displacement outputs.

    iii. Single-input, single-output: input $u_2(t)$ and output $y_3(t)$.

CE1.2a The nonlinear, inherently unstable inverted pendulum is shown in Figure 1.15. The goal is to maintain the pendulum angle $\theta(t) = 0$ by using a feedback controller with a sensor (encoder or potentiometer) for $\theta(t)$ and an actuator to produce an input force $f(t)$. The cart mass is m_1, the pendulum point mass is m_2, and we assume that the pendulum rod is massless. There are two possible outputs, the pendulum angle $\theta(t)$ and the cart displacement $w(t)$. The classical inverted pendulum has only one input, the force $f(t)$. We will consider a second case, using a motor to provide a second input $\tau(t)$ (not shown) at the rotary joint of Figure 1.15. For both cases (they will be very similar), derive the nonlinear model for this system, i.e., draw the free-body diagrams and write the correct number of independent ordinary differential equations. Alternately, you may use the Lagrangian dynamics approach that does not require free-body diagrams. Apply the steps outlined in Section 1.4 to derive a linearized model about the unstable equilibrium condition corresponding to zero angular displacement.

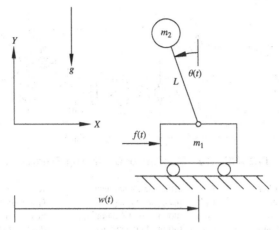

FIGURE 1.15 Diagram for Continuing Exercise 2.

CE1.2b Derive a valid state-space description for the system of Figure 1.15. That is, specify the state variables and derive the coefficient matrices $A, B, C,$ and D. Write out your results in matrix-vector form. Give the system order and matrix-vector dimensions of your result. Consider three distinct cases:

   i. Single-input, single-output: input $f(t)$ and output $\theta(t)$.

  ii. Single-input, multiple-output: one input $f(t)$ and two outputs $w(t)$ and $\theta(t)$.

 iii. Multiple-input, multiple-output: two inputs $f(t)$ and $\tau(t)$ (add a motor to the inverted pendulum rotary joint, traveling with the cart) and two outputs $w(t)$ and $\theta(t)$.

CE1.3a Figure 1.16 shows a single robot joint/link driven through a gear ratio n by an armature-controlled dc servomotor. The input is the dc armature voltage $v_A(t)$ and the output is the load-shaft angle $\theta_L(t)$. Derive the mathematical model for this system; i.e., develop the circuit differential equation, the electromechanical coupling equations, and the rotational mechanical differential equation. Eliminate intermediate variables and simplify; it will be convenient to use a transfer-function approach. Assume the mass-moment of inertia of all outboard links plus any load $J_L(t)$ is a constant (a reasonable assumption when the gear ratio $n = \omega_M/\omega_L$ is much greater than 1, as it is in the case of industrial robots). The parameters in Figure 1.16 are summarized below.

CE1.3b Derive a valid state-space description for the system of Figure 1.16. That is, specify the state variables and derive the

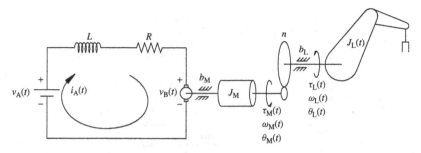

FIGURE 1.16   Diagram for Continuing Exercise 3.

$v_A(t)$	armature voltage	L	armature inductance	R	armature resistance
$i_A(t)$	armature current	$v_B(t)$	back emf voltage	k_B	back emf constant
J_M	motor inertia	b_M	motor viscous damping	$\tau_M(t)$	motor torque
k_T	torque constant	$\omega_M(t)$	motor shaft velocity	$\theta_M(t)$	motor shaft angle
n	gear ratio	$J_L(t)$	load inertia	b_L	load viscous damping
$\tau_L(t)$	load shaft torque	$\omega_L(t)$	load shaft velocity	$\theta_L(t)$	load shaft angle

coefficient matrices A, B, C, and D. Write out your results in matrix-vector form. Give the system order and matrix-vector dimensions of your result. Consider two distinct cases:

   i. Single-input, single-output: armature voltage $v_A(t)$ as the input and robot load shaft angle $\theta_L(t)$ as the output.

   ii. Single-input, single-output: armature voltage $v_A(t)$ as the input and robot load shaft angular velocity $\omega_L(t)$ as the output.

CE1.4 The nonlinear ball and beam apparatus was introduced in Section 1.4, Example 1.7. This system will form the basis for Continuing Exercise 4. Figure 1.8 shows the ball and beam system geometry, where the goal is to control the position of the ball on the slotted rotating beam by applying a torque $\tau(t)$ to the beam. CE1.4a and CE1.4b are already completed for you; i.e., the nonlinear equations of motion have been presented, and a valid state-space realization has been derived and linearized about the given nominal trajectory. Thus the assignment here is to rederive these steps for Continuing Exercise 4. As in Example 1.7, use the single-input, single-output model with input torque $\tau(t)$ and output ball position $p(t)$.

For all ensuing Continuing Exercise 4 assignments, use a special case of the time-varying linear state equation (1.6) to obtain a linear time-invariant state-space realization of the nonlinear model; use zero velocity $v_0 = 0$ and constant nominal ball position $\tilde{p}(t) = p_0$. Derive the linear time-invariant coefficient matrices A, B, C, and D for this special case.

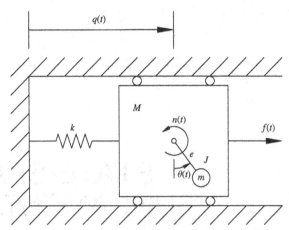

FIGURE 1.17 Diagram for Continuing Exercise 5 (top view).

CE1.5a A nonlinear proof-mass actuator system is shown in Figure 1.17. This system has been proposed as a nonlinear controls benchmark problem (Bupp et al., 1998). However, in this book, the system will be linearized about a nominal trajectory, and the linearization then will be used in all ensuing chapters as Continuing Exercise 5.

This is a vibration-suppression system wherein the control goal is to reject an unknown, unwanted disturbance force $f(t)$ by using the control torque $n(t)$ to drive the unbalanced rotating pendulum (proof mass) to counter these disturbances. The block of mass M is connected to the wall via a spring with spring constant k and is constrained to translate as shown; $q(t)$ is the block displacement. The rotating pendulum has a point mass m at the tip, and the pendulum has mass moment of inertia J. The pendulum length is e and the pendulum angle $\theta(t)$ is measured as shown. Assume that the system is operating in the horizontal plane, so gravity need not be considered. Derive the nonlinear model for this system.

CE1.5b For nominal equilibria corresponding to zero control torque, linearize the nonlinear model from CE1.5a and derive a valid state-space description. That is, follow the procedure of Section 1.4 and derive the linearized coefficient matrices A, B, C, and D. Write out your results in matrix-vector form. Give the system order and matrix-vector dimensions of your result. Consider only the single-input, single-output case with input torque $n(t)$ and output displacement $q(t)$. In ensuing problems, the control objective will be to regulate nonequilibrium initial conditions.

2

STATE-SPACE FUNDAMENTALS

Chapter 1 presented the state-space description for linear time-invariant systems. This chapter establishes several fundamental results that follow from this representation, beginning with a derivation of the state equation solution. In the course of this analysis we encounter the *matrix exponential*, so named because of many similarities with the scalar exponential function. In terms of the state-equation solution, we revisit several familiar topics from linear systems analysis, including decomposition of the complete response into zero-input and zero-state response components, characterizing the system impulse response that permits the zero-state response to be cast as the convolution of the impulse response with the input signal, and the utility of the Laplace transform in computing the state-equation solution and defining the system transfer function.

The chapter continues with a more formal treatment of the state-space realization issue and an introduction to the important topic of state coordinate transformations. As we will see, a linear transformation of the state vector yields a different state equation that also realizes the system's input-output behavior represented by either the associated impulse response or transfer function. This has the interesting consequence that state-space realizations are not unique; beginning with one state-space realization, other realizations of the same system may be derived via a state coordinate transformation. We will see that many topics in the remainder of this book are facilitated by the flexibility afforded by this nonuniqueness. For instance, in this chapter we introduce the so-called diagonal canonical form that specifies a set of decoupled, scalar, first-order ordinary differential equations that may be solved independently.

This chapter also illustrates the use of MATLAB in supporting the computations encountered earlier. As in all chapters, these demonstrations will revisit the MATLAB Continuing Example along with Continuing Examples 1 and 2.

2.1 STATE EQUATION SOLUTION

From Chapter 1, our basic mathematical model for a linear time-invariant system consists of the state differential equation and the algebraic output equation:

$$\dot{x}(t) = Ax(t) + Bu(t) \quad x(t_0) = x_0$$
$$y(t) = Cx(t) + Du(t) \tag{2.1}$$

where we assume that the $n \times n$ system dynamics matrix A, the $n \times m$ input matrix B, the $p \times n$ output matrix C, and the $p \times m$ direct transmission matrix D are known constant matrices. The first equation compactly represents a set of n coupled first-order differential equations that must be solved for the state vector $x(t)$ given the initial state $x(t_0) = x_0$ and input vector $u(t)$. The second equation characterizes a static or instantaneous dependence of the output on the state and input. As we shall see, the real work lies in deriving a solution expression for the state vector. With that in hand, a direct substitution into the second equation yields an expression for the output.

Prior to deriving a closed-form solution of Equation (2.1) for the n–dimensional case as outlined above, we first review the solution of scalar first-order differential equations.

Solution of Scalar First-Order Differential Equations

Consider the one–dimensional system represented by the scalar differential equation

$$\dot{x}(t) = ax(t) + bu(t) \quad x(t_0) = x_0 \tag{2.2}$$

in which a and b are scalar constants, and $u(t)$ is a given scalar input signal. A traditional approach for deriving a solution formula for the scalar state $x(t)$ is to multiply both sides of the differential equation by the *integrating factor* $e^{-a(t-t_0)}$ to yield

$$\frac{d}{dt}(e^{-a(t-t_0)}x(t)) = e^{-a(t-t_0)}\dot{x}(t) - e^{-a(t-t_0)}ax(t)$$

$$= e^{-a(t-t_0)}bu(t)$$

We next integrate from t_0 to t and invoke the fundamental theorem of calculus to obtain

$$e^{-a(t-t_0)}x(t) - e^{-a(t-t_0)}x(t_0) = \int_{t_0}^{t} \frac{d}{dt}(e^{-a(\tau-t_0)}x(\tau))\,d\tau$$

$$= \int_{t_0}^{t} e^{-a(\tau-t_0)}bu(\tau)\,d\tau.$$

After multiplying through by $e^{a(t-t_0)}$ and some manipulation, we get

$$x(t) = e^{a(t-t_0)}x_0 + \int_{t_0}^{t} e^{a(t-\tau)}bu(\tau)\,d\tau \qquad (2.3)$$

which expresses the state response $x(t)$ as a sum of terms, the first owing to the given initial state $x(t_0) = x_0$ and the second owing to the specified input signal $u(t)$. Notice that the first component characterizes the state response when the input signal is identically zero. We therefore refer to the first term as the *zero-input response* component. Similarly, the second component characterizes the state response for zero initial state. We therefore refer to the second term as the *zero-state response* component.

The Laplace transform furnishes an alternate solution strategy. For this, we assume without loss of generality that $t_0 = 0$ and transform the differential equation, using linearity and time-differentiation properties of the Laplace transform, into

$$sX(s) - x_0 = aX(s) + bU(s)$$

in which $X(s)$ and $U(s)$ are the Laplace transforms of $x(t)$ and $u(t)$, respectively. Straightforward algebra yields

$$X(s) = \frac{1}{s-a}x_0 + \frac{b}{s-a}U(s)$$

From the convolution property of the Laplace transform, we obtain

$$x(t) = e^{at}x_0 + e^{at} * bu(t)$$

$$= e^{at}x_0 + \int_0^t e^{a(t-\tau)}bu(\tau)\,d\tau$$

which agrees with the solution (2.3) derived earlier for $t_0 = 0$.

If our first-order system had an associated scalar output signal $y(t)$ defined by the algebraic relationship

$$y(t) = cx(t) + d\,u(t) \tag{2.4}$$

then by simply substituting the state response we obtain

$$y(t) = ce^{at}x_0 + \int_0^t ce^{a(t-\tau)}bu(\tau)\,d\tau + du(t)$$

which also admits a decomposition into zero-input and zero-state response components. In the Laplace domain, we also have

$$Y(s) = \frac{c}{s-a}x_0 + \frac{cb}{s-a}U(s)$$

We recall that the impulse response of a linear time-invariant system is the system's response to an impulsive input $u(t) = \delta(t)$ when the system is initially at rest, which in this setting corresponds to zero initial state $x_0 = 0$. By interpreting the initial time as $t_0 = 0^-$, just prior to when the impulse occurs, the zero-state response component of $y(t)$ yields the system's impulse response, that is,

$$h(t) = \int_{0^-}^t ce^{a(t-\tau)}b\delta(\tau)\,d\tau + d\delta(t)$$

$$= ce^{at}b + d\delta(t) \tag{2.5}$$

where we have used the *sifting property* of the impulse to evaluate the integral. Now, for any input signal $u(t)$, the zero-input response component of $y(t)$ can be expressed as

$$\int_{0^-}^t ce^{a(t-\tau)}bu(\tau)\,d\tau + du(t) = \int_{0^-}^t [ce^{a(t-\tau)}b + d\delta(t-\tau)]u(\tau)\,d\tau$$

$$= \int_{0^-}^t h(t-\tau)u(\tau)\,d\tau$$

$$= h(t) * u(t)$$

which should look familiar to the reader. Alternatively, in the Laplace domain, the system's transfer function $H(s)$ is, by definition,

$$H(s) = \left.\frac{Y(s)}{U(s)}\right|_{\text{zero initial state}} = \left(\frac{cb}{s-a} + d\right)$$

and so the impulse response $h(t)$ and transfer function $H(s)$ form a Laplace transform pair, as we should expect.

Our approach to deriving state-equation solution formulas for the n-dimensional case and discussing various systems-related implications in both the time and Laplace domains is patterned after the preceding development, but greater care is necessary to tackle the underlying matrix-vector computations correctly. Before proceeding, the reader is encouraged to ponder, however briefly, the matrix-vector extensions of the preceding computations.

State Equation Solution

In this subsection we derive a closed-form solution to the n-dimensional linear time invariant state equation (2.1) given a specified initial state $x(t_0) = x_0$ and input vector $u(t)$.

Homogeneous Case   We begin with a related homogeneous matrix differential equation

$$\dot{X}(t) = AX(t) \qquad X(t_0) = I \tag{2.6}$$

where I is the $n \times n$ identity matrix. We assume an infinite power series form for the solution

$$X(t) = \sum_{k=0}^{\infty} X_k(t - t_0)^k \tag{2.7}$$

Each term in the sum involves an $n \times n$ matrix X_k to be determined and depends only on the elapsed time $t - t_0$, reflecting the time-invariance of the state equation. The initial condition for Equation (2.6) yields $X(t_0) = X_0 = I$. Substituting Equation (2.7) into Equation (2.6), formally differentiating term by term with respect to time, and shifting the summation index gives

$$\sum_{k=0}^{\infty} (k+1)X_{k+1}(t-t_0)^k = A \left(\sum_{k=0}^{\infty} X_k(t-t_0)^k \right)$$

$$= \sum_{k=0}^{\infty} A \; X_k(t-t_0)^k$$

By equating like powers of $t - t_0$, we obtain the recursive relationship

$$X_{k+1} = \frac{1}{k+1} A X_k \qquad k \geq 0$$

which, when initialized with $X_0 = I$, leads to

$$X_k = \frac{1}{k!} A^k \qquad k \geq 0$$

Substituting this result into the power series (2.7) yields

$$X(t) = \sum_{k=0}^{\infty} \frac{1}{k!} A^k (t-t_0)^k$$

We note here that the infinite power series (2.7) has the requisite convergence properties so that the infinite power series resulting from term-by-term differentiation converges to $\dot{X}(t)$, and Equation (2.6) is satisfied.

Recall that the scalar exponential function is defined by the following infinite power series

$$e^{at} = 1 + at + \tfrac{1}{2}a^2t^2 + \tfrac{1}{6}a^3t^3 + \cdots$$

$$= \sum_{k=0}^{\infty} \frac{1}{k!} a^k t^k$$

Motivated by this, we define the so-called matrix exponential via

$$e^{At} = I + At + \tfrac{1}{2}A^2t^2 + \tfrac{1}{6}A^3t^3 + \cdots$$

$$= \sum_{k=0}^{\infty} \frac{1}{k!} A^k t^k \tag{2.8}$$

from which the solution to the homogeneous matrix differential equation (2.6) can be expressed compactly as

$$X(t) = e^{A(t-t_0)}$$

It is important to point out that e^{At} is merely notation used to represent the power series in Equation (2.8). Beyond the scalar case, the matrix exponential never equals the matrix of scalar exponentials corresponding to the individual elements in the matrix A. That is,

$$e^{At} \neq [e^{a_{ij}t}]$$

Properties that are satisfied by the matrix exponential are collected in the following proposition.

Proposition 2.1   *For any real $n \times n$ matrix A, the matrix exponential e^{At} has the following properties:*

1. *e^{At} is the unique matrix satisfying*

$$\frac{d}{dt}e^{At} = Ae^{At} \qquad e^{At}\big|_{t=0} = I_n$$

2. *For any t_1 and t_2, $e^{A(t_1+t_2)} = e^{At_1}e^{At_2}$. As a direct consequence, for any t*

$$I = e^{A(0)} = e^{A(t-t)} = e^{At}e^{-At}$$

Thus e^{At} is invertible (nonsingular) for all t with inverse

$$\left[e^{At}\right]^{-1} = e^{-At}$$

3. *A and e^{At} commute with respect to matrix multiplication, that is, $Ae^{At} = e^{At}A$ for all t.*
4. *$[e^{At}]^T = e^{A^T t}$ for all t.*
5. *For any real $n \times n$ matrix B, $e^{(A+B)t} = e^{At}e^{Bt}$ for all t if and only if $AB = BA$, that is, A and B commute with respect to matrix multiplication.*   □

The first property asserts the uniqueness of $X(t) = e^{A(t-t_0)}$ as a solution to Equation (2.6). This property is useful in situations where we must verify whether a given time–dependent matrix $X(t)$ is the matrix exponential for an associated matrix A. To resolve this issue, it is *not* necessary compute e^{At} from scratch via some means. Rather, it suffices to check whether $\dot{X}(t) = AX(t)$ and $X(0) = I$. If a candidate for the matrix exponential is not provided, then it must be computed directly. The defining power series is, except in special cases, not especially useful in this regard. However, there are special cases in which closed-form solutions can be deduced, as shown in the following two examples.

Example 2.1   Consider the 4×4 matrix with ones above the main diagonal and zeros elsewhere:

$$A = \begin{bmatrix} 0 & 1 & 0 & 0 \\ 0 & 0 & 1 & 0 \\ 0 & 0 & 0 & 1 \\ 0 & 0 & 0 & 0 \end{bmatrix}$$

As called for by the power series (2.8), we compute powers of A:

$$A^2 = \begin{bmatrix} 0 & 0 & 1 & 0 \\ 0 & 0 & 0 & 1 \\ 0 & 0 & 0 & 0 \\ 0 & 0 & 0 & 0 \end{bmatrix} \qquad A^3 = \begin{bmatrix} 0 & 0 & 0 & 1 \\ 0 & 0 & 0 & 0 \\ 0 & 0 & 0 & 0 \\ 0 & 0 & 0 & 0 \end{bmatrix}$$

$$A^4 = \begin{bmatrix} 0 & 0 & 0 & 0 \\ 0 & 0 & 0 & 0 \\ 0 & 0 & 0 & 0 \\ 0 & 0 & 0 & 0 \end{bmatrix}$$

from which it follows that $A^k = 0$, for $k \geq 4$, and consequently, the power series (2.8) contains only a finite number of nonzero terms:

$$e^{At} = I + At + \tfrac{1}{2}A^2t^2 + \tfrac{1}{6}A^3t^3$$

$$= \begin{bmatrix} 1 & 0 & 0 & 0 \\ 0 & 1 & 0 & 0 \\ 0 & 0 & 1 & 0 \\ 0 & 0 & 0 & 1 \end{bmatrix} + \begin{bmatrix} 0 & t & 0 & 0 \\ 0 & 0 & t & 0 \\ 0 & 0 & 0 & t \\ 0 & 0 & 0 & 0 \end{bmatrix} + \begin{bmatrix} 0 & 0 & \tfrac{1}{2}t^2 & 0 \\ 0 & 0 & 0 & \tfrac{1}{2}t^2 \\ 0 & 0 & 0 & 0 \\ 0 & 0 & 0 & 0 \end{bmatrix}$$

$$+ \begin{bmatrix} 0 & 0 & 0 & \tfrac{1}{6}t^3 \\ 0 & 0 & 0 & 0 \\ 0 & 0 & 0 & 0 \\ 0 & 0 & 0 & 0 \end{bmatrix} = \begin{bmatrix} 1 & t & \tfrac{1}{2}t^2 & \tfrac{1}{6}t^3 \\ 0 & 1 & t & \tfrac{1}{2}t^2 \\ 0 & 0 & 1 & t \\ 0 & 0 & 0 & 1 \end{bmatrix}$$

Inspired by this result, we claim the following outcome for the $n-$dimensional case:

$$A = \begin{bmatrix} 0 & 1 & 0 & \cdots & 0 \\ 0 & 0 & 1 & \cdots & 0 \\ \vdots & \vdots & \vdots & \ddots & \vdots \\ 0 & 0 & 0 & \cdots & 1 \\ 0 & 0 & 0 & \cdots & 0 \end{bmatrix} \Rightarrow e^{At} = \begin{bmatrix} 1 & t & \frac{1}{2}t^2 & \cdots & \frac{1}{(n-1)!}t^{n-1} \\ 0 & 1 & t & \cdots & \frac{1}{(n-2)!}t^{n-2} \\ \vdots & \vdots & \vdots & \ddots & \vdots \\ 0 & 0 & 0 & \cdots & t \\ 0 & 0 & 0 & \cdots & 1 \end{bmatrix}$$

the veracity of which can be verified by checking that the first property of Proposition 2.1 is satisfied, an exercise left for the reader. □

Example 2.2 Consider the diagonal $n \times n$ matrix:

$$A = \begin{bmatrix} \lambda_1 & 0 & \cdots & 0 & 0 \\ 0 & \lambda_2 & \cdots & 0 & 0 \\ \vdots & \vdots & \ddots & \vdots & \vdots \\ 0 & 0 & \cdots & \lambda_{n-1} & 0 \\ 0 & 0 & \cdots & 0 & \lambda_n \end{bmatrix}$$

Here, the power series (2.8) will contain an infinite number of terms when at least one $\lambda_1 \neq 0$, but since diagonal matrices satisfy

$$A^k = \begin{bmatrix} \lambda_1^k & 0 & \cdots & 0 & 0 \\ 0 & \lambda_2^k & \cdots & 0 & 0 \\ \vdots & \vdots & \ddots & \vdots & \vdots \\ 0 & 0 & \cdots & \lambda_{n-1}^k & 0 \\ 0 & 0 & \cdots & 0 & \lambda_n^k \end{bmatrix}$$

each term in the series is a diagonal matrix, and

$$e^{At} = \sum_{k=0}^{\infty} \frac{1}{k!} \begin{bmatrix} \lambda_1^k & 0 & \cdots & 0 & 0 \\ 0 & \lambda_2^k & \cdots & 0 & 0 \\ \vdots & \vdots & \ddots & \vdots & \vdots \\ 0 & 0 & \cdots & \lambda_{n-1}^k & 0 \\ 0 & 0 & \cdots & 0 & \lambda_n^k \end{bmatrix} t^k$$

$$
= \begin{bmatrix}
\sum_{k=0}^{\infty} \frac{1}{k!}\lambda_1^k t^k & 0 & \cdots & 0 & 0 \\
0 & \sum_{k=0}^{\infty} \frac{1}{k!}\lambda_2^k t^k & \cdots & 0 & 0 \\
\vdots & \vdots & \ddots & \vdots & \vdots \\
0 & 0 & \cdots & \sum_{k=0}^{\infty} \frac{1}{k!}\lambda_{n-1}^k t^k & 0 \\
0 & 0 & \cdots & 0 & \sum_{k=0}^{\infty} \frac{1}{k!}\lambda_n^k t^k
\end{bmatrix}
$$

On observing that each diagonal entry specifies a power series converging to a scalar exponential function, we have

$$
\begin{bmatrix}
e^{\lambda_1 t} & 0 & \cdots & 0 & 0 \\
0 & e^{\lambda_2 t} & \cdots & 0 & 0 \\
\vdots & \vdots & \ddots & \vdots & \vdots \\
0 & 0 & \cdots & e^{\lambda_{n-1} t} & 0 \\
0 & 0 & \cdots & 0 & e^{\lambda_n t}
\end{bmatrix}
\qquad \square
$$

Another useful property of the matrix exponential is that the infinite power series definition (2.8) can be reduced to a finite power series

$$
e^{At} = \sum_{k=0}^{n-1} \alpha_k(t) A^k \tag{2.9}
$$

involving scalar analytic functions $\alpha_0(t), \alpha_1(t), \ldots, \alpha_{n-1}(t)$. As shown in Rugh (1996), the existence of the requisite functions can be verified by equating

$$
\frac{d}{dt} e^{At} = \frac{d}{dt} \left[\sum_{k=0}^{n-1} \alpha_k(t) A^k \right] = \sum_{k=0}^{n-1} \dot{\alpha}_k(t) A^k
$$

and

$$
A \left[\sum_{k=0}^{n-1} \alpha_k(t) A^k \right] = \sum_{k=0}^{n-1} \alpha_k(t) A^{k+1}
$$

We invoke the Cayley-Hamilton theorem (see Appendix B, Section 8) which, in terms of the characteristic polynomial $|\lambda I - A| = \lambda^n$

$+ a_{n-1}\lambda^{n-1} + \cdots + a_1\lambda + a_0$, allows us to write

$$A^n = -a_0 I - a_1 A - \cdots - a_{n-1}A^{n-1}$$

Substituting this identity into the preceding summation yields

$$\sum_{k=0}^{n-1}\alpha_k(t)A^{k+1} = \sum_{k=0}^{n-2}\alpha_k(t)A^{k+1} + \alpha_{n-1}(t)A^n$$

$$= \sum_{k=0}^{n-2}\alpha_k(t)A^{k+1} - \sum_{k=0}^{n-1}a_k\alpha_{n-1}(t)A^k$$

$$= -a_0\alpha_{n-1}(t)I + \sum_{k=1}^{n-1}[\alpha_{k-1}(t) - a_k\alpha_{n-1}(t)]A^k$$

By equating the coefficients of each power of A in the finite series representation for $(d/dt)e^{At}$ and the preceding expression for Ae^{At}, we obtain

$$\dot{\alpha}_0(t) = -a_0\alpha_{n-1}(t)$$

$$\dot{\alpha}_1(t) = \alpha_0(t) - a_1\alpha_{n-1}(t)$$

$$\dot{\alpha}_2(t) = \alpha_1(t) - a_2\alpha_{n-1}(t)$$

$$\vdots$$

$$\dot{\alpha}_{n-1}(t) = \alpha_{n-2}(t) - a_{n-1}\alpha_{n-1}(t)$$

This coupled set of first-order ordinary differential equations can be written in matrix-vector form to yield the homogenous linear state equation

$$\begin{bmatrix} \dot{\alpha}_0(t) \\ \dot{\alpha}_1(t) \\ \dot{\alpha}_2(t) \\ \vdots \\ \dot{\alpha}_{n-1}(t) \end{bmatrix} = \begin{bmatrix} 0 & 0 & \cdots & 0 & -a_0 \\ 1 & 0 & \cdots & 0 & -a_1 \\ 0 & 1 & \cdots & 0 & -a_2 \\ \vdots & \vdots & \ddots & \vdots & \vdots \\ 0 & 0 & \cdots & 1 & -a_{n-1} \end{bmatrix} \begin{bmatrix} \alpha_0(t) \\ \alpha_1(t) \\ \alpha_2(t) \\ \vdots \\ \alpha_{n-1}(t) \end{bmatrix}$$

Using the matrix exponential property $e^{A \cdot 0} = I$, we are led to the initial values $\alpha_0(0) = 1$ and $\alpha_1(0) = \alpha_2(0) = \cdots = \alpha_{n-1}(0) = 0$ which form the initial state vector

$$
\begin{bmatrix}
\alpha_0(0) \\
\alpha_1(0) \\
\alpha_2(0) \\
\vdots \\
\alpha_{n-1}(0)
\end{bmatrix}
=
\begin{bmatrix}
1 \\
0 \\
0 \\
\vdots \\
0
\end{bmatrix}
$$

We have thus characterized coefficient functions that establish the finite power series representation of the matrix exponential in Equation (2.9) as the solution to this homogeneous linear state equation with the specified initial state.

Several technical arguments in the coming chapters are greatly facilitated merely by the existence of the finite power series representation in Equation (2.9) without requiring explicit knowledge of the functions $\alpha_0(t), \alpha_1(t), \ldots, \alpha_{n-1}(t)$. In order to use Equation (2.9) for computational purposes, the preceding discussion is of limited value because it indirectly characterizes these coefficient functions as the solution to a homogeneous linear state equation that, in turn, involves another matrix exponential. Fortunately, a more explicit characterization is available which we now discuss for the case in which the matrix A has distinct eigenvalues $\lambda_1, \lambda_2, \ldots, \lambda_n$. A scalar version of the preceding argument allows us to conclude that the coefficient functions $\alpha_0(t), \alpha_1(t), \ldots, \alpha_{n-1}(t)$ also provide a finite series expansion for the scalar exponential functions $e^{\lambda_i t}$, that is,

$$
e^{\lambda_i t} = \sum_{k=0}^{n-1} \alpha_k(t)\lambda_i^k, \qquad i = 1, \ldots, n
$$

which yields the following system of equations

$$
\begin{bmatrix}
1 & \lambda_1 & \lambda_1^2 & \cdots & \lambda_1^{n-1} \\
1 & \lambda_2 & \lambda_2^2 & \cdots & \lambda_2^{n-1} \\
\vdots & \vdots & \vdots & \ddots & \vdots \\
1 & \lambda_n & \lambda_n^2 & \cdots & \lambda_n^{n-1}
\end{bmatrix}
\begin{bmatrix}
\alpha_0(t) \\
\alpha_1(t) \\
\alpha_2(t) \\
\vdots \\
\alpha_{n-1}(t)
\end{bmatrix}
=
\begin{bmatrix}
e^{\lambda_1 t} \\
e^{\lambda_2 t} \\
\vdots \\
e^{\lambda_n t}
\end{bmatrix}
$$

The $n \times n$ coefficient matrix is called a *Vandermonde matrix* that is nonsingular when and only when the eigenvalues $\lambda_1, \lambda_2, \ldots, \lambda_n$ are distinct. In this case, this system of equations can be solved to uniquely determine the coefficient functions $\alpha_0(t), \alpha_1(t), \ldots, \alpha_{n-1}(t)$.

Example 2.3 Consider the upper-triangular 3×3 matrix

$$A = \begin{bmatrix} 0 & -2 & 1 \\ 0 & -1 & -1 \\ 0 & 0 & -2 \end{bmatrix}$$

with distinct eigenvalues extracted from the main diagonal: $\lambda_1 = 0$, $\lambda_2 = -1$, and $\lambda_3 = -2$. The associated Vandermonde matrix is

$$\begin{bmatrix} 1 & \lambda_1 & \lambda_1^2 \\ 1 & \lambda_2 & \lambda_2^2 \\ 1 & \lambda_3 & \lambda_3^2 \end{bmatrix} = \begin{bmatrix} 1 & 0 & 0 \\ 1 & -1 & 1 \\ 1 & -2 & 4 \end{bmatrix}$$

which has a determinant of -2 is therefore nonsingular. This yields the coefficient functions

$$\begin{bmatrix} \alpha_0(t) \\ \alpha_1(t) \\ \alpha_2(t) \end{bmatrix} = \begin{bmatrix} 1 & 0 & 0 \\ 1 & -1 & 1 \\ 1 & -2 & 4 \end{bmatrix}^{-1} \begin{bmatrix} e^{\lambda_1 t} \\ e^{\lambda_1 t} \\ e^{\lambda_1 t} \end{bmatrix} = \begin{bmatrix} 1 & 0 & 0 \\ \frac{3}{2} & -2 & \frac{1}{2} \\ \frac{1}{2} & -1 & \frac{1}{2} \end{bmatrix} \begin{bmatrix} 1 \\ e^{-t} \\ e^{-2t} \end{bmatrix}$$

$$= \begin{bmatrix} 1 \\ \frac{3}{2} - 2e^{-t} + \frac{1}{2}e^{-2t} \\ \frac{1}{2} - e^{-t} + \frac{1}{2}e^{-2t} \end{bmatrix}$$

The matrix exponential is then

$$e^{At} = \alpha_0(t)I + \alpha_1(t)A + \alpha_2(t)A^2$$

$$= (1) \begin{bmatrix} 1 & 0 & 0 \\ 0 & 1 & 0 \\ 0 & 0 & 1 \end{bmatrix} + \left(\frac{3}{2} - 2e^{-t} + \frac{1}{2}e^{-2t} \right) \begin{bmatrix} 0 & -2 & 1 \\ 0 & -1 & -1 \\ 0 & 0 & -2 \end{bmatrix}$$

$$+ \left(\frac{1}{2} - e^{-t} + \frac{1}{2}e^{-2t} \right) \begin{bmatrix} 0 & 2 & 0 \\ 0 & 1 & 3 \\ 0 & 0 & 4 \end{bmatrix}$$

$$= \begin{bmatrix} 1 & -2 + 2e^{-t} & \frac{3}{2} - 2e^{-t} + \frac{1}{2}e^{-2t} \\ 0 & e^{-t} & -e^{-t} + e^{-2t} \\ 0 & 0 & e^{-2t} \end{bmatrix} \qquad \square$$

The interested reader is referred to Reid (1983) for modifications to this procedure to facilitate the treatment of complex conjugate eigenvalues and to cope with the case of repeated eigenvalues. Henceforth, we will rely mainly on the Laplace transform for computational purposes, and this will pursued later in this chapter.

The linear time-invariant state equation (2.1) in the unforced case [$u(t) \equiv 0$] reduces to the homogeneous state equation

$$\dot{x}(t) = Ax(t) \quad x(t_0) = x_0 \tag{2.10}$$

For the specified initial state, the unique solution is given by

$$x(t) = e^{A(t-t_0)}x_0 \tag{2.11}$$

which is easily verified using the first two properties of the matrix exponential asserted in Proposition 2.1.

A useful interpretation of this expression is that the matrix exponential $e^{A(t-t_0)}$ characterizes the transition from the initial state x_0 to the state $x(t)$ at any time $t \geq t_0$. As such, the matrix exponential $e^{A(t-t_0)}$ is often referred to as the *state-transition matrix* for the state equation and is denoted by $\Phi(t, t_0)$. The component $\phi_{ij}(t, t_0)$ of the state-transition matrix is the time response of the ith state variable resulting from an initial condition of 1 on the jth state variable with zero initial conditions on all other state variables (think of the definitions of linear superposition and matrix multiplication).

General Case   Returning to the general forced case, we derive a solution formula for the state equation (2.1). For this, we define

$$z(t) = e^{-A(t-t_0)}x(t)$$

from which $z(t_0) = e^{-A(t_0-t_0)}x(t_0) = x_0$ and

$$\dot{z}(t) = \frac{d}{dt}[e^{-A(t-t_0)}]x(t) + e^{-A(t-t_0)}\dot{x}(t)$$

$$= (-A)e^{-A(t-t_0)}x(t) + e^{-A(t-t_0)}[Ax(t) + Bu(t)]$$

$$= e^{-A(t-t_0)}Bu(t)$$

Since the right-hand side above does not involve $z(t)$, we may solve for $z(t)$ by applying the fundamental theorem of calculus

$$z(t) = z(t_0) + \int_{t_0}^{t} \dot{z}(\tau)d\tau$$

$$= z(t_0) + \int_{t_0}^{t} e^{-A(\tau-t_0)}Bu(\tau)d\tau$$

From our original definition, we can recover $x(t) = e^{A(t-t_0)}z(t)$ and $x(t_0) = z(t_0)$ so that

$$x(t) = e^{A(t-t_0)}\left[z(t_0) + \int_{t_0}^{t} e^{-A(\tau-t_0)}Bu(\tau)\,d\tau\right]$$

$$= e^{A(t-t_0)}x(t_0) + \int_{t_0}^{t} e^{A(t-t_0)}e^{-A(\tau-t_0)}Bu(\tau)\,d\tau \tag{2.12}$$

$$= e^{A(t-t_0)}x(t_0) + \int_{t_0}^{t} e^{A(t-\tau)}Bu(\tau)\,d\tau$$

This constitutes a closed-form expression (the so-called variation-of-constants formula) for the *complete* solution to the linear time-invariant state equation that we observe is decomposed into two terms. The first term is due to the initial state $x(t_0) = x_0$ and determines the solution in the case where the input is identically zero. As such, we refer to it as the *zero-input response* $x_{zi}(t)$ and often write

$$x_{zi}(t) = e^{A(t-t_0)}x(t_0) \tag{2.13}$$

The second term is due to the input signal and determines the solution in the case where the initial state is the zero vector. Accordingly, we refer to it as the *zero-state response* $x_{zs}(t)$ and denote

$$x_{zs}(t) = \int_{t_0}^{t} e^{A(t-\tau)}Bu(\tau)\,d\tau \tag{2.14}$$

so that by the principle of linear superposition, the complete response is given by $x(t) = x_{zi}(t) + x_{zs}(t)$. Having characterized the complete state response, a straightforward substitution yields the complete output response

$$y(t) = Ce^{A(t-t_0)}x(t_0) + \int_{t_0}^{t} Ce^{A(t-\tau)}Bu(\tau)\,d\tau + Du(t) \tag{2.15}$$

which likewise can be decomposed into zero-input and zero-state response components, namely,

$$y_{zi}(t) = Ce^{A(t-t_0)}x(t_0) \quad y_{zs}(t) = \int_{t_0}^{t} Ce^{A(t-\tau)}Bu(\tau)\,d\tau + Du(t) \tag{2.16}$$

2.2 IMPULSE RESPONSE

We assume without loss of generality that $t_0 = 0$ and set the initial state $x(0) = 0$ to focus on the zero-state response, that is,

$$y_{zs}(t) = \int_{0^-}^{t} Ce^{A(t-\tau)} Bu(\tau)\, d\tau + Du(t)$$

$$= \int_{0^-}^{t} [Ce^{A(t-\tau)} B + D\delta(t - \tau)]u(\tau)\, d\tau$$

Also, we partition coefficient matrices B and D column-wise

$$B = \begin{bmatrix} b_1 & b_2 & \cdots & b_m \end{bmatrix} \quad D = \begin{bmatrix} d_1 & d_2 & \cdots & d_m \end{bmatrix}$$

With $\{e_1, \ldots, e_m\}$ denoting the standard basis for $\mathbb{R}^m$, an impulsive input on the ith input $u(t) = e_i \delta(t)$, $i = 1, \ldots, m$, yields the response

$$\int_0^t Ce^{A(t-\tau)} Be_i\, \delta(\tau)\, d\tau + De_i \delta(t) = Ce^{At} b_i \delta(t) + d_i \delta(t) \quad t \geq 0$$

This forms the ith column of the $p \times m$ *impulse response matrix*

$$h(t) = Ce^{At} B + D\delta(t) \quad t \geq 0 \tag{2.17}$$

in terms of which the zero-state response component has the familiar characterization

$$y_{zs}(t) = \int_0^t h(t - \tau)u(\tau)\, d\tau \tag{2.18}$$

2.3 LAPLACE DOMAIN REPRESENTATION

Taking Laplace transforms of the state equation (2.1) with $t_0 = 0$, using linearity and time-differentiation properties, yields

$$sX(s) - x_0 = AX(s) + BU(s)$$

Grouping $X(s)$ terms on the left gives

$$(sI - A)X(s) = x_0 + BU(s)$$

Now, from basic linear algebra, the determinant $|sI - A|$ is a degree-n monic polynomial (i.e., the coefficient of s^n is 1), and so it is not the zero polynomial. Also, the adjoint of $(sI - A)$ is an $n \times n$ matrix of polynomials having degree at most $n - 1$. Consequently,

$$(sI - A)^{-1} = \frac{\text{adj}(sI - A)}{|sI - A|}$$

is an $n \times n$ matrix of rational functions in the complex variable s. Moreover, each element of $(sI - A)^{-1}$ has numerator degree that is guaranteed to be strictly less than its denominator degree and therefore is a *strictly proper* rational function.

The preceding equation now can be solved for $X(s)$ to obtain

$$X(s) = (sI - A)^{-1}x_0 + (sI - A)^{-1}BU(s) \qquad (2.19)$$

As in the time domain, we can decompose $X(s)$ into zero-input and zero-state response components $X(s) = X_{zi}(s) + X_{zs}(s)$ in which

$$X_{zi}(s) = (sI - A)^{-1}x_0 \quad X_{zs}(s) = (sI - A)^{-1}BU(s) \qquad (2.20)$$

Denoting a Laplace transform pair by $f(t) \leftrightarrow F(s)$, we see that since

$$x_{zi}(t) = e^{At}x_0 \leftrightarrow X_{zi}(s) = (sI - A)^{-1}x_0$$

holds for any initial state, we can conclude that

$$e^{At} \leftrightarrow (sI - A)^{-1} \qquad (2.21)$$

This relationship suggests an approach for computing the matrix exponential by first forming the matrix inverse $(sI - A)^{-1}$, whose elements are guaranteed to be rational functions in the complex variable s, and then applying partial fraction expansion element by element to determine the inverse Laplace transform that yields e^{At}.

Taking Laplace transforms through the output equation and substituting for $X(s)$ yields

$$\begin{aligned} Y(s) &= CX(s) + DU(s) \\ &= C(sI - A)^{-1}x_0 + [C(sI - A)^{-1}B + D]U(s) \end{aligned} \qquad (2.22)$$

from which zero-input and zero-state response components are identified as follows:

$$Y_{zi}(s) = C(sI - A)^{-1}x_0 \quad Y_{zs}(s) = [C(sI - A)^{-1}B + D]U(s) \quad (2.23)$$

Focusing on the zero-state response component, the convolution property of the Laplace transform indicates that

$$y_{zs}(t) = \int_0^t h(t - \tau)u(\tau)\,d\tau \;\leftrightarrow\; Y_{zs}(s) = [C(sI - A)^{-1}B + D]U(s)$$

$$(2.24)$$

yielding the familiar relationship between the impulse response and transfer function

$$h(t) = Ce^{At}B + D\delta(t) \;\leftrightarrow\; H(s) = C(sI - A)^{-1}B + D \quad (2.25)$$

where, in the multiple-input, multiple-output case, the transfer function is a $p \times m$ matrix of rational functions in s.

Example 2.4   In this example we solve the linear second-order ordinary differential equation $\ddot{y}(t) + 7\dot{y}(t) + 12y(t) = u(t)$, given that the input $u(t)$ is a step input of magnitude 3 and the initial conditions are $y(0) = 0.10$ and $\dot{y}(0) = 0.05$. The system characteristic polynomial is $s^2 + 7s + 12 = (s + 3)(s + 4)$, and the system eigenvalues are $s_{1,2} = -3, -4$. These eigenvalues are distinct, negative real roots, so the system is overdamped. Using standard solution techniques, we find the solution is

$$y(t) = 0.25 - 0.55e^{-3t} + 0.40e^{-4t} \quad t \geq 0$$

We now derive this same solution using the techniques of this chapter. First, we must derive a valid state-space description. We define the state vector as

$$x(t) = \begin{bmatrix} x_1(t) \\ x_2(t) \end{bmatrix} = \begin{bmatrix} y(t) \\ \dot{y}(t) \end{bmatrix}$$

Then the state differential equation is

$$\begin{bmatrix} \dot{x}_1(t) \\ \dot{x}_2(t) \end{bmatrix} = \begin{bmatrix} 0 & 1 \\ -12 & -7 \end{bmatrix} \begin{bmatrix} x_1(t) \\ x_2(t) \end{bmatrix} + \begin{bmatrix} 0 \\ 1 \end{bmatrix} u(t)$$

We are given that $u(t)$ is a step input of magnitude 3, and the initial state is $x(0) = [y(0), \dot{y}(0)]^T = [0.10, 0.05]^T$. We can solve this problem in the Laplace domain, by first writing

$$X(s) = (sI - A)^{-1}X(0) + (sI - A)^{-1}BU(s)$$

$$(sI - A) = \begin{bmatrix} s & -1 \\ 12 & s+7 \end{bmatrix}$$

$$(sI - A)^{-1} = \frac{1}{s^2 + 7s + 12} \begin{bmatrix} s+7 & 1 \\ -12 & s \end{bmatrix}$$

where the denominator $s^2 + 7s + 12 = (s+3)(s+4)$ is $|sI - A|$, the characteristic polynomial. Substituting these results into the preceding expression we obtain:

$$X(s) = \frac{1}{s^2 + 7s + 12} \begin{bmatrix} s+7 & 1 \\ -12 & s \end{bmatrix} \begin{bmatrix} 0.10 \\ 0.05 \end{bmatrix}$$

$$+ \frac{1}{s^2 + 7s + 12} \begin{bmatrix} s+7 & 1 \\ -12 & s \end{bmatrix} \begin{bmatrix} 0 \\ 1 \end{bmatrix} \frac{3}{s}$$

where the Laplace transform of the unit step function is $\frac{1}{s}$. Simplifying, we find the state solution in the Laplace domain:

$$X(s) = \begin{bmatrix} X_1(s) \\ X_2(s) \end{bmatrix} = \frac{1}{(s+3)(s+4)} \begin{bmatrix} 0.10s + 0.75 + \dfrac{3}{s} \\ 0.05s + 1.80 \end{bmatrix}$$

$$= \begin{bmatrix} \dfrac{0.10s^2 + 0.75s + 3}{s(s+3)(s+4)} \\ \dfrac{0.05s + 1.80}{(s+3)(s+4)} \end{bmatrix}$$

A partial fraction expansion of the first state variable yields the residues

$$C_1 = 0.25$$

$$C_2 = -0.55$$

$$C_3 = 0.40$$

The output $y(t)$ equals the first state variable $x_1(t)$, found by the inverse Laplace transform:

$$y(t) = x_1(t) = L^{-1}\{X_1(s)\}$$
$$= L^{-1}\left\{\frac{0.25}{s} - \frac{0.55}{s+3} + \frac{0.40}{s+4}\right\}$$
$$= 0.25 - 0.55e^{-3t} + 0.40e^{-4t} \qquad t \ge 0$$

which agrees with the stated solution. Now we find the solution for the second state variable $x_2(t)$ in a similar manner. A partial fraction expansion of the second state variable yields the residues:

$$C_1 = 1.65$$
$$C_2 = -1.60$$

Then the second state variable $x_2(t)$ is found from the inverse Laplace transform:

$$x_2(t) = L^{-1}\{X_2(s)\}$$
$$= L^{-1}\left\{\frac{1.65}{s+3} - \frac{1.60}{s+4}\right\}$$
$$= 1.65e^{-3t} - 1.60e^{-4t} \qquad t \ge 0$$

We can check this result by verifying that $x_2(t) = \dot{x}_1(t)$:

$$\dot{x}_1(t) = -(-3)0.55e^{-3t} + (-4)0.40e^{-4t} = 1.65e^{-3t} - 1.60e^{-4t}$$

which agrees with the $x_2(t)$ solution. Figure 2.1 shows the state response versus time for this example.

We can see that the initial state $x(0) = [0.10, 0.05]^T$ is satisfied and that the steady state values are 0.25 for $x_1(t)$ and 0 for $x_2(t)$. $\qquad \square$

Example 2.5 Consider the two-dimensional state equation

$$\begin{bmatrix} \dot{x}_1(t) \\ \dot{x}_2(t) \end{bmatrix} = \begin{bmatrix} 0 & 1 \\ -2 & -3 \end{bmatrix} \begin{bmatrix} x_1(t) \\ x_2(t) \end{bmatrix} + \begin{bmatrix} 0 \\ 1 \end{bmatrix} u(t) \qquad x(0) = x_0 = \begin{bmatrix} -1 \\ 1 \end{bmatrix}$$

$$y(t) = \begin{bmatrix} 1 & 1 \end{bmatrix} \begin{bmatrix} x_1(t) \\ x_2(t) \end{bmatrix}$$

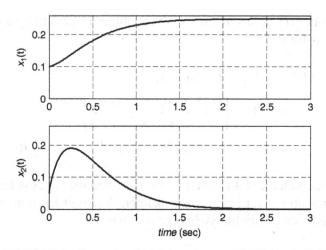

FIGURE 2.1   Second-order state responses for Example 2.4.

From

$$sI - A = \begin{bmatrix} s & 0 \\ 0 & s \end{bmatrix} - \begin{bmatrix} 0 & 1 \\ -2 & -3 \end{bmatrix} = \begin{bmatrix} s & -1 \\ 2 & s+3 \end{bmatrix}$$

we find, by computing the matrix inverse and performing partial fraction expansion on each element,

$$(sI - A)^{-1} = \frac{\text{adj}(sI - A)}{|sI - A|} = \frac{\begin{bmatrix} s+3 & 1 \\ -2 & s \end{bmatrix}}{s^2 + 3s + 2}$$

$$= \begin{bmatrix} \dfrac{s+3}{(s+1)(s+2)} & \dfrac{1}{(s+1)(s+2)} \\[2ex] \dfrac{-2}{(s+1)(s+2)} & \dfrac{s}{(s+1)(s+2)} \end{bmatrix}$$

$$= \begin{bmatrix} \dfrac{2}{s+1} + \dfrac{-1}{s+2} & \dfrac{1}{s+1} + \dfrac{-1}{s+2} \\[2ex] \dfrac{-2}{s+1} + \dfrac{2}{s+2} & \dfrac{-1}{s+1} + \dfrac{2}{s+2} \end{bmatrix}$$

It follows directly that

$$e^{At} = L^{-1}[(sI - A)^{-1}]$$

$$= \begin{bmatrix} 2e^{-t} - e^{-2t} & e^{-t} - e^{-2t} \\ -2e^{-t} + 2e^{-2t} & -e^{-t} + 2e^{-2t} \end{bmatrix} \quad t \geq 0$$

For the specified initial state, the zero-input response component of the state and output are

$$x_{zi}(t) = e^{At}x_0 = \begin{bmatrix} -e^{-t} \\ e^{-t} \end{bmatrix} \quad y_{zi}(t) = Ce^{At}x_0 = Cx_{zi}(t) = 0 \quad t \geq 0$$

For a unit-step input signal, the Laplace domain representation of the zero-state components of the state and output response are

$$X_{zs}(s) = (sI - A)^{-1}BU(s)$$

$$= \frac{\begin{bmatrix} s+3 & 1 \\ -2 & s \end{bmatrix}}{s^2 + 3s + 2} \begin{bmatrix} 0 \\ 1 \end{bmatrix} \frac{1}{s}$$

$$= \begin{bmatrix} \dfrac{1}{(s+1)(s+2)} \\ \dfrac{s}{(s+1)(s+2)} \end{bmatrix} \frac{1}{s}$$

$$= \begin{bmatrix} \dfrac{1}{s(s+1)(s+2)} \\ \dfrac{1}{(s+1)(s+2)} \end{bmatrix}$$

$$= \begin{bmatrix} \dfrac{1/2}{s} - \dfrac{1}{s+1} + \dfrac{1/2}{s+2} \\ \dfrac{1}{s+1} - \dfrac{1}{s+2} \end{bmatrix}$$

$$Y_{zs}(s) = CX_{zs}(s) + DU(s)$$

$$= \begin{bmatrix} 1 & 1 \end{bmatrix} \begin{bmatrix} \dfrac{1/2}{s} - \dfrac{1}{s+1} + \dfrac{1/2}{s+2} \\ \dfrac{1}{s+1} - \dfrac{1}{s+2} \end{bmatrix} + [0]\dfrac{1}{s}$$

$$= \dfrac{1/2}{s} - \dfrac{1/2}{s+2}$$

from which

$$x_{zs}(t) = \begin{bmatrix} 1/2 - e^{-t} + 1/2e^{-2t} \\ e^{-t} - e^{-2t} \end{bmatrix} \quad y_{zs}(t) = \frac{1}{2}(1 - e^{-2t}) \quad t \geq 0$$

and complete state and output responses then are

$$x(t) = \begin{bmatrix} 1/2 - 2e^{-t} + 1/2e^{-2t} \\ 2e^{-t} - e^{-2t} \end{bmatrix} \quad y(t) = \frac{1}{2}(1 - e^{-2t}) \quad t \geq 0$$

Finally, the transfer function is given as

$$H(s) = C(sI - A)^{-1}B + D$$

$$= \begin{bmatrix} 1 & 1 \end{bmatrix} \frac{\begin{bmatrix} s+3 & 1 \\ -2 & s \end{bmatrix}}{s^2 + 3s + 2} \begin{bmatrix} 0 \\ 1 \end{bmatrix} + 0$$

$$= \frac{s+1}{s^2 + 3s + 2} = \frac{s+1}{(s+1)(s+2)} = \frac{1}{s+2}$$

with associated impulse response

$$h(t) = e^{-2t} \quad t \geq 0 \qquad \qquad \square$$

2.4   STATE-SPACE REALIZATIONS REVISITED

Recall that at the conclusion of Section 1.3 we called a state equation a *state-space realization* of a linear time-invariant system's input-output behavior if, loosely speaking, it corresponds to the Laplace domain relationship $Y(s) = H(s)U(s)$ involving the system's transfer function. Since this pertains to the zero-state response component, we see that Equation (2.25) implies that a state-space realization is required to satisfy

$$C(sI - A)^{-1}B + D = H(s)$$

or, equivalently, in terms of the impulse response in the time domain,

$$Ce^{At}B + D\delta(t) = h(t)$$

Example 2.6   Here we extend to arbitrary dimensions the state-space realization analysis conducted for the three–dimensional system of Example 1.6. Namely, we show that the single-input, single-output n–dimensional strictly proper transfer function

$$H(s) = \frac{b(s)}{a(s)} = \frac{b_{n-1}s^{n-1} + \cdots + b_1 s + b_0}{s^n + a_{n-1}s^{n-1} + \cdots + a_1 s + a_0} \qquad (2.26)$$

has a state-space realization specified by coefficient matrices

$$
A = \begin{bmatrix}
0 & 1 & 0 & \cdots & 0 & 0 \\
0 & 0 & 1 & \cdots & 0 & 0 \\
\vdots & \vdots & \vdots & \ddots & \vdots & \vdots \\
0 & 0 & 0 & \cdots & 1 & 0 \\
0 & 0 & 0 & \cdots & 0 & 1 \\
-a_0 & -a_1 & -a_2 & \cdots & -a_{n-2} & -a_{n-1}
\end{bmatrix}
$$

$$
B = \begin{bmatrix}
0 \\ 0 \\ \vdots \\ 0 \\ 0 \\ 1
\end{bmatrix}
$$

$$
C = \begin{bmatrix} b_0 & b_1 & b_2 & \cdots & b_{n-2} & b_{n-1} \end{bmatrix} \quad D = 0
$$

(2.27)

Moreover, by judiciously ordering the calculations, we avoid the unpleasant prospect of symbolically rendering $(sI - A)^{-1}$, which is seemingly at the heart of the identity we must verify. First, observe that

$$
(sI - A) \begin{bmatrix} 1 \\ s \\ s^2 \\ \vdots \\ s^{n-2} \\ s^{n-1} \end{bmatrix}
= \begin{bmatrix}
s & -1 & 0 & \cdots & 0 & 0 \\
0 & s & -1 & \cdots & 0 & 0 \\
\vdots & \vdots & \vdots & \ddots & \vdots & \vdots \\
0 & 0 & 0 & \cdots & -1 & 0 \\
0 & 0 & 0 & \cdots & s & -1 \\
a_0 & a_1 & a_2 & \cdots & a_{n-2} & s + a_{n-1}
\end{bmatrix}
\begin{bmatrix} 1 \\ s \\ s^2 \\ \vdots \\ s^{n-2} \\ s^{n-1} \end{bmatrix}
$$

$$
= \begin{bmatrix}
s \cdot 1 + (-1) \cdot s \\
s \cdot s + (-1) \cdot s^2 \\
\vdots \\
s \cdot s^{n-3} + (-1) \cdot s^{n-2} \\
s \cdot s^{n-2} + (-1) \cdot s^{n-1} \\
a_0 \cdot 1 + a_1 \cdot s + a_2 \cdot s^2 + \cdots + a_{n-2} \cdot s^{n-2} + (s + a_{n-1})s^{n-1}
\end{bmatrix}
$$

$$
= \begin{bmatrix}
0 \\
0 \\
\vdots \\
0 \\
0 \\
s^n + a_{n-1}s^{n-1} + \cdots + a_1 s + a_0
\end{bmatrix}
$$

$$
= Ba(s)
$$

Rearranging to solve for $(sI - A)^{-1}B$ and substituting into $C(sI - A)^{-1}B + D$ yields

$$C(sI - A)^{-1}B + D = \begin{bmatrix} b_0 & b_1 & b_2 & \cdots & b_{n-2} & b_{n-1} \end{bmatrix} \dfrac{\begin{bmatrix} 1 \\ s \\ s^2 \\ \vdots \\ s^{n-2} \\ s^{n-1} \end{bmatrix}}{a(s)} + 0$$

$$= \frac{b_0 \cdot 1 + b_1 \cdot s + b_2 \cdot s^2 + \cdots + b_{n-2} \cdot s^{n-2} + b_{n-1}s^{n-1}}{a(s)}$$

$$= \frac{b(s)}{a(s)} = H(s)$$

as required. $\square$

2.5 COORDINATE TRANSFORMATIONS

Here we introduce the concept of a state coordinate transformation and study the impact that such a transformation has on the state equation itself, as well as various derived quantities. Especially noteworthy is what does *not* change with the application of a state coordinate transformation. The reader is referred to Appendix B (Sections B.4 and B.6) for an overview of linear algebra topics (change of basis and linear transformations) that underpin our development here. Once having discussed state coordinate transformations in general terms, we consider a particular application: transforming a given state equation into the so-called diagonal canonical form.

General Coordinate Transformations

For the n-dimensional linear time-invariant state equation (2.1), any *non-singular* $n \times n$ matrix T defines a coordinate transformation via

$$x(t) = Tz(t) \quad z(t) = T^{-1}x(t) \tag{2.28}$$

Associated with the transformed state vector $z(t)$ is the transformed state equation

$$\dot{z}(t) = T^{-1}\dot{x}(t)$$

$$= T^{-1}[Ax(t) + Bu(t)]$$

$$= T^{-1}ATz(t) + T^{-1}Bu(t)$$

$$y(t) = CTz(t) + Du(t)$$

That is, the coefficient matrices are transformed according to

$$\hat{A} = T^{-1}AT \qquad \hat{B} = T^{-1}B \qquad \hat{C} = CT \qquad \hat{D} = D \qquad (2.29)$$

and we henceforth write

$$\dot{z}(t) = \hat{A}z(t) + \hat{B}u(t)$$
$$y(t) = \hat{C}z(t) + \hat{D}u(t) \qquad z(t_0) = z_0 \qquad (2.30)$$

where, in addition, the initial state is transformed via $z(t_0) = T^{-1}x(t_0)$.

For the system dynamics matrix, the coordinate transformation yields $\hat{A} = T^{-1}AT$. This is called a *similarity transformation*, so the new system dynamics matrix $\hat{A}$ has the same characteristic polynomial and eigenvalues as A. However, the eigenvectors are different but are linked by the coordinate transformation.

The impact that state coordinate transformations have on various quantities associated with linear time-invariant state equations is summarized in the following proposition.

Proposition 2.2 *For the linear time-invariant state equation (2.1) and coordinate transformation (2.28):*

1. $|sI - \hat{A}| = |sI - A|$
2. $(sI - \hat{A})^{-1} = T^{-1}(sI - A)^{-1}T$
3. $e^{\hat{A}t} = T^{-1}e^{At}T$
4. $\hat{C}(sI - \hat{A})^{-1}\hat{B} + \hat{D} = C(sI - A)^{-1}B + D$
5. $\hat{C}e^{\hat{A}t}\hat{B} + \hat{D}\delta(t) = Ce^{At}B + D\delta(t)$ ☐

Item 1 states that the system characteristic polynomial and thus the system eigenvalues are invariant under the coordinate transformation (2.28). This proposition is proved using determinant properties as follows:

$$|sI - \hat{A}| = |sI - T^{-1}AT|$$
$$= |sT^{-1}T - T^{-1}AT|$$
$$= |T^{-1}(sI - A)T|$$
$$= |T^{-1}||sI - A||T|$$
$$= |T^{-1}||T||sI - A|$$
$$= |sI - A|$$

where the last step follows from $|T^{-1}||T| = |T^{-1}T| = |I| = 1$. Therefore, since a nonsingular matrix and its inverse have reciprocal determinants, $\hat{A}$ and A have the same characteristic polynomial and eigenvalues.

Items 4 and 5 indicate that the transfer function and impulse response are unaffected by (or invariant with respect to) any state coordinate transformation. Consequently, given one state-space realization of a transfer function or impulse response, there are infinitely many others (of the same dimension) because there are infinitely many ways to specify a state coordinate transformation.

Diagonal Canonical Form

There are some special realizations that can be obtained by applying the state coordinate transformation (2.28) to a given linear state equation. In this subsection we present the diagonal canonical form for the single-input, single-output case.

Diagonal canonical form is also called modal form because it yields a decoupled set of n first-order ordinary differential equations. This is clearly convenient because in this form, n scalar first-order differential equation solutions may be formulated independently, instead of using coupled system solution methods. Any state-space realization with a diagonalizable A matrix can be transformed to diagonal canonical form (DCF) by

$$x(t) = T_{\text{DCF}}\, z(t)$$

where the diagonal canonical form coordinate transformation matrix $T_{\text{DCF}} = [\, v_1 \quad v_2 \quad \cdots \quad v_n \,]$ consists of eigenvectors v_i of A arranged column-wise (see Appendix B, Section B.8 for an overview of eigenvalues and eigenvectors). Because A is assumed to be diagonalizable, the n eigenvectors are linearly independent, yielding a nonsingular T_{DCF}. The diagonal canonical form is characterized by a diagonal A matrix with eigenvalues appearing on the diagonal, where eigenvalue λ_i is associated with the eigenvector $v_i, i = 1, 2, \ldots, n$:

$$A_{\text{DCF}} = T_{\text{DCF}}^{-1} A T_{\text{DCF}} = \begin{bmatrix} \lambda_1 & 0 & 0 & \cdots & 0 \\ 0 & \lambda_2 & 0 & \cdots & 0 \\ 0 & 0 & \lambda_3 & \cdots & 0 \\ \vdots & \vdots & \vdots & \ddots & \vdots \\ 0 & 0 & 0 & 0 & \lambda_n \end{bmatrix} \qquad (2.31)$$

$B_{\text{DCF}} = T_{\text{DCF}}^{-1} B$, $C_{\text{DCF}} = C T_{\text{DCF}}$, and $D_{\text{DCF}} = D$ have no particular form.

Example 2.7 We now compute the diagonal canonical form via a state coordinate transformation for the linear time-invariant state equation in Example 2.5. For this example, the state coordinate transformation (2.28) given by

$$T_{DCF} = \begin{bmatrix} 1 & -1 \\ -1 & 2 \end{bmatrix} \quad T_{DCF}^{-1} = \begin{bmatrix} 2 & 1 \\ 1 & 1 \end{bmatrix}$$

yields transformed coefficient matrices

$$
\begin{aligned}
A_{DCF} &= T_{DCF}^{-1} A T_{DCF} & B_{DCF} &= T_{DCF}^{-1} B \\
&= \begin{bmatrix} 2 & 1 \\ 1 & 1 \end{bmatrix} \begin{bmatrix} 0 & 1 \\ -2 & -3 \end{bmatrix} \begin{bmatrix} 1 & -1 \\ -1 & 2 \end{bmatrix} & &= \begin{bmatrix} 2 & 1 \\ 1 & 1 \end{bmatrix} \begin{bmatrix} 0 \\ 1 \end{bmatrix} \\
&= \begin{bmatrix} -1 & 0 \\ 0 & -2 \end{bmatrix} & &= \begin{bmatrix} 1 \\ 1 \end{bmatrix}
\end{aligned}
$$

$$
\begin{aligned}
C_{DCF} &= C T_{DCF} \\
&= \begin{bmatrix} 1 & 1 \end{bmatrix} \begin{bmatrix} 1 & -1 \\ -1 & 2 \end{bmatrix} \qquad D_{DCF} = D = 0 \\
&= \begin{bmatrix} 0 & 1 \end{bmatrix} \quad D_{DCF}
\end{aligned}
$$

Note that this yields a diagonal A_{DCF} matrix so that the diagonal canonical form represents two decoupled first-order ordinary differential equations, that is,

$$\dot{z}_1(t) = -z_1(t) + u(t) \quad \dot{z}_2(t) = -2z_2(t) + u(t)$$

which therefore can be solved independently to yield complete solutions

$$z_1(t) = e^{-t} z_1(0) + \int_0^t e^{-(t-\tau)} u(\tau) \, d\tau$$

$$z_2(t) = e^{-t} z_2(0) + \int_0^t e^{-2(t-\tau)} u(\tau) \, d\tau$$

We also must transform the initial state given in Example 2.5 using $z(0) = T_{DCF}^{-1} x(0)$:

$$\begin{bmatrix} z_1(0) \\ z_2(0) \end{bmatrix} = \begin{bmatrix} 2 & 1 \\ 1 & 1 \end{bmatrix} \begin{bmatrix} -1 \\ 1 \end{bmatrix} = \begin{bmatrix} -1 \\ 0 \end{bmatrix}$$

which together with a unit step input yields

$$z_1(t) = 1 - 2e^{-t} \quad z_2(t) = \tfrac{1}{2}(1 - e^{-2t}) \quad t \geq 0$$

The complete solution in the original coordinates then can be recovered from the relationship $x(t) = T_{\mathrm{DCF}} z(t)$:

$$\begin{bmatrix} x_1(t) \\ x_2(t) \end{bmatrix} = \begin{bmatrix} 1 & -1 \\ -1 & 2 \end{bmatrix} \begin{bmatrix} 1 - 2e^{-t} \\ \tfrac{1}{2}(1 - e^{-2t}) \end{bmatrix} = \begin{bmatrix} \tfrac{1}{2} - 2e^{-t} + \tfrac{1}{2}e^{-2t} \\ 2e^{-t} - e^{-2t} \end{bmatrix}$$

which agrees with the result in Example 2.5. Note also from C_{DCF} that $y(t) = z_2(t)$, which directly gives

$$y_{zs}(t) = \int_0^t e^{-2(t-\tau)} u(\tau) \, d\tau$$

from which we identify the impulse response and corresponding transfer function as

$$h(t) = e^{-2t} \quad t \geq 0 \quad \leftrightarrow \quad H(s) = \frac{1}{s+2}$$

This agrees with the outcome observed in Example 2.5, in which a pole-zero cancellation resulted in the first-order transfer function above. Here, a first-order transfer function arises because the transformed state equation is decomposed into two decoupled first-order subsystems, each associated with one of the state variables $z_1(t)$ and $z_2(t)$. Of these two subsystems, the first is disconnected from the output $y(t)$ so that the system's input-output behavior is directly governed by the z_2 subsystem alone.

The previously calculated zero-state output responses in both Laplace and time domains are verified easily. It is interesting to note that the preceding pole-zero cancellation results in a first-order transfer function having the one–dimensional state space realization

$$\dot{z}_2(t) = -2z_2(t) + u(t)$$
$$y(t) = z_2(t)$$

We can conclude that not only do there exist different state-space realizations of the same dimension for a given transfer function, but there also may exist realizations of different and possibly lower dimension (this will be discussed in Chapter 5). □

Example 2.8   Given the three–dimensional single-input, single-output linear time-invariant state equation specified below, we calculate the diagonal canonical form.

$$A = \begin{bmatrix} 8 & -5 & 10 \\ 0 & -1 & 1 \\ -8 & 5 & -9 \end{bmatrix} \quad B = \begin{bmatrix} -1 \\ 0 \\ 1 \end{bmatrix} \quad C = \begin{bmatrix} 1 & -2 & 4 \end{bmatrix} \quad D = 0$$

The characteristic polynomial is $s^3 + 2s^2 + 4s + 8$; the eigenvalues of A are the roots of this characteristic polynomial, $\pm 2i, -2$. The diagonal canonical form transformation matrix T_{DCF} is constructed from three eigenvectors v_i arranged column-wise.

$$T_{\text{DCF}} = \begin{bmatrix} v_1 & v_2 & v_3 \end{bmatrix}$$

$$= \begin{bmatrix} 5 & 5 & -3 \\ 2i & -2i & -2 \\ -4+2i & -4-2i & 4 \end{bmatrix}$$

The resulting diagonal canonical form state-space realization is

$$A_{\text{DCF}} = T_{\text{DCF}}^{-1} A T_{\text{DCF}} \qquad B_{DCF} = T_{\text{DCF}}^{-1} B$$

$$= \begin{bmatrix} 2i & 0 & 0 \\ 0 & -2i & 0 \\ 0 & 0 & -2 \end{bmatrix} \qquad = \begin{bmatrix} -0.0625 - 0.0625i \\ -0.0625 + 0.0625i \\ 0.125 \end{bmatrix}$$

$$C_{\text{DCF}} = C T_{\text{DCF}} \qquad\qquad D_{\text{DCF}} = D$$
$$= \begin{bmatrix} -11+4i & -11-4i & 9 \end{bmatrix} \qquad = 0$$

If one were to start with a valid diagonal canonical form realization, $T_{\text{DCF}} = I_n$ because the eigenvectors can be taken to be the standard basis vectors.   □

As seen in Example 2.8, when A has complex eigenvalues occurring in a conjugate pair, the associated eigenvectors can be chosen to form a conjugate pair. The coordinate transformation matrix T_{DCF} formed from linearly independent eigenvectors of A will consequently contain complex elements. Clearly, the diagonal matrix A_{DCF} will also contain complex elements, namely, the complex eigenvalues. In addition, the matrices B_{DCF} and C_{DCF} computed from T_{DCF} will also have complex entries in general. To avoid a state-space realization with complex coefficient matrices, we can modify the construction of the coordinate transformation matrix T_{DCF} as follows. We assume for simplicity that $\lambda_1 = \sigma + j\omega$ and $\lambda_2 = \sigma - j\omega$ are the only complex eigenvalues of A with associated eigenvectors $v_1 = u + jw$ and $v_2 = u - jw$. It is not difficult to show that linear independence of the complex eigenvectors v_1 and v_2 is equivalent to linear independence of the real vectors $u = Re(v_1)$ and $w = Im(v_1)$. Letting $\lambda_3, \ldots, \lambda_n$ denote the remaining real eigenvalues of A with associated real eigenvectors $v_3, \ldots, v_n$, the matrix

$$T_{\text{DCF}} = \begin{bmatrix} u & w & v_3 & \cdots & v_n \end{bmatrix}$$

is real and nonsingular. Using this to define a state coordinate transformation, we obtain

$$A_{\text{DCF}} = T_{\text{DCF}}^{-1} A T_{\text{DCF}} = \begin{bmatrix} \sigma & \omega & 0 & \cdots & 0 \\ -\omega & \sigma & 0 & \cdots & 0 \\ 0 & 0 & \lambda_3 & \cdots & 0 \\ \vdots & \vdots & \vdots & \ddots & \vdots \\ 0 & 0 & 0 & 0 & \lambda_n \end{bmatrix}$$

which is a real matrix but no longer diagonal. However, A_{DCF} is a block diagonal matrix that contains a 2×2 block displaying the real and imaginary parts of the complex conjugate eigenvalues λ_1, λ_2. Also, because T_{DCF} now is a real matrix, $B_{\text{DCF}} = T_{\text{DCF}}^{-1} B$ and $C_{\text{DCF}} = C T_{\text{DCF}}$ are guaranteed to be real yielding a state-space realization with purely real coefficient matrices. This process can be generalized to handle a system dynamics matrix A with any combination of real and complex-conjugate eigenvalues. The real A_{DCF} matrix that results will have a block diagonal structure with each real eigenvalue appearing directly on the main diagonal and a 2×2 matrix displaying the real and imaginary part of each complex conjugate pair. The reader is invited to revisit Example 2.8 and instead apply the state coordinate transformation specified by

$$T_{\text{DCF}} = \begin{bmatrix} 5 & 0 & -3 \\ 0 & 2 & -2 \\ -4 & 2 & 2 \end{bmatrix}$$

2.6 MATLAB FOR SIMULATION AND COORDINATE TRANSFORMATIONS

MATLAB and the accompanying Control Systems Toolbox provide many useful functions for the analysis, simulation, and coordinate transformations of linear time-invariant systems described by state equations. A subset of these MATLAB functions is discussed in this section.

MATLAB for Simulation of State-Space Systems

Some MATLAB functions that are useful for analysis and simulation of state-space systems are

eig(A)	Find the eigenvalues of **A**.
poly(A)	Find the system characteristic polynomial coefficients from **A**.
roots(den)	Find the roots of the characteristic polynomial.
damp(A)	Calculate the second-order system damping ratio and undamped natural frequency (for each mode if $n > 2$) from the system dynamics matrix **A**.
damp(den)	Calculate the second-order system damping ratio and undamped natural frequency (for each mode if $n > 2$) from the coefficients **den** of the system characteristic polynomial.
impulse(SysName)	Determine the unit impulse response for a system numerically.
step(SysName)	Determine the unit step response for a system numerically.
lsim(SysName,u,t,x0)	General linear simulation; calculate the output $y(t)$ and state $x(t)$ numerically given the system data structure.
expm(A*t)	Evaluate the state transition matrix at time **t** seconds.
plot(x,y)	Plot dependent variable **y** versus independent variable **x**.

One way to invoke the MATLAB function **lsim** with left-hand side arguments is

```
[y,t,x] = lsim(SysName,u,t,x0)
```

The **lsim** function inputs are the state-space data structure **SysName**, the input matrix **u** [**length(t)** rows by number of inputs m columns], an evenly spaced time vector **t** supplied by the user, and the $n \times 1$ initial state vector **x0**. No plot is generated, but the **lsim** function yields the system output solution **y** [a matrix of **length(t)** rows by number of outputs p columns], the same time vector **t**, and the system state solution **x** [a matrix of **length(t)** rows by number of states n columns]. The matrices **y**, **x**, and **u** all have time increasing along rows, with one column for each component, in order. The user then can plot the desired output and state components versus time after the **lsim** command has been executed.

MATLAB for Coordinate Transformations and Diagonal Canonical Form

Some MATLAB functions that are useful for coordinate transformations and the diagonal canonical form realization are

  canon MATLAB function for canonical forms (use the **modal** switch for diagonal canonical form)

  ss2ss Coordinate transformation of one state-space realization to another.

The **canon** function with the **modal** switch handles complex conjugate eigenvalues using the approach described following Example 2.8 and returns a states-space realization with purely real coefficient matrices.

Continuing MATLAB Example

State-Space Simulation   For the Continuing MATLAB Example [single-input, single-output rotational mechanical system with input torque $\tau(t)$ and output angular displacement $\theta(t)$], we now simulate the open-loop system response given zero input torque $\tau(t)$ and initial state $x(0) = [0.4, 0.2]^T$. We invoke the **lsim** command which numerically solves for the state vector $x(t)$ from $\dot{x}(t) = Ax(t) + Bu(t)$ given the zero input $u(t)$ and the initial state $x(0)$. Then **lsim** also yields the output $y(t)$ from $y(t) = Cx(t) + Du(t)$. The following MATLAB code, in combination with that in Chapter 1, performs the open-loop system simulation for this example. Appendix C summarizes the entire Continuing MATLAB Example m-file.

```
%-------------------------------------------------
% Chapter 2. Simulation of State-Space Systems
%-------------------------------------------------
t = [0:.01:4];              % Define array of time
                            % values
U = [zeros(size(t))];       % Zero single input of
                            % proper size to go with t
x0 = [0.4; 0.2];            % Define initial state
                            % vector [x10; x20]

CharPoly = poly(A)          % Find characteristic
                            % polynomial from A
```

```
Poles = roots(CharPoly)        % Find the system poles

EigsO  = eig(A);               % Calculate open-loop
                               % system eigenvalues
damp(A);                       % Calculate eigenvalues,
                               % zeta, and wn from ABCD

[Yo,t,Xo] = lsim(JbkR,U,t,xO);% Open-loop response
                               % (zero input, given ICs)

Xo(101,:);                     % State vector value at
                               % t=1 sec
X1 = expm(A*1)*XO;             % Compare with state
                               % transition matrix
                               % method

figure;                        % Open-loop State Plots
subplot(211), plot(t,Xo(:,1)); grid;
   axis([0 4 -0.2 0.5]);
set(gca,'FontSize',18);
ylabel('{\itx}_1 (\itrad)')
subplot(212), plot(t,Xo(:,2)); grid; axis([0 4 -2 1]);
set(gca,'FontSize',18);
xlabel('\ittime (sec)'); ylabel('{\itx}_2 (\itrad/s)');
```

This m-file, combined with the m-file from Chapter 1, generates the following results for the open-loop characteristic polynomial, poles, eigenvalues, damping ratio ξ and undamped natural frequency ω_n, and the value of the state vector at 1 second. The eigenvalues of A agree with those from the **damp** command, and also with **roots** applied to the characteristic polynomial.

```
CharPoly =
   1.0000 4.0000  40.0000

Poles =
   -2.0000 + 6.0000i
   -2.0000 - 6.0000i

EigsO =
   -2.0000 + 6.0000i
   -2.0000 - 6.0000i
```

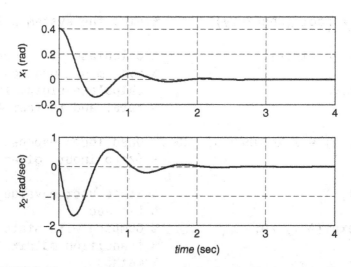

FIGURE 2.2 Open-loop state responses for the Continuing MATLAB Example.

```
Eigenvalue                      Damping      Freq. (rad/s)
-2.00e + 000 + 6.00e + 000i   3.16e - 001   6.32e + 000
-2.00e + 000 - 6.00e + 000i   3.16e - 001   6.32e + 000

X1 =
    0.0457
    0.1293
```

The m-file also generates the open-loop state response of Figure 2.2.

Coordinate Transformations and Diagonal Canonical Form For
the Continuing MATLAB Example, we now compute the diagonal canonical
form state-space realization for the given open-loop system. The following
MATLAB code, along with the MATLAB code from Chapter 1, which also
appears in Appendix C, performs this computation.

```
%-----------------------------------------------------
% Chapter 2. Coordinate Transformations and Diagonal
% Canonical Form
%-----------------------------------------------------

[Tdcf,E] = eig(A);              % Transform to DCF
                                % via formula
Adcf = inv(Tdcf)*A*Tdcf;
Bdcf = inv(Tdcf)*B;
```

```
Cdcf = C*Tdcf;
Ddcf = D;

[JbkRm,Tm] = canon(JbkR,'modal');   % Calculate DCF
                                    % using MATLAB canon
Am = JbkRm.a
Bm = JbkRm.b
Cm = JbkRm.c
Dm = JbkRm.d
```

This m-file, combined with the Chapter 1 m-file, produces the diagonal canonical form realization for the Continuing MATLAB Example:

```
Tdcf =
  -0.0494 - 0.1482i    -0.0494 + 0.1482i
   0.9877     0.9877

Adcf =
  -2.0000 + 6.0000i     0 - 0.0000i
   0.0000 - 0.0000i    -2.0000 - 6.0000i

Bdcf =
   0.5062 + 0.1687i
   0.5062 - 0.1687i

Cdcf =
  -0.0494 - 0.1482i     -0.0494 + 0.1482i

Ddcf =
    0

Tm =
    0       1.0124
  -6.7495  -0.3375

Am =
  -2.0000     6.0000
  -6.0000    -2.0000

Bm =
   1.0124
  -0.3375
```

```
Cm =
   -0.0494  -0.1482
```

```
Dm =
      0
```

We observe that **Am** is a real 2×2 matrix that displays the real and imaginary parts of the complex conjugate eigenvalues $-2 \pm 6i$. The MAT-LAB modal transformation matrix **Tm** above is actually the inverse of our coordinate transformation matrix given in Equation (2.28). Therefore, the inverse of this matrix, for use in our coordinate transformation, is

```
inv(Tm) =
   -0.0494  -0.1482
    0.9877           0
```

The first column of **inv(Tm)** is the real part of the first column of **Tdcf**, which is an eigenvector corresponding to the eigenvalue $-2 + 6i$. The second column of **inv(Tm)** is the imaginary part of this eigenvector.

2.7   CONTINUING EXAMPLES FOR SIMULATION AND COORDINATE TRANSFORMATIONS

Continuing Example 1: Two-Mass Translational Mechanical System

Simulation   The constant parameters in Table 2.1 are given for Continuing Example 1 (two-mass translational mechanical system):

For *case a*, we simulate the open-loop system response given zero initial state and step inputs of magnitudes 20 and 10 N, respectively, for $u_1(t)$ and $u_2(t)$.

For *case b*, we simulate the open-loop system response given zero input $u_2(t)$ and initial state $x(0) = [0.1, 0, 0.2, 0]^T$ [initial displacements of $y_1(0) = 0.1$ and $y_2(0) = 0.2$ m, with zero initial velocities].

TABLE 2.1   Numerical Parameters for Continuing Example 1

i	m_i (kg)	c_i (Ns/m)	k_i (N/m)
1	40	20	400
2	20	10	200

Case a.   For case a, we invoke **lsim** to numerically solve for the state vector $x(t)$ given the inputs $u(t)$ and the zero initial state $x(0)$; **lsim** also yields the output $y(t)$. The state-space coefficient matrices, with parameter values from Table 2.1 above, are

$$
A = \begin{bmatrix} 0 & 1 & 0 & 0 \\ -15 & -0.75 & 5 & 0.25 \\ 0 & 0 & 0 & 1 \\ 10 & 0.5 & -10 & -0.5 \end{bmatrix} \qquad B = \begin{bmatrix} 0 & 0 \\ 0.025 & 0 \\ 0 & 0 \\ 0 & 0.05 \end{bmatrix}
$$

$$
C = \begin{bmatrix} 1 & 0 & 0 & 0 \\ 0 & 0 & 1 & 0 \end{bmatrix} \qquad D = \begin{bmatrix} 0 & 0 \\ 0 & 0 \end{bmatrix}
$$

The plots of outputs $y_1(t)$ and $y_2(t)$ versus time are given in Figure 2.3. We see from Figure 2.3 that this system is lightly damped; there is significant overshoot, and the masses are still vibrating at 40 seconds. The vibratory motion is an underdamped second-order transient response, settling to final nonzero steady-state values resulting from the step inputs. The four open-loop system eigenvalues of A are $s_{1,2} = -0.5 \pm 4.44i$ and $s_{3,4} = -0.125 \pm 2.23i$. The fourth-order system characteristic polynomial is

$$
s^4 + 1.25s^3 + 25.25s^2 + 10s + 100
$$

This characteristic polynomial was found using the MATLAB function **poly(A)**; the roots of this polynomial are identical to the system

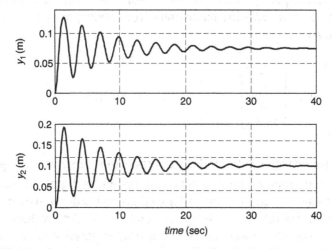

FIGURE 2.3   Open-loop output response for Continuing Example 1, case a.

eigenvalues. There are two modes of vibration in this two-degrees-of-freedom system; both are underdamped with $\xi_1 = 0.112$ and $\omega_{n1} = 4.48$ rad/s for $s_{1,2}$ and $\xi_2 = 0.056$ and $\omega_{n2} = 2.24$ rad/s for $s_{3,4}$. Note that each mode contributes to both $y_1(t)$ and $y_2(t)$ in Figure 2.3. The steady-state values are found by setting $\dot{x}(t) = 0$ in $\dot{x}(t) = Ax(t) + Bu(t)$ to yield $x_{ss} = -A^{-1}Bu$. As a result of the step inputs, the output displacement components do not return to zero in steady-state, as the velocities do: $x_{ss} = [0.075, 0, 0.125, 0]^T$.

Although we focus on state-space techniques, for completeness, the matrix of transfer functions $H(s)$ $[Y(s) = H(s)U(s)]$ is given below for Continuing Example 1, Case a (found from the MATLAB function **tf**):

$$H(s) =$$

$$\begin{bmatrix} \dfrac{0.025s^2 + 0.0125s + 0.25}{s^4 + 1.25s^3 + 25.25s^2 + 10s + 100} & \dfrac{0.0125s + 0.25}{s^4 + 1.25s^3 + 25.25s^2 + 10s + 100} \\[4mm] \dfrac{0.0125s + 0.25}{s^4 + 1.25s^3 + 25.25s^2 + 10s + 100} & \dfrac{0.05s^2 + 0.0375s + 0.75}{s^4 + 1.25s^3 + 25.25s^2 + 10s + 100} \end{bmatrix}$$

Note that the denominator polynomial in every element of H(s) is the same and agrees with the system characteristic polynomial derived from the A matrix and presented earlier. Consequently, the roots of the system characteristic polynomial are identical to the eigenvalues of A.

Case b. For Case b, we again use **lsim** to solve for the state vector $x(t)$ given the zero input $u_2(t)$ and the initial state $x(0)$. The state-space coefficient matrices, with specific parameters from above, are (A is unchanged from Case a):

$$B = \begin{bmatrix} 0 \\ 0 \\ 0 \\ 0.05 \end{bmatrix} \qquad C = [1 \quad 0 \quad 0 \quad 0] \qquad D = 0$$

The plots for states $x_1(t)$ through $x_4(t)$ versus time are given in Figure 2.4. We again see from Figure 2.4 that this system is lightly damped. The vibratory motion is again an underdamped second-order transient response to the initial conditions, settling to zero steady-state values for zero input $u_2(t)$. The open-loop system characteristic polynomial, eigenvalues,

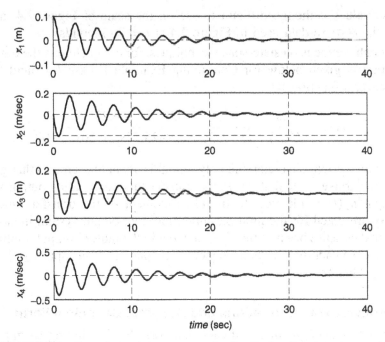

FIGURE 2.4   Open-loop state response for Continuing Example 1, case *b*.

damping ratios, and undamped natural frequencies are all identical to the Case *a* results.

In Figure 2.4, we see that states $x_1(t)$ and $x_3(t)$ start from the given initial displacement values of 0.1 and 0.2, respectively. The given initial velocities are both zero. Note that in this Case *b* example, the final values are all zero because after the transient response has died out, each spring returns to its equilibrium position referenced as zero displacement.

When focusing on the zero-input response, we can calculate the state vector at any desired time by using the state transition matrix $\Phi(t, t_0) = e^{A(t-t_0)}$. For instance, at time $t = 20$ sec:

$$x(20) = \Phi(20, 0)x(0)$$

$$= e^{A(20)}x(0)$$

$$= \begin{bmatrix} 0.0067 \\ -0.0114 \\ 0.0134 \\ -0.0228 \end{bmatrix}$$

These values, although difficult to see on the scale of Figure 2.4, agree with the MATLAB data used in Figure 2.4 at $t = 20$ seconds.

Although we focus on state-space techniques, for completeness the transfer function is given below for Continuing Example 1, case b (found from the MATLAB function **tf**):

$$H(s) = \frac{Y_1(s)}{U_2(s)} = \frac{0.0125s + 0.25}{s^4 + 1.25s^3 + 25.25s^2 + 10s + 100}$$

where the system characteristic polynomial is again the same as that given in case a above. Note that this scalar transfer function relating output $y_1(t)$ to input $u_2(t)$ is identical to the $(1,2)$ element of the transfer function matrix presented for the multiple-input, multiple-output system in case a. This makes sense because the $(1,2)$ element in the multiple-input, multiple-output case captures the dependence of output $y_1(t)$ on input $u_2(t)$ with $u_1(t)$ set to zero.

Coordinate Transformations and Diagonal Canonical Form

We now calculate the diagonal canonical form for Continuing Example 1, case b. If we allow complex numbers in our realization, the transformation matrix to diagonal canonical form is composed of eigenvectors of A arranged column-wise:

$$T_{\text{DCF}} = \begin{bmatrix} 0.017 + 0.155i & 0.017 - 0.155i & -0.010 - 0.182i & -0.010 + 0.182i \\ -0.690 & -0.690 & 0.408 & 0.408 \\ -0.017 - 0.155i & -0.017 + 0.155i & -0.020 - 0.365i & -0.020 + 0.365i \\ 0.690 & 0.690 & 0.817 & 0.817 \end{bmatrix}$$

Applying this coordinate transformation, we obtain the diagonal canonical form:

$$A_{\text{DCF}} = T_{\text{DCF}}^{-1} A T_{\text{DCF}}$$

$$= \begin{bmatrix} -0.50 + 4.44i & 0 & 0 & 0 \\ 0 & -0.50 - 4.44i & 0 & 0 \\ 0 & 0 & -0.125 + 2.23i & 0 \\ 0 & 0 & 0 & -0.125 - 2.23i \end{bmatrix}$$

$$B_{\text{DCF}} = T_{DCF}^{-1} B = \begin{bmatrix} 0.0121 + 0.0014i \\ 0.0121 - 0.0014i \\ 0.0204 + 0.0011i \\ 0.0204 - 0.0011i \end{bmatrix}$$

$$C_{\mathrm{DCF}} = CT_{\mathrm{DCF}} = \begin{bmatrix} 0.017 + 0.155i & 0.017 - 0.155i & -0.010 - 0.182i \end{bmatrix}$$
$$- 0.010 + 0.182i]$$

$$D_{\mathrm{DCF}} = D = 0$$

Note that, as expected, the eigenvalues of the system appear on the diagonal of A_{DCF}.

The MATLAB **canon** function with the switch **modal** yields

$$A_{\mathrm{m}} = \begin{bmatrix} -0.50 & 4.44 & 0 & 0 \\ -4.44 & -0.50 & 0 & 0 \\ 0 & 0 & -0.125 & 2.23 \\ 0 & 0 & -2.23 & -0.125 \end{bmatrix}$$

$$B_{\mathrm{m}} = \begin{bmatrix} 0.024 \\ -0.003 \\ 0.041 \\ -0.002 \end{bmatrix}$$

$$C_{\mathrm{m}} = \begin{bmatrix} 0.017 & 0.155 & -0.010 & -0.182 \end{bmatrix}$$

$$D_{\mathrm{m}} = D = 0$$

which is consistent with our preceding discussions.

Continuing Example 2: Rotational Electromechanical System

Simulation The constant parameters in Table 2.2 are given for Continuing Example 2 [single-input, single-output rotational electromechanical system with input voltage $v(t)$ and output angular displacement $\theta(t)$].

We now simulate the open-loop system response given zero initial state and unit step voltage input. We use the **lsim** function to solve for the state

TABLE 2.2 Numerical Parameters for Continuing Example 2

Parameter	Value	Units	Name
L	1	H	Circuit inductance
R	2	Ω	Circuit resistance
J	1	kg-m^2	Motor shaft polar inertia
b	1	N-m-s	Motor shaft damping constant
k_T	2	N-m/A	Torque constant

vector $x(t)$ given the unit step input $u(t)$ and zero initial state $x(0)$; $y(t)$ also results from the **lsim** function. The state-space coefficient matrices, with parameter values from Table 2.2 above, are

$$A = \begin{bmatrix} 0 & 1 & 0 \\ 0 & 0 & 1 \\ 0 & -2 & -3 \end{bmatrix} \quad B = \begin{bmatrix} 0 \\ 0 \\ 2 \end{bmatrix} \quad C = [1 \quad 0 \quad 0] \quad D = 0$$

Plots for the three state variables versus time are given in Figure 2.5.

We see from Figure 2.5 (top) that the motor shaft angle $\theta(t) = x_1(t)$ increases linearly in the steady state, after the transient response has died out. This is to be expected: If a constant voltage is applied, the motor angular displacement should continue to increase linearly because there is no torsional spring. Then a constant steady-state current and torque result. The steady-state linear slope of $x_1(t)$ in Figure 2.5 is the steady-state value of $\dot{\theta}(t) = x_2(t)$, namely, 1 rad/s. This $x_2(t)$ response is an overdamped second-order response. The third state response, $\ddot{\theta}(t) = x_3(t)$, rapidly rises from its zero initial condition value to a maximum of 0.5 rad/s^2; in steady state, $\ddot{\theta}(t)$ is zero owing to the constant angular velocity $\dot{\theta}(t)$ of the motor shaft. The three open-loop system eigenvalues of A are $s_{1,2,3} = 0, -1, -2$. The third-order system characteristic polynomial is

$$s^3 + 3s^2 + 2s = s(s^2 + 3s + 2)$$
$$= s(s+1)(s+2)$$

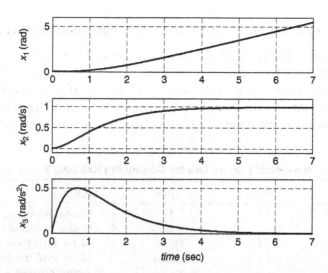

FIGURE 2.5 Open-loop response for Continuing Example 2.

This was found using the MATLAB function **poly(A)**; the roots of this polynomial are identical to the system eigenvalues. The zero eigenvalue corresponds to the rigid-body rotation of the motor shaft; the remaining two eigenvalues -1, -2 led to the conclusion that the shaft angular velocity $\dot{\theta}(t) = \omega(t)$ response is overdamped. Note that we cannot calculate steady state values from $x_{ss} = -A^{-1}Bu$ as in Continuing Example 1 because the system dynamics matrix A is singular because of the zero eigenvalue.

For completeness the scalar transfer function is given below for this example (found via the MATLAB function **tf**):

$$H(s) = \frac{\theta(s)}{V(s)} = \frac{2}{s^3 + 3s^2 + 2s} = \frac{2}{s(s+1)(s+2)}$$

Note the same characteristic polynomial results as reported earlier. The roots of the denominator polynomial are the same as the eigenvalues of A. The preceding transfer function $H(s)$ relates output motor angular displacement $\theta(t)$ to the applied voltage $v(t)$. If we wish to consider the motor shaft angular velocity $\omega(t)$ as the output instead, we must differentiate $\theta(t)$, which is equivalent to multiplying by s, yielding the overdamped second-order system discussed previously:

$$H_2(s) = \frac{\omega(s)}{V(s)} = \frac{2}{(s+1)(s+2)}$$

We could develop an associated two-dimensional state-space realization if we wished to control $\omega(t)$ rather than $\theta(t)$ as the output:

$$x_1(t) = \omega(t)$$

$$x_2(t) = \dot{\omega}(t) = \dot{x}_1(t)$$

$$\begin{bmatrix} \dot{x}_1(t) \\ \dot{x}_2(t) \end{bmatrix} = \begin{bmatrix} 0 & 1 \\ \dfrac{-Rb}{LJ} & \dfrac{-(Lb+RJ)}{LJ} \end{bmatrix} \begin{bmatrix} x_1(t) \\ x_2(t) \end{bmatrix} + \begin{bmatrix} 0 \\ \dfrac{k_T}{LJ} \end{bmatrix} v(t)$$

$$= \begin{bmatrix} 0 & 1 \\ -2 & -3 \end{bmatrix} \begin{bmatrix} x_1(t) \\ x_2(t) \end{bmatrix} + \begin{bmatrix} 0 \\ 2 \end{bmatrix} v(t)$$

$$\omega(t) = \begin{bmatrix} 1 & 0 \end{bmatrix} \begin{bmatrix} x_1(t) \\ x_2(t) \end{bmatrix} + [0]v(t)$$

Coordinate Transformations and Diagonal Canonical Form

We now present the diagonal canonical form for Continuing Example 2. The coordinate transformation matrix for diagonal canonical form is composed of eigenvectors of A arranged column-wise:

$$T_{\text{DCF}} = \begin{bmatrix} 1 & -0.577 & 0.218 \\ 0 & 0.577 & -0.436 \\ 0 & -0.577 & 0.873 \end{bmatrix}$$

Applying this coordinate transformation, we obtain the diagonal canonical form:

$$A_{\text{DCF}} = T_{\text{DCF}}^{-1} A T_{\text{DCF}} = \begin{bmatrix} 0 & 0 & 0 \\ 0 & -1 & 0 \\ 0 & 0 & -2 \end{bmatrix}$$

$$B_{\text{DCF}} = T_{\text{DCF}}^{-1} B = \begin{bmatrix} 1 \\ 3.464 \\ 4.583 \end{bmatrix}$$

$$C_{\text{DCF}} = C T_{\text{DCF}} = \begin{bmatrix} 1 & -0.577 & 0.218 \end{bmatrix}$$

$$D_{\text{DCF}} = D = 0$$

Note that the system eigenvalues appear on the main diagonal of diagonal matrix A_{DCF}, as expected.

The MATLAB **canon** function with the **modal** switch yields identical results because the system eigenvalues are real.

2.8   HOMEWORK EXERCISES

Numerical Exercises

NE2.1 Solve $2\dot{x}(t) + 5x(t) = u(t)$ for $x(t)$, given that $u(t)$ is the unit step function and initial state $x(0) = 0$. Calculate the time constant and plot your $x(t)$ result versus time.

NE2.2 For the following systems described by the given state equations, derive the associated transfer functions.

a. $$\begin{bmatrix} \dot{x}_1(t) \\ \dot{x}_2(t) \end{bmatrix} = \begin{bmatrix} -3 & 0 \\ 0 & -4 \end{bmatrix} \begin{bmatrix} x_1(t) \\ x_2(t) \end{bmatrix} + \begin{bmatrix} 1 \\ 1 \end{bmatrix} u(t)$$

$$y(t) = \begin{bmatrix} 1 & 1 \end{bmatrix} \begin{bmatrix} x_1(t) \\ x_2(t) \end{bmatrix} + [0]u(t)$$

b.
$$\begin{bmatrix} \dot{x}_1(t) \\ \dot{x}_2(t) \end{bmatrix} = \begin{bmatrix} 0 & 1 \\ -3 & -2 \end{bmatrix} \begin{bmatrix} x_1(t) \\ x_2(t) \end{bmatrix} + \begin{bmatrix} 0 \\ 1 \end{bmatrix} u(t)$$

$$y(t) = \begin{bmatrix} 1 & 0 \end{bmatrix} \begin{bmatrix} x_1(t) \\ x_2(t) \end{bmatrix} + [0]u(t)$$

c.
$$\begin{bmatrix} \dot{x}_1(t) \\ \dot{x}_2(t) \end{bmatrix} = \begin{bmatrix} 0 & -2 \\ 1 & -12 \end{bmatrix} \begin{bmatrix} x_1(t) \\ x_2(t) \end{bmatrix} + \begin{bmatrix} 1 \\ 0 \end{bmatrix} u(t)$$

$$y(t) = \begin{bmatrix} 0 & 1 \end{bmatrix} \begin{bmatrix} x_1(t) \\ x_2(t) \end{bmatrix} + [0]u(t)$$

d.
$$\begin{bmatrix} \dot{x}_1(t) \\ \dot{x}_2(t) \end{bmatrix} = \begin{bmatrix} 1 & 2 \\ 3 & 4 \end{bmatrix} \begin{bmatrix} x_1(t) \\ x_2(t) \end{bmatrix} + \begin{bmatrix} 5 \\ 6 \end{bmatrix} u(t)$$

$$y(t) = \begin{bmatrix} 7 & 8 \end{bmatrix} \begin{bmatrix} x_1(t) \\ x_2(t) \end{bmatrix} + [9]u(t)$$

NE2.3 Determine the characteristic polynomial and eigenvalues for the systems represented by the following system dynamics matrices.

a. $A = \begin{bmatrix} -1 & 0 \\ 0 & -2 \end{bmatrix}$

b. $A = \begin{bmatrix} 0 & 1 \\ -10 & -20 \end{bmatrix}$

c. $A = \begin{bmatrix} 0 & 1 \\ -10 & 0 \end{bmatrix}$

d. $A = \begin{bmatrix} 0 & 1 \\ 0 & -20 \end{bmatrix}$

NE2.4 For the given homogeneous system below, subject only to the initial state $x(0) = [2, \ 1]^T$, calculate the matrix exponential and the state vector at time $t = 4$ seconds.

$$\begin{bmatrix} \dot{x}_1(t) \\ \dot{x}_2(t) \end{bmatrix} = \begin{bmatrix} 0 & 1 \\ -6 & -12 \end{bmatrix} \begin{bmatrix} x_1(t) \\ x_2(t) \end{bmatrix}$$

NE2.5 Solve the two-dimensional state equation below for the state vector $x(t)$, given that the input $u(t)$ is a unit step input and zero initial state. Plot both state components versus time.

$$\begin{bmatrix} \dot{x}_1(t) \\ \dot{x}_2(t) \end{bmatrix} = \begin{bmatrix} 0 & 1 \\ -8 & -6 \end{bmatrix} \begin{bmatrix} x_1(t) \\ x_2(t) \end{bmatrix} + \begin{bmatrix} 0 \\ 1 \end{bmatrix} u(t)$$

NE2.6 Solve the two-dimensional state equation

$$\begin{bmatrix} \dot{x}_1(t) \\ \dot{x}_2(t) \end{bmatrix} = \begin{bmatrix} 0 & 1 \\ -2 & -2 \end{bmatrix} \begin{bmatrix} x_1(t) \\ x_2(t) \end{bmatrix} + \begin{bmatrix} 0 \\ 1 \end{bmatrix} u(t)$$

$$\begin{bmatrix} x_1(0) \\ x_2(0) \end{bmatrix} = \begin{bmatrix} 1 \\ -1 \end{bmatrix}$$

for a unit step $u(t)$.

NE2.7 Solve the two-dimensional state equation

$$\begin{bmatrix} \dot{x}_1(t) \\ \dot{x}_2(t) \end{bmatrix} = \begin{bmatrix} 0 & 1 \\ -5 & -2 \end{bmatrix} \begin{bmatrix} x_1(t) \\ x_2(t) \end{bmatrix} + \begin{bmatrix} 0 \\ 1 \end{bmatrix} u(t)$$

$$\begin{bmatrix} x_1(0) \\ x_2(0) \end{bmatrix} = \begin{bmatrix} 1 \\ -1 \end{bmatrix}$$

for a unit step $u(t)$.

NE2.8 Solve the two-dimensional state equation

$$\begin{bmatrix} \dot{x}_1(t) \\ \dot{x}_2(t) \end{bmatrix} = \begin{bmatrix} 0 & 1 \\ -1 & -2 \end{bmatrix} \begin{bmatrix} x_1(t) \\ x_2(t) \end{bmatrix} + \begin{bmatrix} 0 \\ 1 \end{bmatrix} u(t)$$

$$\begin{bmatrix} x_1(0) \\ x_2(0) \end{bmatrix} = \begin{bmatrix} 1 \\ -2 \end{bmatrix}$$

$$y(t) = \begin{bmatrix} 2 & 1 \end{bmatrix} \begin{bmatrix} x_1(t) \\ x_2(t) \end{bmatrix}$$

for $u(t) = e^{-2t}$, $t \geq 0$.

NE2.9 Calculate the complete response $y(t)$ for the state equation

$$\begin{bmatrix} \dot{x}_1(t) \\ \dot{x}_2(t) \end{bmatrix} = \begin{bmatrix} -2 & 0 \\ 0 & -3 \end{bmatrix} \begin{bmatrix} x_1(t) \\ x_2(t) \end{bmatrix} + \begin{bmatrix} 1 \\ 1 \end{bmatrix} u(t)$$

$$x(0) = \begin{bmatrix} \frac{2}{3} \\ \frac{1}{2} \end{bmatrix}$$

$$y(t) = \begin{bmatrix} -3 & 4 \end{bmatrix} \begin{bmatrix} x_1(t) \\ x_2(t) \end{bmatrix}$$

for the input signal $u(t) = 2e^t$, $t \geqslant 0$. Identify zero-input and zero-state response components.

NE2.10 Diagonalize the following system dynamics matrices A using coordinate transformations.

a. $A = \begin{bmatrix} 0 & 1 \\ -8 & -10 \end{bmatrix}$

b. $A = \begin{bmatrix} 0 & 1 \\ 10 & 6 \end{bmatrix}$

c. $A = \begin{bmatrix} 0 & -10 \\ 1 & -1 \end{bmatrix}$

d. $A = \begin{bmatrix} 0 & 10 \\ 1 & 0 \end{bmatrix}$

Analytical Exercises

AE2.1 If A and B are $n \times n$ matrices, show that

$$e^{(A+B)t} = e^{At} + \int_0^t e^{A(t-\tau)} B e^{(A+B)\tau} \, d\tau$$

You may wish to use the Leibniz rule for differentiating an integral:

$$\frac{d}{dt} \int_{a(t)}^{b(t)} X(t, \tau) \, d\tau = X[t, b(t)] \dot{b}(t)$$

$$- X[t, a(t)] \dot{a}(t) + \int_{a(t)}^{b(t)} \frac{\partial X(t, \tau)}{\partial t} \, d\tau$$

AE2.2 Show that for any $n \times n$ matrix A and any scalar γ

$$e^{(\gamma I + A)t} = e^{\gamma t} e^{At}$$

AE2.3 A real $n \times n$ matrix A is called *skew symmetric* if $A^T = -A$. A real $n \times n$ matrix R is called *orthogonal* if $R^{-1} = R^T$. Given a skew symmetric $n \times n$ matrix A, show that the matrix exponential e^{At} is an orthogonal matrix for every t.

AE2.4 Show that the upper block triangular matrix

$$A = \begin{bmatrix} A_{11} & A_{12} \\ 0 & A_{22} \end{bmatrix}$$

has the matrix exponential

$$e^{At} = \begin{bmatrix} e^{A_{11}t} & \int_0^t e^{A_{11}(t-\tau)} A_{12} e^{A_{22}\tau} \, d\tau \\ 0 & e^{A_{22}t} \end{bmatrix}$$

AE2.5 Show that the matrix exponential satisfies

$$e^{At} = I + A \int_0^t e^{A\tau} \, d\tau$$

Use this to derive an expression for $\int_0^t e^{A\tau} d\tau$ in the case where A is nonsingular.

AE2.7 For $n \times n$ matrices A and Q, show that the matrix differential equation

$$\dot{W}(t, t_0) = A\, W(t, t_0) + W(t, t_0) A^T + Q \quad W(t_0, t_0) = 0$$

has the solution

$$W(t, t_0) = \int_{t_0}^t e^{A(t-\tau)} \, Q \, e^{A^T(t-\tau)} \, d\tau$$

AE2.8 Verify that the three–dimensional state equation

$$\dot{x}(t) = Ax(t) + Bu(t)$$

specified by

$$A = \begin{bmatrix} 0 & 1 & 0 \\ 0 & 0 & 1 \\ -a_0 & -a_1 & -a_2 \end{bmatrix} \quad B = \begin{bmatrix} 1 & 0 & 0 \\ a_2 & 1 & 0 \\ a_1 & a_2 & 1 \end{bmatrix}^{-1} \begin{bmatrix} b_2 \\ b_1 \\ b_0 \end{bmatrix}$$

$$C = \begin{bmatrix} 1 & 0 & 0 \end{bmatrix}$$

is a state-space realization of the transfer function

$$H(s) = \frac{b_2 s^2 + b_1 s + b_0}{s^3 + a_2 s^2 + a_1 s + a_0}$$

AE2.9 Verify that the three–dimensional state equation

$$\dot{x}(t) = Ax(t) + Bu(t)$$
$$y(t) = C_x(t)$$

specified by

$$A = \begin{bmatrix} 0 & 0 & -a_0 \\ 1 & 0 & -a_1 \\ 0 & 1 & -a_2 \end{bmatrix} \quad B = \begin{bmatrix} 1 \\ 0 \\ 0 \end{bmatrix}$$

$$C = \begin{bmatrix} b_2 & b_1 & b_0 \end{bmatrix} \begin{bmatrix} 1 & a_2 & a_1 \\ 0 & 1 & a_2 \\ 0 & 0 & 1 \end{bmatrix}^{-1}$$

is a state-space realization of the transfer function

$$H(s) = \frac{b_2 s^2 + b_1 s + b_0}{s^3 + a_2 s^2 + a_1 s + a_0}$$

AE2.10 Show that if the multiple-input, multiple-output state equation

$$\dot{x}(t) = Ax(t) + Bu(t)$$
$$y(t) = Cx(t) + Du(t)$$

is a state-space realization of the transfer function matrix $H(s)$, then the so-called dual state equation

$$\dot{z}(t) = -A^T z(t) - C^T v(t)$$
$$w(t) = B^T z(t) + D^T v(t)$$

is a state-space realization of $H^T(-s)$.

AE2.11 Let the single-input, single-output state equation

$$\dot{x}(t) = Ax(t) + Bu(t) \quad x(0) = x_0$$
$$y(t) = Cx(t) + Du(t)$$

be a state-space realization of the transfer function $H(s)$. Suppose that is $z_0 \in \mathbb{C}$ not an eigenvalue of A for which

$$\begin{bmatrix} z_0 I - A & -B \\ C & D \end{bmatrix}$$

is singular. Show that z_0 is a zero of the transfer function $H(s)$. Furthermore, show that there exist nontrivial $x_0 \in \mathbb{C}^n$ and $u. \in \mathbb{C}$ such that $x(0) = x_0$ and $u(t) = u_0 \, e^{z_0 t}$ yield $y(t) = 0$ for all $t \geqslant 0$.

AE2.12 For the single-input, single-output state equation

$$\dot{x}(t) = Ax(t) + Bu(t) \quad x(0) = x_0$$
$$y(t) = Cx(t) + Du(t)$$

with $D \neq 0$, show that the related state equation

$$\dot{z}(t) = (A - BD^{-1}C)z(t) + BD^{-1}v(t) \quad z(0) = z_0$$
$$w(t) = -D^{-1}Cz(t) + D^{-1}v(t)$$

has the property that if $z_0 = x_0$ and $v(t) = y(t)$, then $w(t) = u(t)$.

AE2.13 For the m-input, m-output state equation

$$\dot{x}(t) = Ax(t) + Bu(t) \quad x(0) = x_0$$
$$y(t) = Cx(t)$$

with the $m \times m$ matrix CB nonsingular, show that the related state equation

$$\dot{z}(t) = (A - B(CB)^{-1}CA)z(t) + B(CB)^{-1}v(t) \quad z(0) = z_0$$
$$w(t) = -(CB)^{-1}CAz(t) + (CB)^{-1}v(t)$$

has the property that if $z_0 = x_0$ and $v(t) = \dot{y}(t)$, then $w(t) = u(t)$.

Continuing MATLAB Exercises

CME2.1 For the system given in CME1.1:
   a. Determine and plot the impulse response.

b. Determine and plot the unit step response for zero initial state.

c. Determine and plot the zero input response given $x_0 = [1, -1]^T$.

d. Calculate the diagonal canonical form.

CME2.2 For the system given in CME1.2:

a. Determine and plot the impulse response.

b. Determine and plot the unit step response for zero initial state.

c. Determine and plot the zero input response given $x_0 = [1, 2, 3]^T$.

d. Calculate the diagonal canonical form.

CME2.3 For the system given in CME1.3:

a. Determine and plot the impulse response.

b. Determine and plot the unit step response for zero initial state.

c. Determine and plot the zero input response given $x_0 = [4, 3, 2, 1]^T$.

d. Calculate the diagonal canonical form.

CME2.4 For the system given in CME1.4:

a. Determine and plot the impulse response.

b. Determine and plot the unit step response for zero initial state.

c. Determine and plot the zero input response given $x_0 = [1, 2, 3, 4]^T$.

d. Calculate the diagonal canonical form.

Continuing Exercises

CE2.1a Use the numerical parameters in Table 2.3 for this and all ensuing CE1 assignments (see Figure 1.14).

Simulate and plot the resulting open-loop output displacements for three cases (for this problem, use the state-space realizations of CE1.1b):

   i. Multiple-input, multiple-output: three inputs $u_i(t)$ and three outputs $y_i(t)$, $i = 1, 2, 3$.

TABLE 2.3 Numerical Parameters for CE1 System

i	m_i (kg)	c_i (Ns/m)	k_i (N/m)
1	1	0.4	10
2	2	0.8	20
3	3	1.2	30
4		1.6	40

(a) Step inputs of magnitudes $u_1(t) = 3$, $u_2(t) = 2$, and $u_3(t) = 1$ (N). Assume zero initial state.

(b) Zero inputs $u(t)$. Assume initial displacements $y_1(0) = 0.005$, $y_2(0) = 0.010$, and $y_3(0) = 0.015$ (m); Assume zero initial velocities. Plot all six state components.

ii. Multiple-input, multiple-output: two unit step inputs $u_1(t)$ and $u_3(t)$, three displacement outputs $y_i(t), i = 1, 2, 3$. Assume zero initial state.

iii. Single-input, single-output: unit step input $u_2(t)$ and output $y_3(t)$. Assume zero initial state. Plot all six state components.

For each case, simulate long enough to demonstrate the steady-state behavior. For all plots, use the MATLAB **subplot** function to plot each variable on separate axes, aligned vertically with the same time range. What are the system eigenvalues? These define the nature of the system transient response. For case i(b) only, check your state vector results at $t = 10$ seconds using the matrix exponential.

Since this book does not focus on modeling, the solution for CE1.1a is given below:

$$m_1\ddot{y}_1(t) + (c_1 + c_2)\dot{y}_1(t) + (k_1 + k_2)y_1(t)$$
$$- c_2\dot{y}_2(t) - k_2y_2(t) = u_1(t)$$
$$m_2\ddot{y}_2(t) + (c_2 + c_3)\dot{y}_2(t) + (k_2 + k_3)y_2(t) - c_2\dot{y}_1(t) - k_2y_1(t)$$
$$- c_3\dot{y}_3(t) - k_3y_3(t) = u_2(t)$$
$$m_3\ddot{y}_3(t) + (c_3 + c_4)\dot{y}_3(t) + (k_3 + k_4)y_3(t)$$
$$- c_3\dot{y}_2(t) - k_3y_2(t) = u_3(t)$$

One possible solution for CE1.1b (system dynamics matrix A only) is given below. This A matrix is the same for all input-output cases, while $B, C,$ and D will be different for each case. First, the state variables associated with this realization are

$$x_1(t) = y_1(t) \qquad x_3(t) = y_2(t) \qquad x_5(t) = y_3(t)$$
$$x_2(t) = \dot{y}_1(t) = \dot{x}_1(t) \quad x_4(t) = \dot{y}_2(t) = \dot{x}_3(t) \quad x_6(t) = \dot{y}_3(t) = \dot{x}_5(t)$$

$$A = \begin{bmatrix} 0 & 1 & 0 & 0 & 0 & 0 \\ -\dfrac{(k_1 + k_2)}{m_1} & -\dfrac{(c_1 + c_2)}{m_1} & \dfrac{k_2}{m_1} & \dfrac{c_2}{m_1} & 0 & 0 \\ 0 & 0 & 0 & 1 & 0 & 0 \\ \dfrac{k_2}{m_2} & \dfrac{c_2}{m_2} & -\dfrac{(k_2 + k_3)}{m_2} & -\dfrac{(c_2 + c_3)}{m_2} & \dfrac{k_3}{m_2} & \dfrac{c_3}{m_2} \\ 0 & 0 & 0 & 0 & 0 & 1 \\ 0 & 0 & \dfrac{k_3}{m_3} & \dfrac{c_3}{m_3} & -\dfrac{(k_3 + k_4)}{m_3} & -\dfrac{(c_3 + c_4)}{m_3} \end{bmatrix}$$

CE2.1b Calculate the diagonal canonical form realization for the case iii CE1 system. Comment on the structure of the results.

CE2.2a Use the numerical parameters in Table 2.4 for this and all ensuing CE2 assignments (see Figure 1.15).
Simulate and plot the open-loop state variable responses for three cases (for this problem use the state-space realizations of CE1.2b); assume zero initial state for all cases [except Case i(b) below]:

    i. Single-input, single-output: input $f(t)$ and output $\theta(t)$.
      (a) unit impulse input $f(t)$ and zero initial state.
      (b) zero input $f(t)$ and an initial condition of $\theta(0) = 0.1$ rad (zero initial conditions on all other state variables).
    ii. Single-input, multiple-output: impulse input $f(t)$ and two outputs $w(t)$ and $\theta(t)$.
    iii. Multiple-input, multiple-output: two unit step inputs $f(t)$ and $\tau(t)$ and two outputs $w(t)$ and $\theta(t)$.

Simulate long enough to demonstrate the steady-state behavior. What are the system eigenvalues? Based on these eigenvalues and the physical system, explain the system responses.

TABLE 2.4   Numerical Parameters for CE2 System

Parameter	Value	Units	Name
m_1	2	kg	cart mass
m_2	1	kg	pendulum mass
L	0.75	m	pendulum length
g	9.81	m/s^2	gravitational acceleration

Since this book does not focus on modeling, the solution for CE1.2a is given below:

Coupled Nonlinear Differential Equations

$$(m_1 + m_2)\ddot{w}(t) - m_2 L \cos\theta(t)\ddot{\theta}(t) + m_2 L \sin\theta(t)\dot{\theta}(t)^2 = f(t)$$

$$m_2 L^2\ddot{\theta}(t) - m_2 L \cos\theta(t)\ddot{w}(t) - m_2 g L \sin\theta(t) = 0$$

Coupled Linearized Differential Equations

$$(m_1 + m_2)\ddot{w}(t) - m_2 L\ddot{\theta}(t) = f(t)$$

$$-m_2\ddot{w}(t) + m_2 L\ddot{\theta}(t) - m_2 g\theta(t) = 0$$

Coupled Linearized Differential Equations with Torque Motor Included

$$(m_1 + m_2)\ddot{w}(t) - m_2 L\ddot{\theta}(t) = f(t)$$

$$-m_2 L\ddot{w}(t) + m_2 L^2\ddot{\theta}(t) - m_2 g L\theta(t) = \tau(t)$$

Note that the coupled nonlinear differential equations could have been converted first to state-space form and then linearized about a nominal trajectory, as described in Section 1.4; a natural choice for the nominal trajectory is zero pendulum angle and rate, plus zero cart position (center of travel) and rate. Consider this as an alternate solution to CE1.2b—you will get the same A, B, C, and D matrices.

One possible solution for CE1.2b (system dynamics matrix A only) is given below. This A matrix is the same for all input-output cases, whereas B, C, and D will be different for each case. First, the state variables associated with this realization are

$$x_1(t) = w(t) \qquad\qquad x_3(t) = \theta(t)$$

$$x_2(t) = \dot{w}(t) = \dot{x}_1(t) \qquad x_4(t) = \dot{\theta}(t) = \dot{x}_3(t)$$

$$A = \begin{bmatrix} 0 & 1 & 0 & 0 \\ 0 & 0 & \dfrac{m_2 g}{m_1} & 0 \\ 0 & 0 & 0 & 1 \\ 0 & 0 & \dfrac{(m_1 + m_2)g}{m_1 L} & 0 \end{bmatrix}$$

TABLE 2.5 Numerical Parameters for CE3 System

Parameter	Value	Units	Name
L	0.0006	H	armature inductance
R	1.40	Ω	armature resistance
k_B	0.00867	V/deg/s	motor back emf constant
J_M	0.00844	lb$_f$-in-s^2	motor shaft polar inertia
b_M	0.00013	lb$_f$-in/deg/s	motor shaft damping constant
k_T	4.375	lb$_f$-in/A	torque constant
n	200	unitless	gear ratio
J_L	1	lb$_f$-in-s^2	load shaft polar inertia
b_L	0.5	lb$_f$-in/deg/s	load shaft damping constant

CE2.2b For the case i CE2 system, try to calculate the diagonal canonical form realization (diagonal canonical form cannot be found—why?).

CE2.3a Use the numerical parameters in Table 2.5 for this and all ensuing CE3 assignments (see Figure 1.16).

Simulate and plot the open-loop state variable responses for two cases (for this problem, use the state-space realizations of CE1.3b):

i. Single-input, single-output: input armature voltage $v_A(t)$ and output robot load shaft angle $\theta_L(t)$.

   (a) Unit step input armature voltage $v_A(t)$; plot all three state variables given zero initial state.

   (b) Zero input armature voltage $v_A(t)$; plot all three state variables given initial state $\theta_L(0) = 0, \dot{\theta}_L(0) = 1$, and $\ddot{\theta}_L(0) = 0$.

ii. Single-input, single-output: unit step input armature voltage $v_A(t)$ and output robot load shaft angular velocity $\omega_L(t)$; plot both state variables. For this case, assume zero initial state.

Simulate long enough to demonstrate the steady-state behavior. What are the system eigenvalues? Based on these eigenvalues and the physical system, explain the system responses.

Since this book does not focus on modeling, the solution for CE1.3a is given below; the overall transfer function is

$$G(s) = \frac{\Theta_L(s)}{V_A(s)} = \frac{k_T/n}{LJs^3 + (Lb + RJ)s^2 + (Rb + k_Tk_B)s}$$

where $J = J_M + \dfrac{J_L}{n^2}$ and $b = b_M + \dfrac{b_L}{n^2}$ are the effective polar inertia and viscous damping coefficient reflected to the motor shaft. The associated single-input, single-output ordinary differential equation is

$$LJ\ddot{\theta}_L(t) + (Lb + RJ)\ddot{\theta}_L(t) + (Rb + k_Tk_B)\dot{\theta}_L(t) = \frac{k_T}{n}v_A(t)$$

One possible solution for CE1.3b (case i) is given below. The state variables and output associated with the solution below are:

$$x_1(t) = \theta_L(t)$$
$$x_2(t) = \dot{\theta}_L(t) = \dot{x}_1(t) \qquad y(t) = \theta_L(t) = x_1(t)$$
$$x_3(t) = \ddot{\theta}_L(t) = \ddot{x}_1(t) = \dot{x}_2(t)$$

The state differential and algebraic output equations are

$$\begin{bmatrix} \dot{x}_1(t) \\ \dot{x}_2(t) \\ \dot{x}_3(t) \end{bmatrix} = \begin{bmatrix} 0 & 1 & 0 \\ 0 & 0 & 1 \\ 0 & -\dfrac{(Rb + k_Tk_B)}{LJ} & -\dfrac{(Lb + RJ)}{LJ} \end{bmatrix} \begin{bmatrix} x_1(t) \\ x_2(t) \\ x_3(t) \end{bmatrix}$$

$$+ \begin{bmatrix} 0 \\ 0 \\ \dfrac{k_T}{LJn} \end{bmatrix} v_A(t)$$

$$y(t) = \begin{bmatrix} 1 & 0 & 0 \end{bmatrix} \begin{bmatrix} x_1(t) \\ x_2(t) \\ x_3(t) \end{bmatrix} + [0]v_A(t)$$

The solution for case ii is similar, but of reduced (second) order:

$$x_1(t) = \omega_L(t)$$
$$x_2(t) = \dot{\omega}_L(t) = \dot{x}_1(t) \qquad y(t) = \omega_L(t) = x_1(t)$$

$$\begin{bmatrix} \dot{x}_1(t) \\ \dot{x}_2(t) \end{bmatrix} = \begin{bmatrix} 0 & 1 \\ -\dfrac{(Rb + k_Tk_B)}{LJ} & -\dfrac{(Lb + RJ)}{LJ} \end{bmatrix} \begin{bmatrix} x_1(t) \\ x_2(t) \end{bmatrix}$$

$$+ \begin{bmatrix} 0 \\ \dfrac{k_T}{LJn} \end{bmatrix} v_A(t)$$

$$y(t) = \begin{bmatrix} 1 & 0 \end{bmatrix} \begin{bmatrix} x_1(t) \\ x_2(t) \end{bmatrix} + [0]v_A(t)$$

CE2.3b Calculate the diagonal canonical form realization for the case i CE3 system. Comment on the structure of the results.

CE2.4a Use the numerical parameters in Table 2.6 for this and all ensuing CE4 assignments (see Figure 1.8).

Simulate and plot the open-loop state variables in response to an impulse torque input $\tau(t) = \delta(t)$ Nm and $p(0) = 0.25$ m with zero initial conditions on all other state variables. Simulate long enough to demonstrate the steady-state behavior. What are the system eigenvalues? Based on these eigenvalues and the physical system, explain the system response to these initial conditions.

A valid state-space realization for this system is given in Example 1.7, linearized about the nominal trajectory discussed there. This linearization was performed for a horizontal beam with a ball trajectory consisting of an initial ball position and constant ball translational velocity. However, the realization in Example 1.7 is time varying because it depends on the nominal ball position $\tilde{p}(t)$. Therefore, for CE4, place a further constraint on this system linearization to obtain a linear time-invariant system: Set the constant ball velocity to zero ($v_0 = 0$) and set $\tilde{p}(t) = p_0 = 0.25$ m. Discuss the likely real-world impact of this linearization and constraint.

CE2.4b Calculate the diagonal canonical form realization for the CE4 system.

CE2.5a Use the numerical parameters in Table 2.7 (Bupp et al., 1998) for this and all ensuing CE5 assignments (see Figure 1.17).

TABLE 2.6 Numerical Parameters for CE4 System

Parameter	Value	Units	Name
L	1	m	beam length (rotates about center)
J	0.0676	kg-m^2	beam mass moment of inertia
m	0.9048	kg	ball mass
r	0.03	m	ball radius
J_b	0.000326	kg-m^2	ball mass moment of inertia
g	9.81	m/s^2	gravitational acceleration

TABLE 2.7   Numerical Parameters for CE5 System

Parameter	Value	Units	Name
M	1.3608	kg	cart mass
k	186.3	N/m	spring stiffness constant
m	0.096	kg	pendulum-end point mass
J	0.0002175	kg-m^2	pendulum mass moment of inertia
e	0.0592	m	pendulum length

Simulate and plot the open-loop state variables in response to the initial conditions $q(0) = 0.05$ m, $\dot{q}(0) = 0$, $\theta(0) = \pi/4$ rad, and $\dot{\theta}(0) = 0$ rad/s. Simulate long enough to demonstrate the steady-state behavior. What are the system eigenvalues? Based on these eigenvalues and the physical system, explain the system response to these initial conditions.

Since this book does not focus on modeling, the solution for CE1.5a is given below:

$$(M + m)\ddot{q}(t) + kq(t) + me(\ddot{\theta}(t)\cos\theta(t) - \dot{\theta}^2(t)\sin\theta(t)) = 0$$

$$(J + me^2)\ddot{\theta}(t) + me\ddot{q}(t)\cos\theta(t) = n(t)$$

A valid state-space realization for this system is given below:

$$x_1(t) = q(t) \qquad x_3(t) = \theta(t)$$
$$x_2(t) = \dot{q}(t) = \dot{x}_1(t) \qquad x_4(t) = \dot{\theta}(t) = \dot{x}_3(t)$$

$$A = \begin{bmatrix} 0 & 1 & 0 & 0 \\ \dfrac{-k(J + me^2)}{d(\tilde{\theta})} & 0 & 0 & 0 \\ 0 & 0 & 0 & 1 \\ \dfrac{kme\cos(\tilde{\theta})}{d(\tilde{\theta})} & 0 & 0 & 0 \end{bmatrix} \qquad B = \begin{bmatrix} 0 \\ \dfrac{-me\cos(\tilde{\theta})}{d(\tilde{\theta})} \\ 0 \\ \dfrac{M + m}{d(\tilde{\theta})} \end{bmatrix}$$

$$C = \begin{bmatrix} 1 & 0 & 0 & 0 \end{bmatrix} \qquad\qquad D = 0$$

where $d(\tilde{\theta}) = (M + m)(J + me^2) - (me\cos(\tilde{\theta}))^2$. Note that this linearized state-space realization depends on the zero-torque equilibrium for which the linearization was performed. For CE5, place a further constraint on this system linearization to obtain a linear time-invariant system: Set the nominal pendulum angle to $\bar{\theta} = \pi/4$. Discuss the likely impact of this linearization and constraint.

Note: The original system of CE1.5 is nonlinear (because the pendulum can rotate without limit); in order to control it properly, one must use nonlinear techniques that are beyond the scope of this book. Please see Bernstein (1998) a special nonlinear control journal issue with seven articles that survey different nonlinear control approaches applied to this benchmark problem.

CE2.5b Calculate the diagonal canonical form realization for the CE5 system.

3

CONTROLLABILITY

In this chapter we explore the input-to-state interaction of the n-dimensional linear time-invariant state equation, seeking to characterize the extent to which state trajectories can be controlled by the input signal. Specifically, we derive conditions under which, starting anywhere in the state space, the state trajectory can be driven to the origin by piecewise continuous input signals over a finite time interval. More generally, it turns out that if it is possible to do this, then it is also possible to steer the state trajectory to *any* final state in finite time via a suitable input signal.

While the controllability analysis presented in this chapter has a decidedly open-loop flavor involving input signals that have a prescribed effect on the state trajectory without any mention of feedback, there is also an important connection between controllability and the design of feedback-control laws that will be pursued in Chapter 7.

This chapter initiates a strategy that will recur throughout the remainder of this book. Specifically, our aim is to characterize important properties of *dynamic systems* described by our state-space model via analysis of the state-equation data, namely the coefficient matrices A, B, C, and D. Fundamental questions pertaining to the complex interaction of input, state, and output can be answered using the tools of linear algebra. Much of what we present in what follows either originated with or was inspired by the pioneering work of R. E. Kalman, unquestionably the father of state-space

Linear State-Space Control Systems, by Robert L. Williams II and Douglas A. Lawrence
Copyright © 2007 John Wiley & Sons, Inc.

methods in linear systems and control theory. In our investigation of controllability, we identify connections between input-to-state behavior, as characterized by the state-equation solution derived in Chapter 2, and linear algebraic properties of the state equation's coefficient matrices.

Once the basic notion of controllability is defined, analyzed, and illustrated with several examples, we study the relationship between controllability and state coordinate transformations and present additional algebraic characterizations of controllability that are of utility not only in the analysis of particular systems but also in establishing further general results. We also illustrate the use of MATLAB for controllability analysis and once again return to the MATLAB Continuing Example along with Continuing Examples 1 and 2.

3.1 FUNDAMENTAL RESULTS

We consider the linear time-invariant state differential equation

$$\dot{x}(t) = Ax(t) + Bu(t) \quad x(t_0) = x_0 \tag{3.1}$$

in which the algebraic output equation has been omitted because it will play no role in the ensuing analysis. Our point of departure is the following definition.

Definition 3.1 A state $x \in \mathbb{R}^n$ is *controllable to the origin* if for a given initial time t_0 there exists a finite final time $t_f > t_0$ and a piecewise continuous input signal $u(\cdot)$ defined on $[t_0, t_f]$ such that with initial state $x(t_0) = x$, the final state satisfies

$$x(t_f) = e^{A(t_f - t_0)}x + \int_{t_0}^{t_f} e^{A(t_f - \tau)} Bu(\tau)d\tau$$

$$= 0 \in \mathbb{R}^n$$

The state equation (3.1) is *controllable* if every state $x \in \mathbb{R}^n$ is controllable to the origin.

Based on this definition alone, determining whether or not a particular state equation is controllable appears to be a daunting task because it is not immediately clear how to characterize input signals that have the prescribed effect on the state trajectory. Our immediate goal is to overcome this difficulty by translating this controllability property of state equations into an equivalent linear algebraic property of the state equation's coefficient matrices A and B.

Theorem 3.2   *The linear state equation* (3.1) *is controllable if and only if*

$$\text{rank}[\, B \quad AB \quad A^2 B \quad \cdots \quad A^{n-1} B \,] = n$$

We refer to this matrix as the *controllability matrix* and often denote it by P to save considerable writing. We note that this controllability matrix is constructed directly from the state equation's coefficient matrices A and B. The theorem asserts that controllability of the state equation (3.1) is equivalent to P having full-row rank, thereby yielding a linear algebraic test for controllability. This equivalence affords us a measure of pithiness because we will henceforth take controllability of a matrix pair (A, B) to mean controllability of the linear state equation with coefficient matrices A and B.

For the general multiple-input case, we see that P consists of n matrix blocks $B, AB, A^2 B, \ldots, A^{n-1} B$, each with dimension $n \times m$, stacked side by side. Thus P has dimension $n \times (n\, m)$ and therefore has more columns than rows in the multiple-input case. Furthermore, the rank condition in the theorem requires that the n rows of P are linearly independent when viewed as row vectors of dimension $1 \times (nm)$. Consequently, P satisfying the preceding rank condition is frequently referred to as having full-row rank. An alternative interpretation of the rank condition is that of the $n\, m$ columns of P written out as

$$P = [\, b_1 \quad b_2 \quad \cdots \quad b_m \,|\, Ab_1 \quad Ab_2 \quad \cdots \quad Ab_m \,|$$
$$\cdots \,|\, A^{n-1} b_1 \quad A^{n-1} b_2 \quad \cdots \quad A^{n-1} b_m \,]$$

there must be at least one way to select n linearly independent columns.

For the single-input case, B consists of a single column, as do AB, $A^2 B$, $\ldots, A^{n-1} B$, yielding a square $n \times n$ controllability matrix P. Therefore, a single-input linear state equation is controllable if and only if the associated controllability matrix is nonsingular. We can check that P is nonsingular by verifying that P has a nonzero determinant.

In order to prove Theorem 3.2, we introduce the so-called controllability Gramian, defined as follows for any initial time t_0 and any finite final time $t_f > t_0$:

$$W(t_0, t_f) = \int_{t_0}^{t_f} e^{A(t_0 - \tau)} B B^T e^{A^T(t_0 - \tau)} d\tau$$

The controllability Gramian is a square $n \times n$ matrix. Also, it is straightforward to check that it is a *symmetric* matrix, that is, $W(t_0, t_f) = W^T(t_0, t_f)$. Finally, it is a *positive semidefinite* matrix because the associated real-valued *quadratic form* $x^T W(t_0, t_f)x$ satisfies the following for all vectors $x \in \mathbb{R}^n$

$$x^T W(t_0, t_f)x = x^T \int_{t_0}^{t_f} e^{A(t_0-\tau)} B B^T e^{A^T(t_0-\tau)} d\tau \; x$$

$$= \int_{t_0}^{t_f} \left\| B^T e^{A^T(t_0-\tau)} x \right\|^2 d\tau$$

$$\geq 0$$

The asserted inequality follows from the fact that integrating a nonnegative integral forward in time from t_0 to t_f must produce a nonnegative result.

□

Lemma 3.3

$$\text{rank}[\, B \quad AB \quad A^2 B \quad \cdots \quad A^{n-1}B \,] = n$$

if and only if for any initial time t_0 and any finite final time $t_f > t_0$, the controllability Gramian $W(t_0, t_f)$ is nonsingular.

Proof. The lemma involves two implications:

If rank $P = n$, then $W(t_0, t_f)$ is nonsingular for any t_0 and any finite $t_f > t_0$.

If $W(t_0, t_f)$ is nonsingular for any t_0 and any finite $t_f > t_0$, then rank $P = n$.

We begin with a proof of the contrapositive of the first implication:

If $W(t_0, t_f)$ is singular for some t_0 and finite $t_f > t_0$, then rank $P < n$.

Assuming that $W(t_0, t_f)$ is singular for some t_0 and $t_f > t_0$, there exists a nonzero vector $x \in \mathbb{R}^n$ for which $W(t_0, t_f)x = 0 \in \mathbb{R}^n$ and therefore $x^T W(t_0, t_f)x = 0 \in \mathbb{R}$. Using the controllability Gramian definition, we have

$$0 = x^T \int_{t_0}^{t_f} e^{A(t_0-\tau)} B B^T e^{A^T(t_0-\tau)} d\tau \; x$$

$$= \int_{t_0}^{t_f} \left\| B^T e^{A^T(t_0-\tau)} x \right\|^2 d\tau$$

The integrand in this expression involves the Euclidean norm of the $m \times 1$ vector quantity $B^T e^{A^T(t_0-\tau)} x$ which we view as an analytic function of the integration variable $\tau \in [t_0, t_f]$. Since this integrand can never be negative, the only way the integral over a positive time interval can evaluate to zero is if the integrand is identically zero for all $\tau \in [t_0, t_f]$.

Because only the zero vector has a zero Euclidean norm, we must have

$$B^T e^{A^T(t_0-\tau)} x = 0 \in \mathbb{R}^m \quad \text{for all} \quad \tau \in [t_0, t_f]$$

This means that the derivative of this expression of any order with respect to τ also must be the zero vector for all $\tau \in [t_0, t_f]$ and in particular at $\tau = t_0$. Going one step further, the same must hold for the transpose of this expression. In other words,

$$0 = \frac{d^k}{d\tau^k} x^T e^{A(t_0-\tau)} B \bigg|_{\tau=t_0} = (-1)^k x^T A^k e^{A(t_0-\tau)} B|_{\tau=t_0}$$

$$= (-1)^k x^T A^k B \quad \text{for all} \quad k \geq 0$$

The alternating sign does not interfere with the conclusion that

$$[0 \quad 0 \quad 0 \quad \cdots \quad 0] = [x^T B \quad x^T A B \quad x^T A^2 B \quad \cdots \quad x^T A^{n-1} B]$$

$$= x^T [B \quad AB \quad A^2 B \quad \cdots \quad A^{n-1} B]$$

The components of the vector $x \in \mathbb{R}^n$ specify a linear combination of the rows of the controllability matrix P that yields a $1 \times (nm)$ zero vector. Since we initially assumed that x is not the zero vector, at least one of its components is nonzero. Thus we conclude that the controllability matrix P has linearly dependent rows and, consequently, less than full-row rank. In other words, rank $P < n$.

Next, we prove the contrapositive of the second implication:

If rank $P < n$, then $W(t_0, t_f)$ is singular for some t_0 and finite $t_f > t_0$.

Assuming that P has less than full-row rank, there exists a nonzero vector $x \in \mathbb{R}^n$ whose components specify a linear combination of the rows of P that yield a $1 \times (nm)$ zero vector. In other words,

$$x^T [B \quad AB \quad A^2 B \quad \cdots \quad A^{n-1} B]$$

$$= [x^T B \quad x^T A B \quad x^T A^2 B \quad \cdots \quad x^T A^{n-1} B]$$

$$= [0 \quad 0 \quad 0 \quad \cdots \quad 0]$$

and this shows that $x^T A^k B = 0$, for $k = 0, 1, \ldots, n-1$. We next use the fact that there exist scalar analytic functions $\alpha_0(t), \alpha_1(t), \ldots, \alpha_{n-1}(t)$ yielding the finite series representation for the matrix exponential

$$e^{At} = \sum_{k=0}^{n-1} \alpha_k(t) A^k$$

This allows us to write

$$x^T e^{At} B = x^T \left(\sum_{k=0}^{n-1} \alpha_k(t) A^k \right) B = \sum_{k=0}^{n-1} \alpha_k(t) x^T A^k B = 0 \quad \text{for all } t$$

in which the zero vector has dimension $1 \times m$. The transpose of this identity yields $B^T e^{A^T t} x = 0$ for all t. This enables us to show that for any t_0 and $t_f > t_0$,

$$W(t_0, t_f)x = \int_{t_0}^{t_f} e^{A(t_0-\tau)} B B^T e^{A^T(t_0-\tau)} d\tau \ x$$

$$= \int_{t_0}^{t_f} e^{A(t_0-\tau)} B (B^T e^{A^T(t_0-\tau)} x) d\tau$$

$$= 0$$

We conclude that $W(t_0, t_f)$ has a null space of dimension at least one. Consequently, by Sylvester's law, rank $W(t_0, t_f) < n$, and therefore, $W(t_0, t_f)$ is singular.   □

The preceding lemma is instrumental in showing that the rank condition of Theorem 3.2 implies controllability of the linear state equation (3.1) in the sense of Definition 3.1. In particular, Lemma 3.3 allows us to easily specify input signals that steer the state trajectory from any initial state to the origin in finite time.

Proof of Theorem 3.2.   The theorem involves the implications
    If the state equation (3.1) is controllable, then rank $P = n$.
    If rank $P = n$, then the linear state equation (3.1) is controllable.
For the first implication, we assume that the linear state equation (3.1) is controllable so that, by definition, every initial state is controllable to the origin. That is, for any $x \in \mathbb{R}^n$, there exists a finite $t_f > t_0$ and a piecewise continuous input signal $u(\cdot)$ defined on $[t_0, t_f]$ such that with $x(t_0) = x$,

$$0 = e^{A(t_f-t_0)} x + \int_{t_0}^{t_f} e^{A(t_f-\tau)} Bu(\tau) d\tau.$$

This expression can be rearranged as follows, again using the finite series expansion of the matrix exponential:

$$x = -e^{-A(t_0-t_f)} \int_{t_0}^{t_f} e^{A(t_f-\tau)} Bu(\tau) d\tau$$

$$= -\int_{t_0}^{t_f} e^{A(t_0-\tau)} Bu(\tau)d\tau$$

$$= -\int_{t_0}^{t_f} \left(\sum_{k=0}^{n-1} \alpha_k(t_0 - \tau)A^k\right) Bu(\tau)d\tau$$

$$= -\sum_{k=0}^{n-1} A^k B \int_{t_0}^{t_f} \alpha_k(t_0 - \tau)u(\tau)d\tau$$

We observe that the definite integral appearing in each term of the sum indexed by k evaluates to a constant $m \times 1$ vector that we denote by u_k. This allows us to write

$$x = -\sum_{k=0}^{n-1} A^k B u_k$$

$$= -[B \quad AB \quad A^2 B \quad \cdots \quad A^{n-1}B] \begin{bmatrix} u_0 \\ u_1 \\ u_2 \\ \vdots \\ u_{n-1} \end{bmatrix}$$

This expresses $x \in \mathbb{R}^n$ as a linear combination of the columns of the controllability matrix. Under the assumption that the linear state equation (3.1) is controllable, such a relationship must hold for any $x \in \mathbb{R}^n$. This implies that the image or range space of the controllability matrix must be all of $\mathbb{R}^n$. It follows directly that rank $P = n$.

To prove the second implication, we assume that rank $P = n$, which by Lemma 3.3 implies that for any t_0 and finite $t_f > t_0$ the controllability Gramian $W(t_0, t_f)$ is nonsingular. We must show that every state $x \in \mathbb{R}^n$ is controllable to the origin in the sense of Definition 3.1. With $x \in \mathbb{R}^n$ fixed but arbitrary, consider the input signal defined on $[t_0, t_f]$ via

$$u(t) = -B^T e^{A^T(t_0-t)} W^{-1}(t_0, t_f)x$$

The resulting state trajectory initialized with $x(t_0) = x$ satisfies

$$x(t_f) = e^{A(t_f-t_0)}x + \int_{t_0}^{t_f} e^{A(t_f-\tau)} Bu(\tau)d\tau$$

$$= e^{A(t_f-t_0)}\left[x + \int_{t_0}^{t_f} e^{A(t_0-\tau)} B(-B^T e^{A^T(t_0-\tau)} W^{-1}(t_0, t_f)x)d\tau\right]$$

$$= e^{A(t_f-t_0)} \left[x - \int_{t_0}^{t_f} e^{A(t_0-\tau)} B B^T e^{A^T(t_0-\tau)} d\tau \ W^{-1}(t_0, t_f)x \right]$$

$$= e^{A(t_f-t_0)} \left[x - W(t_0, t_f) W^{-1}(t_0, t_f)x \right]$$

$$= 0$$

thereby verifying that the arbitrarily selected state is controllable to the origin, which concludes the proof.   □

Returning to the claim made at the beginning of this section, given a controllable state equation, a straightforward calculation left to the reader shows that the input signal defined by

$$u(t) = B^T e^{A^T(t_0-t)} W^{-1}(t_0, t_f)(e^{A(t_0-t_f)}x_f - x_0) \quad \text{for} \quad t_0 \le t \le t_f$$

steers the state trajectory from the initial state $x(t_0) = x_0$ to the final state $x(t_f) = x_f$, with x_0 and x_f lying anywhere in the state space. For the special case in which $x(t_0) = 0 \in \mathbb{R}^n$, the final state $x(t_f) = x_f$ is often referred to as *reachable from the origin*. Consequently, a controllable state equation is also sometimes referred to as a *reachable* state equation.

3.2 CONTROLLABILITY EXAMPLES

Example 3.1   Given the following single-input two–dimensional linear state equation, we now assess its controllability.

$$\begin{bmatrix} \dot{x}_1(t) \\ \dot{x}_2(t) \end{bmatrix} = \begin{bmatrix} 1 & 5 \\ 8 & 4 \end{bmatrix} \begin{bmatrix} x_1(t) \\ x_2(t) \end{bmatrix} + \begin{bmatrix} -2 \\ 2 \end{bmatrix} u(t)$$

The controllability matrix P is found as follows:

$$B = \begin{bmatrix} -2 \\ 2 \end{bmatrix} \quad AB = \begin{bmatrix} 8 \\ -8 \end{bmatrix} \quad P = \begin{bmatrix} -2 & 8 \\ 2 & -8 \end{bmatrix}$$

Clearly, $|P| = 0$, so the state equation is not controllable. To see why this is true, consider a different state definition

$$\begin{bmatrix} z_1(t) \\ z_2(t) \end{bmatrix} = \begin{bmatrix} x_1(t) \\ x_1(t) + x_2(t) \end{bmatrix}$$

The associated coordinate transformation (see Section 2.5) is

$$x(t) = Tz(t)$$

$$\begin{bmatrix} x_1(t) \\ x_2(t) \end{bmatrix} = \begin{bmatrix} 1 & 0 \\ -1 & 1 \end{bmatrix} \begin{bmatrix} z_1(t) \\ z_2(t) \end{bmatrix}$$

Applying this coordinate transformation yields the transformed state equation

$$\begin{bmatrix} \dot{z}_1(t) \\ \dot{z}_2(t) \end{bmatrix} = \begin{bmatrix} -4 & 5 \\ 0 & 9 \end{bmatrix} \begin{bmatrix} z_1(t) \\ z_2(t) \end{bmatrix} + \begin{bmatrix} -2 \\ 0 \end{bmatrix} u(t)$$

We see that $\dot{z}_2(t)$ does not depend on the input $u(t)$, so this state variable is not controllable. $\qquad\square$

Example 3.2   Given the following three-dimensional single-input state equation, that is,

$$\begin{bmatrix} \dot{x}_1(t) \\ \dot{x}_2(t) \\ \dot{x}_3(t) \end{bmatrix} = \begin{bmatrix} 0 & 1 & 0 \\ 0 & 0 & 1 \\ -6 & -11 & -6 \end{bmatrix} \begin{bmatrix} x_1(t) \\ x_2(t) \\ x_3(t) \end{bmatrix} + \begin{bmatrix} 0 \\ 1 \\ -3 \end{bmatrix} u(t)$$

we construct the controllability matrix P using

$$B = \begin{bmatrix} 0 \\ 1 \\ -3 \end{bmatrix}$$

$$AB = \begin{bmatrix} 0 & 1 & 0 \\ 0 & 0 & 1 \\ -6 & -11 & -6 \end{bmatrix} \begin{bmatrix} 0 \\ 1 \\ -3 \end{bmatrix} = \begin{bmatrix} 1 \\ -3 \\ 7 \end{bmatrix}$$

$$A^2B = A(AB) = \begin{bmatrix} 0 & 1 & 0 \\ 0 & 0 & 1 \\ -6 & -11 & -6 \end{bmatrix} \begin{bmatrix} 1 \\ -3 \\ 7 \end{bmatrix} = \begin{bmatrix} -3 \\ 7 \\ -15 \end{bmatrix}$$

This yields

$$P = \begin{bmatrix} B & AB & A^2B \end{bmatrix}$$

$$= \begin{bmatrix} 0 & 1 & -3 \\ 1 & -3 & 7 \\ -3 & 7 & -15 \end{bmatrix}$$

To check controllability, we calculate

$$|P| = |B \quad AB \quad A^2B|$$

$$= \begin{vmatrix} 0 & 1 & -3 \\ 1 & -3 & 7 \\ -3 & 7 & -15 \end{vmatrix}$$

$$= [0 + (-21) + (-21)] - [(-27) + 0 + (-15)]$$

$$= -42 - (-42)$$
$$= 0$$

and thus rank $P < 3$. This indicates that the state equation is *not* controllable. The upper left 2×2 submatrix

$$\begin{bmatrix} 0 & 1 \\ 1 & -3 \end{bmatrix}$$

has nonzero determinant, indicating that rank $P = 2$.   □

Example 3.3   We investigate the controllability of the three–dimensional state equation

$$\begin{bmatrix} \dot{x}_1(t) \\ \dot{x}_2(t) \\ \dot{x}_3(t) \end{bmatrix} = \begin{bmatrix} 0 & 1 & 0 \\ 0 & 0 & 1 \\ -a_0 & -a_1 & -a_2 \end{bmatrix} \begin{bmatrix} x_1(t) \\ x_2(t) \\ x_3(t) \end{bmatrix} + \begin{bmatrix} 0 \\ 0 \\ 1 \end{bmatrix} u(t)$$

$$y(t) = \begin{bmatrix} b_0 & b_1 & b_2 \end{bmatrix} \begin{bmatrix} x_1(t) \\ x_2(t) \\ x_3(t) \end{bmatrix}$$

which the reader will recall is a state-space realization of the transfer function

$$H(s) = \frac{b_2 s^2 + b_1 s + b_0}{s^3 + a_2 s^2 + a_1 s + a_0}$$

The controllability matrix P is found as follows:

$$B = \begin{bmatrix} 0 \\ 0 \\ 1 \end{bmatrix} \quad AB = \begin{bmatrix} 0 \\ 1 \\ -a_2 \end{bmatrix}$$

$$A^2 B = \begin{bmatrix} 1 \\ -a_2 \\ a_2^2 - a_1 \end{bmatrix}$$

$$P = \begin{bmatrix} B & AB & A^2 B \end{bmatrix}$$

$$= \begin{bmatrix} 0 & 0 & 1 \\ 0 & 1 & -a_2 \\ 1 & -a_2 & a_2^2 - a_1 \end{bmatrix}$$

The controllability matrix P is independent of the transfer function numerator coefficients b_0, b_1, and b_2. The determinant of the controllability matrix is $|P| = -1 \neq 0$, so the state equation is controllable. Note that this result, i.e. the determinant of P, is independent of the characteristic polynomial coefficients a_0, a_1, and a_2, so a state-space realization in this form is always controllable. This is also true for any system order n, a claim that we will verify shortly. $\square$

Example 3.4 Given the five-dimensional, two-input state equation

$$
\begin{bmatrix} \dot{x}_1(t) \\ \dot{x}_2(t) \\ \dot{x}_3(t) \\ \dot{x}_4(t) \\ \dot{x}_5(t) \end{bmatrix} = \begin{bmatrix} 0 & 1 & 0 & 0 & 0 \\ 0 & 0 & 0 & 0 & 0 \\ 0 & 0 & 0 & 1 & 0 \\ 0 & 0 & 0 & 0 & 1 \\ 0 & 0 & 0 & 0 & 0 \end{bmatrix} \begin{bmatrix} x_1(t) \\ x_2(t) \\ x_3(t) \\ x_4(t) \\ x_5(t) \end{bmatrix} + \begin{bmatrix} 0 & 0 \\ 1 & 0 \\ 0 & 0 \\ 0 & 0 \\ 0 & 1 \end{bmatrix} \begin{bmatrix} u_1(t) \\ u_2(t) \end{bmatrix}
$$

the controllability matrix is

$$
P = \begin{bmatrix} b_1 & b_2 \,|\, Ab_1 & Ab_2 \,|\, A^2b_1 & A^2b_2 \,|\, A^3b_1 & A^3b_2 \,|\, A^4b_1 & A^4b_2 \end{bmatrix}
$$

$$
= \begin{bmatrix} 0 & 0 & 1 & 0 & 0 & 0 & 0 & 0 & 0 & 0 \\ 1 & 0 & 0 & 0 & 0 & 0 & 0 & 0 & 0 & 0 \\ 0 & 0 & 0 & 0 & 0 & 1 & 0 & 0 & 0 & 0 \\ 0 & 0 & 0 & 1 & 0 & 0 & 0 & 0 & 0 & 0 \\ 0 & 1 & 0 & 0 & 0 & 0 & 0 & 0 & 0 & 0 \end{bmatrix}
$$

This state equation is controllable because P has full-row rank, as can be seen from the pattern of ones and zeros. Also, columns 1, 2, 3, 4, and 6 form a linearly independent set. Since the remaining columns are each zero vectors, this turns out to be the only way to select five linearly independent columns from P. $\square$

Example 3.5 Now consider the following five-dimensional, two-input state equation

$$
\begin{bmatrix} \dot{x}_1(t) \\ \dot{x}_2(t) \\ \dot{x}_3(t) \\ \dot{x}_4(t) \\ \dot{x}_5(t) \end{bmatrix} = \begin{bmatrix} 0 & 1 & 0 & 0 & 0 \\ 1 & 0 & -1 & 0 & 1 \\ 0 & 0 & 0 & 1 & 0 \\ 0 & 0 & 0 & 0 & 1 \\ 0 & -1 & 0 & -1 & 0 \end{bmatrix} \begin{bmatrix} x_1(t) \\ x_2(t) \\ x_3(t) \\ x_4(t) \\ x_5(t) \end{bmatrix} + \begin{bmatrix} 0 & 0 \\ 1 & -1 \\ 0 & 0 \\ 0 & 0 \\ 1 & 1 \end{bmatrix} \begin{bmatrix} u_1(t) \\ u_2(t) \end{bmatrix}
$$

This equation differs from the preceding example only in the second and fifth rows of both A and B. These adjustments yield the more interesting controllability matrix

$$P = \begin{bmatrix} b_1 & b_2 \,|\, Ab_1 & Ab_2 \,|\, A^2b_1 & A^2b_2 \,|\, A^3b_1 & A^3b_2 \,|\, A^4b_1 & A^4b_2 \end{bmatrix}$$

$$= \begin{bmatrix} 0 & 0 & 1 & -1 & 1 & 1 & 0 & 0 & -2 & -2 \\ 1 & -1 & 1 & 1 & 0 & 0 & -2 & -2 & 2 & -2 \\ 0 & 0 & 0 & 0 & 1 & 1 & -1 & 1 & -2 & -2 \\ 0 & 0 & 1 & 1 & -1 & 1 & -2 & -2 & 1 & -1 \\ 1 & 1 & -1 & 1 & -2 & -2 & 1 & -1 & 4 & 4 \end{bmatrix}$$

which also has rank equal to 5. If we search from left to right for five linearly independent columns, we find that the first five qualify. However, there are many more ways to select five linearly independent columns from this controllability matrix, as the reader may verify with the aid of MATLAB. □

3.3 COORDINATE TRANSFORMATIONS AND CONTROLLABILITY

The linear time-invariant state equation (3.1) together with the state coordinate transformation $x(t) = Tz(t)$ (see Section 2.5) yields the transformed state equation

$$\dot{z}(t) = \hat{A}z(t) + \hat{B}u(t) \quad z(t_0) = z_0 \tag{3.2}$$

in which

$$\hat{A} = T^{-1}AT \quad \hat{B} = T^{-1}B \quad z_0 = T^{-1}x_0$$

Proceeding directly from Definition 3.1, we see that if a state x is controllable to the origin for the original state equation (3.1), then the state $z = T^{-1}x$ is controllable to the origin for the transformed state equation (3.2) over the same time interval using the same control signal, for if

$$0 = e^{A(t_f-t_0)}x + \int_{t_0}^{t_f} e^{A(t_f-\tau)}Bu(\tau)d\tau$$

then with $z(t_0) = z$,

$$z(t_f) = e^{\hat{A}(t_f-t_0)}z + \int_{t_0}^{t_f} e^{\hat{A}(t_f-\tau)}\hat{B}u(\tau)d\tau$$

$$= (T^{-1}e^{A(t_f-t_0)}T)(T^{-1}x) + \int_{t_0}^{t_f} (T^{-1}e^{A(t_f-\tau)}T)(T^{-1}B)u(\tau)d\tau$$

$$= T^{-1} \left[e^{A(t_f - t_0)} x + \int_{t_0}^{t_f} e^{A(t_f - \tau)} Bu(\tau) d\tau \right]$$

$$= 0$$

Conversely, if a state z is controllable to the origin for the state equation (3.2), then the state $x = T z$ is controllable to the origin for the state equation (3.1). We immediately conclude that the transformed state equation (3.2) is controllable if and only if the original state equation (3.1) is controllable. In short, we say that controllability is invariant with respect to state coordinate transformations.

Any equivalent characterization of controllability must reflect this outcome. For instance, we expect that the rank test for controllability provided by Theorem 3.2 should yield consistent results across all state equations related by a coordinate transformation. To see this, we relate the controllability matrix for the state equation (3.2) to the controllability matrix for the state equation (3.1). Using the fact that $\hat{A}^k = T^{-1} A^k T$ for any positive integer k, we have

$$\hat{P} = \begin{bmatrix} \hat{B} & \hat{A}\hat{B} & \cdots & \hat{A}^{n-1}\hat{B} \end{bmatrix}$$

$$= \begin{bmatrix} T^{-1}B | (T^{-1}AT)(T^{-1}B) | \cdots | (T^{-1}A^{n-1}T)(T^{-1}B) \end{bmatrix}$$

$$= \begin{bmatrix} T^{-1}B | T^{-1}AB | \cdots | T^{-1}A^{n-1}B \end{bmatrix}$$

$$= T^{-1} \begin{bmatrix} B & AB & \cdots & A^{n-1}B \end{bmatrix}$$

$$= T^{-1}P$$

Since either pre- or post-multiplication by a square nonsingular matrix does not affect matrix rank, we see that

$$\text{rank } \hat{P} = \text{rank } P$$

In addition, the $n \times n$ controllability Gramians for Equations (3.2) and (3.1) are related according to

$$\hat{W}(t_0, t_f) = \int_{t_0}^{t_f} e^{\hat{A}(t_0 - \tau)} \hat{B} \hat{B}^T e^{\hat{A}^T(t_0 - \tau)} d\tau$$

$$= \int_{t_0}^{t_f} (T^{-1} e^{A(t_0 - \tau)} T)(T^{-1}B)(T^{-1}B)^T (T^{-1} e^{A(t_0 - \tau)} T)^T d\tau$$

$$= T^{-1} \int_{t_0}^{t_f} e^{A(t_0 - \tau)} B B^T e^{A(t_0 - \tau)} d\tau T^{-T}$$

$$= T^{-1} W(t_0, t_f) T^{-T}$$

where T^{-T} is shorthand notation for $(T^{-1})^T = (T^T)^{-1}$. We conclude that $\hat{W}(t_0, t_f)$ is nonsingular for some initial time t_0 and finite final time $t_f > t_0$ if and only if $W(t_0, t_f)$ is nonsingular for the same t_0 and t_f.

Various system properties are often revealed by a judiciously chosen state-space realization. Here we examine the role that the diagonal canonical form plays in controllability analysis. To do so, we assume that the state equation (3.1) has a single input and a diagonalizable system dynamics matrix A. As discussed in Chapter 2, this ensures that the state equation (3.1) can be transformed into diagonal canonical form. Using the identity:

$$
A_{\text{DCF}}^k B_{\text{DCF}} =
\begin{bmatrix}
\lambda_1^k & 0 & \cdots & 0 \\
0 & \lambda_2^k & \cdots & 0 \\
\vdots & \vdots & \ddots & \vdots \\
0 & 0 & \cdots & \lambda_n^k
\end{bmatrix}
\begin{bmatrix}
b_1 \\
b_2 \\
\vdots \\
b_n
\end{bmatrix}
=
\begin{bmatrix}
\lambda_1^k b_1 \\
\lambda_2^k b_2 \\
\vdots \\
\lambda_n^k b_n
\end{bmatrix}
$$

for any integer $k \geq 0$, we see that the controllability matrix for the diagonal canonical form is given by

$$
P_{\text{DCF}} =
\begin{bmatrix}
b_1 & \lambda_1 b_1 & \lambda_1^2 b_1 & \cdots & \lambda_1^{n-1} b_1 \\
b_2 & \lambda_2 b_2 & \lambda_2^2 b_2 & \cdots & \lambda_2^{n-1} b_2 \\
\vdots & \vdots & \vdots & \ddots & \vdots \\
b_n & \lambda_n b_n & \lambda_n^2 b_n & \cdots & \lambda_n^{n-1} b_n
\end{bmatrix}
$$

$$
=
\begin{bmatrix}
b_1 & 0 & \cdots & 0 \\
0 & b_2 & \cdots & 0 \\
\vdots & \vdots & \ddots & \vdots \\
0 & 0 & \cdots & b_n
\end{bmatrix}
\begin{bmatrix}
1 & \lambda_1 & \lambda_1^2 & \cdots & \lambda_1^{n-1} \\
1 & \lambda_2 & \lambda_2^2 & \cdots & \lambda_2^{n-1} \\
\vdots & \vdots & \vdots & \ddots & \vdots \\
1 & \lambda_n & \lambda_n^2 & \cdots & \lambda_n^{n-1}
\end{bmatrix}
$$

which is nonsingular if and only if each factor on the right-hand is nonsingular. The diagonal left factor is nonsingular if and only if every element of B_{DCF} is nonzero. The right factor is the Vandermonde matrix encountered in Chapter 2, which is nonsingular if and only if the eigenvalues $\lambda_1, \lambda_2, \ldots, \lambda_n$ are distinct. Since controllability is invariant with respect to state coordinate transformations, we conclude that controllability of a single-input state equation can be assessed by inspection of its diagonal canonical form (when it exists). Specifically, a necessary and sufficient condition for controllability is that the eigenvalues of A displayed on the diagonal of A_{DCF} must be distinct and every element of B_{DCF} must be nonzero.

The condition on B_{DCF} is relatively easy to interpret; because of the decoupled nature of the diagonal canonical form, if any component of B_{DCF} is zero, the corresponding state variable is disconnected from the input signal, and therefore cannot be steered between prescribed initial and final values. The eigenvalue condition is a bit more subtle. Suppose that A has a repeated eigenvalue, say, $\lambda_i = \lambda_j = \lambda$. Given zero initial conditions on the corresponding state variables $x_i(t)$ and $x_j(t)$, and any input signal $u(t)$, we have

$$
b_j x_i(t) = b_j \left[\int_0^t e^{\lambda(t-\tau)} b_i u(\tau) d\tau \right] = \int_0^t e^{\lambda(t-\tau)} (b_j b_i) u(\tau) d\tau
$$

$$
= b_i \left[\int_0^t e^{\lambda(t-\tau)} b_j u(\tau) d\tau \right] = b_i x_j(t)
$$

for all $t \geq 0$. That these state variable responses are constrained to satisfy this relationship regardless of the input signal indicates that not every state can be reached from the origin in finite time. Hence when A has a repeated eigenvalue, the state equation in diagonal canonical form is not controllable, implying the same for the original state equation.

This section concludes with three results that, in different ways, relate coordinate transformations and controllability. The first two apply only to the single-input, single-output case. The third holds in the general multiple-input, multiple-output case.

Controllable Realizations of the Same Transfer Function

Consider two single-input, single-output state equations

$$
\dot{x}_1(t) = A_1 x_1(t) + B_1 u(t) \qquad \dot{x}_2(t) = A_2 x_2(t) + B_2 u(t)
$$

$$
y(t) = C_1 x_1(t) + D_1 u(t) \qquad y(t) = C_2 x_2(t) + D_2 u(t) \qquad (3.3)
$$

that are both n-dimensional realizations of the same transfer function. We show that when both realizations are controllable, they are related by a uniquely and explicitly defined coordinate transformation.

Lemma 3.4 *Two controllable n-dimensional single-input, single-output realizations* (3.3) *of the same transfer function are related by the unique* *state coordinate transformation $x_1(t) = T x_2(t)$, where*

$$T = \begin{bmatrix} B_1 & A_1 B_1 & \cdots & A_1^{n-1} B_1 \end{bmatrix} \begin{bmatrix} B_2 & A_2 B_2 & \cdots & A_2^{n-1} B_2 \end{bmatrix}^{-1}$$
$$= P_1 P_2^{-1}$$

Proof. We must verify the identities

$$A_2 = T^{-1} A_1 T \quad B_2 = T^{-1} B_1 \quad C_2 = C_1 T$$

The first identity can be recast as

$$\begin{bmatrix} B_2 & A_2 B_2 & \cdots & A_2^{n-1} B_2 \end{bmatrix}^{-1} A_2 \begin{bmatrix} B_2 & A_2 B_2 & \cdots & A_2^{n-1} B_2 \end{bmatrix}$$
$$= \begin{bmatrix} B_1 & A_1 B_1 & \cdots & A_1^{n-1} B_1 \end{bmatrix}^{-1} A_1 \begin{bmatrix} B_1 & A_1 B_1 & \cdots & A_1^{n-1} B_1 \end{bmatrix}$$

$$(3.4)$$

Now, for any matrix pair (A, B) describing a controllable state equation with $|sI - A| = s^n + a_{n-1} s^{n-1} + \cdots + a_1 s + a_0$, the Cayley-Hamilton theorem gives

$$\begin{bmatrix} B & AB & \cdots & A^{n-1} B \end{bmatrix}^{-1} A \begin{bmatrix} B & AB & \cdots & A^{n-1} B \end{bmatrix}$$

$$= \begin{bmatrix} B & AB & \cdots & A^{n-1} B \end{bmatrix}^{-1} \begin{bmatrix} AB & A^2 B & \cdots & A^n B \end{bmatrix}$$

$$= \begin{bmatrix} B & AB & \cdots & A^{n-1} B \end{bmatrix}^{-1}$$

$$\times \begin{bmatrix} AB & A^2 B & \cdots & (-a_0 B - a_1 AB - \cdots - a_{n-1} A^{n-1} B) \end{bmatrix}$$

$$= \begin{bmatrix} 0 & 0 & \cdots & 0 & 0 & -a_0 \\ 1 & 0 & \cdots & 0 & 0 & -a_1 \\ 0 & 1 & \cdots & 0 & 0 & -a_2 \\ \vdots & \vdots & \ddots & \vdots & \vdots & \vdots \\ 0 & 0 & \cdots & 1 & 0 & -a_{n-2} \\ 0 & 0 & \cdots & 0 & 1 & -a_{n-1} \end{bmatrix}$$

Since the state equations (3.3) are each n-dimensional realizations of the same transfer function $H(s) = b(s)/a(s)$ with deg $a(s) = n$, we necessarily have $|sI - A_1| = |sI - A_2|$. Thus both matrix pairs (A_1, B_1) and (A_2, B_2) satisfy an identity analogous to the preceding one, with an identical outcome in each case. Hence Equation (3.4) holds, yielding $A_2 = T^{-1} A_1 T$.

Next, it is straightforward to check that

$$
\begin{bmatrix} B_2 & A_2 B_2 & \cdots & A_2^{n-1} B_2 \end{bmatrix}^{-1} B_2 =
\begin{bmatrix} 1 \\ 0 \\ 0 \\ \vdots \\ 0 \\ 0 \end{bmatrix}
$$

$$
= \begin{bmatrix} B_1 & A_1 B_1 & \cdots & A_1^{n-1} B_1 \end{bmatrix}^{-1} B_1
$$

which can be repackaged to give $B_2 = T^{-1} B_1$.

Finally, since the state equations (3.3) are each n-dimensional realizations of the same impulse response and $D_2 = D_1$, it follows that

$$
C_2 e^{A_2 t} B_2 = C_1 e^{A_1 t} B_1 \quad t \geq 0
$$

Repeatedly differentiating this identity and evaluating at $t = 0$ gives

$$
C_2 A_2^k B_2 = C_1 A_1^k B_1 \quad k = 0, 1, \ldots, n - 1
$$

which implies that

$$
C_2 \begin{bmatrix} B_2 & A_2 B_2 & \cdots & A_2^{n-1} B_2 \end{bmatrix} = C_1 \begin{bmatrix} B_1 & A_1 B_1 & \cdots & A_1^{n-1} B_1 \end{bmatrix}
$$

which can be rearranged to yield the third identity $C_2 = C_1 T$.

Uniqueness is a consequence of the fact that any state coordinate transformation $x_1(t) = T x_2(t)$ linking the state equations (3.3) necessarily must satisfy

$$
\begin{bmatrix} B_1 & A_1 B_1 & \cdots & A_1^{n-1} B_1 \end{bmatrix} = T \begin{bmatrix} B_2 & A_2 B_2 & \cdots & A_2^{n-1} B_2 \end{bmatrix}
$$

along with the nonsingularity of each controllability matrix.   □

Controller Canonical Form

As noted earlier, state coordinate transformations permit the construction of special state-space realizations that facilitate a particular type of analysis. For instance, the diagonal canonical form realization describes decoupled first-order scalar equations that are easier to solve than a general coupled state equation and controllability can be determined by inspection. We have previously encountered another highly structured realization of the scalar transfer function

$$H(s) = \frac{b(s)}{a(s)} = \frac{b_{n-1}s^{n-1} + \cdots + b_1 s + b_0}{s^n + a_{n-1}s^{n-1} + \cdots + a_1 s + a_0}$$

that was determined by inspection from the coefficients of the numerator and denominator polynomials and was referred to as either the *phase-variable canonical form* or the *controller canonical form* (CCF). Here we justify the latter terminology by introducing the following notation for this realization:

$$\dot{x}_{CCF} = A_{CCF} \, x_{CCF}(t) + B_{CCF} \, u(t)$$

$$y(t) = C_{CCF} \, x_{CCF}(t)$$

in which

$$A_{CCF} = \begin{bmatrix} 0 & 1 & 0 & \cdots & 0 & 0 \\ 0 & 0 & 1 & \cdots & 0 & 0 \\ \vdots & \vdots & \vdots & \ddots & \vdots & \vdots \\ 0 & 0 & 0 & \cdots & 1 & 0 \\ 0 & 0 & 0 & \cdots & 0 & 1 \\ -a_0 & -a_1 & -a_2 & \cdots & -a_{n-2} & -a_{n-1} \end{bmatrix} \qquad B_{CCF} = \begin{bmatrix} 0 \\ 0 \\ \vdots \\ 0 \\ 0 \\ 1 \end{bmatrix}$$

$$C_{CCF} = \begin{bmatrix} b_0 & b_1 & b_2 & \cdots & b_{n-2} & b_{n-1} \end{bmatrix}$$

Not surprisingly, the controller canonical form defines a controllable state equation (see Example 3.3). This can be verified by deriving an explicit representation for the associated controllability matrix.

Lemma 3.5 *The n-dimensional Controller Canonical Form has the controllability matrix*

$$P_{CCF} = \begin{bmatrix} B_{CCF} & A_{CCF}B_{CCF} & A_{CCF}^2 B_{CCF} & \cdots & A_{CCF}^{n-1}B_{CCF} \end{bmatrix}$$

$$= \begin{bmatrix} a_1 & a_2 & \cdots & a_{n-1} & 1 \\ a_2 & a_3 & \cdots & 1 & 0 \\ \vdots & \vdots & \ddots & \vdots & \vdots \\ a_{n-1} & 1 & \cdots & 0 & 0 \\ 1 & 0 & \cdots & 0 & 0 \end{bmatrix}^{-1} \qquad (3.5)$$

Proof. It suffices to check that $P_{CCF} P_{CCF}^{-1} = I$, that is,

$$
\begin{bmatrix} B_{CCF} & A_{CCF} B_{CCF} & A_{CCF}^2 B_{CCF} & \cdots & A_{CCF}^{n-1} B_{CCF} \end{bmatrix}
$$

$$
\times \begin{bmatrix}
a_1 & a_2 & \cdots & a_{n-1} & 1 \\
a_2 & a_3 & \cdots & 1 & 0 \\
\vdots & \vdots & \ddots & \vdots & \vdots \\
a_{n-1} & 1 & \cdots & 0 & 0 \\
1 & 0 & \cdots & 0 & 0
\end{bmatrix} = I
$$

Proceeding column-wise through this identity from last to first, we have by definition $B_{CCF} = e_n$ (see above), and using the structure of the right factor, we must show in addition that

$$
A_{CCF}^{n-j} B_{CCF} + \sum_{k=1}^{n-j} a_{n-k} A_{CCF}^{n-j-k} B_{CCF} = e_j \quad j = 1, 2, \ldots, n-1 \quad (3.6)
$$

We establish Equation (3.6) by induction on j in reverse order, starting with $j = n - 1$. The structure of A_{CCF} allows us to write

$$
A_{CCF} e_j = e_{j-1} - a_{j-1} e_n, \quad j = 2, \ldots, n-1
$$

For $j = n - 1$, Equation (3.6) reduces to

$$
A_{CCF} B_{CCF} + a_{n-1} B_{CCF} = e_{n-1}
$$

which holds by virtue of the preceding relationship $B_{CCF} = e_n$. Next, suppose that Equation (3.6) holds for arbitrary $j \leq n - 1$. Then, for $j - 1$, we have

$$
A_{CCF}^{n-(j-1)} B_{CCF} + \sum_{k=1}^{n-(j-1)} a_{n-k} A_{CCF}^{n-(j-1)-k} B_{CCF}
$$

$$
= A_{CCF} \left(A_{CCF}^{n-j} B_{CCF} + \sum_{k=1}^{n-j} a_{n-k} A_{CCF}^{n-j-k} B_{CCF} \right) + a_{j-1} B_{CCF}
$$

$$
= A_{CCF} e_j + a_{j-1} e_n
$$

$$
= e_{j-1}
$$

which concludes the proof. $\qquad \qquad \square$

We note that P_{CCF} depends explicitly on the characteristic polynomial coefficients $a_1, \ldots, a_{n-1}$ (excluding a_0). We further observe that the inverse of the controller canonical form controllability matrix P_{CCF}, namely,

$$
P_{CCF}^{-1} = \begin{bmatrix}
a_1 & a_2 & \cdots & a_{n-1} & 1 \\
a_2 & a_3 & \cdots & 1 & 0 \\
\vdots & \vdots & \ddots & \vdots & \vdots \\
a_{n-1} & 1 & \cdots & 0 & 0 \\
1 & 0 & \cdots & 0 & 0
\end{bmatrix}
$$

is symmetric. The reader should check that because matrix inversion and matrix transposition are interchangeable operations, P_{CCF} is symmetric as well.

It follows from Lemma 3.5 that beginning with an arbitrary controllable realization of a given transfer function with state vector $x(t)$, the associated controller canonical form is obtained by applying the coordinate transformation $x(t) = T_{CCF}z(t)$, where $T_{CCF} = P \ P_{CCF}^{-1}$, or more explicitly,

$$
T_{CCF} = \begin{bmatrix} B & AB & A^2B & \cdots & A^{n-1}B \end{bmatrix} \begin{bmatrix}
a_1 & a_2 & \cdots & a_{n-1} & 1 \\
a_2 & a_3 & \cdots & 1 & 0 \\
\vdots & \vdots & \ddots & \vdots & \vdots \\
a_{n-1} & 1 & \cdots & 0 & 0 \\
1 & 0 & \cdots & 0 & 0
\end{bmatrix}
$$

$$(3.7)$$

Based on our earlier discussion regarding controllability of the diagonal canonical form, we see that the controller canonical form can be transformed to diagonal canonical form when and only when A_{CCF} has distinct eigenvalues. The eigenvalue condition is in general sufficient for diagonalizability as noted in Appendix B, Section 8. Conversely, if the controller canonical form can be transformed to diagonal canonical form, then the latter is necessarily controllable, which implies that the eigenvalues of A_{CCF} must be distinct as noted earlier. Alternatively, the necessity of the eigenvalue condition can be argued using the eigenstructure results for A_{CCF} established in AE3.1. It also follows from AE3.1, that when A_{CCF} has distinct eigenvalues, a coordinate transformation from controller canonical form to diagonal canonical form is given by the transposed

Vandermonde matrix

$$
T_{\text{DCF}} = \begin{bmatrix}
1 & 1 & 1 & \cdots & 1 \\
\lambda_1 & \lambda_2 & \lambda_3 & \cdots & \lambda_n \\
\lambda_1^2 & \lambda_2^2 & \lambda_3^2 & \cdots & \lambda_n^2 \\
\vdots & \vdots & \vdots & \ddots & \vdots \\
\lambda_1^{n-1} & \lambda_2^{n-1} & \lambda_3^{n-1} & \cdots & \lambda_n^{n-1}
\end{bmatrix} \tag{3.8}
$$

which is nonsingular.

Example 3.6 Given the three-dimensional single-input, single-output state equation specified by the coefficient matrices given below, we now compute the controller canonical form. This is the same system given in Example 2.8, for which the diagonal canonical form was computed.

$$
A = \begin{bmatrix} 8 & -5 & 10 \\ 0 & -1 & 1 \\ -8 & 5 & -9 \end{bmatrix} \quad B = \begin{bmatrix} -1 \\ 0 \\ 1 \end{bmatrix} \quad C = [1 \quad -2 \quad 4]
$$

$$
D = 0
$$

The system characteristic polynomial is again $|sI - A| = s^3 + 2s^2 + 4s + 8$, and the eigenvalues are $\pm 2i, -2$. The controller canonical form transformation matrix $T_{\text{CCF}} = P \, P_{\text{CCF}}^{-1}$ is computed as follows.

$$
T_{\text{CCF}} = \begin{bmatrix} B & AB & A^2B \end{bmatrix} \begin{bmatrix} a_1 & a_2 & 1 \\ a_2 & 1 & 0 \\ 1 & 0 & 0 \end{bmatrix}
$$

$$
= \begin{bmatrix} -1 & 2 & 1 \\ 0 & 1 & -2 \\ 1 & -1 & -2 \end{bmatrix} \begin{bmatrix} 4 & 2 & 1 \\ 2 & 1 & 0 \\ 1 & 0 & 0 \end{bmatrix}
$$

$$
= \begin{bmatrix} 1 & 0 & -1 \\ 0 & 1 & 0 \\ 0 & 1 & 1 \end{bmatrix}
$$

The resulting controller canonical form state-space realization is given by

$$
A_{\text{CCF}} = T_{\text{CCF}}^{-1} A T_{\text{CCF}} \qquad B_{\text{CCF}} = T_{\text{CCF}}^{-1} B
$$

$$
= \begin{bmatrix} 0 & 1 & 0 \\ 0 & 0 & 1 \\ -8 & -4 & -2 \end{bmatrix} \qquad = \begin{bmatrix} 0 \\ 0 \\ 1 \end{bmatrix}
$$

$$C_{CCF} = CT_{CCF} \qquad\qquad D_{CCF} = D$$
$$= \begin{bmatrix} 1 & 2 & 3 \end{bmatrix} \qquad\qquad = 0$$

Because A_{CCF} also has the distinct eigenvalues $\pm 2i, -2$, the matrix

$$T_{DCF} = \begin{bmatrix} 1 & 1 & 1 \\ 2i & -2i & -2 \\ -4 & -4 & 4 \end{bmatrix}$$

is nonsingular and yields the diagonal canonical form obtained in Example 2.8

$$A_{DCF} = T_{DCF}^{-1} A_{CCF} T_{DCF} \qquad\qquad B_{DCF} = T_{DCF}^{-1} B_{CCF}$$

$$= \begin{bmatrix} 2i & 0 & 0 \\ 0 & -2i & 0 \\ 0 & 0 & -2 \end{bmatrix} \qquad\qquad = \begin{bmatrix} -0.06 - 0.06i \\ -0.06 + 0.06i \\ 0.13 \end{bmatrix}$$

$$C_{DCF} = C_{CCF} T_{DCF} \qquad\qquad D_{DCF} = D$$

$$= \begin{bmatrix} -11 + 4i & -11 - 4i & 9 \end{bmatrix} \qquad\qquad = 0$$

Here A_{DCF} displays the distinct eigenvalues, and all components of B_{DCF} are nonzero, as we expect.   □

Uncontrollable State Equations

The following result provides a useful characterization of state equations that are not controllable. We refer to this as the *standard form for uncontrollable state equations*.

Theorem 3.6 *Suppose that*

$$\text{rank} \begin{bmatrix} B & AB & A^2 B & \cdots & A^{n-1} B \end{bmatrix} = q < n$$

Then there exists a state coordinate transformation $x(t) = T\, z(t)$ *such that the transformed state equation has*

$$\hat{A} = T^{-1} A T \qquad\qquad \hat{B} = T^{-1} B$$

$$= \begin{bmatrix} A_{11} & A_{12} \\ 0 & A_{22} \end{bmatrix} \qquad\qquad = \begin{bmatrix} B_1 \\ 0 \end{bmatrix}$$

where the pair (A_{11}, B_1) *defines a controllable q-dimensional state equation.*

Proof. First select q linearly independent columns labeled $t_1, t_2, \ldots, t_q$ from the n m columns of the controllability matrix P. Then let t_{q+1}, $t_{q+2}, \ldots, t_n$ be additional $n \times 1$ vectors such that $\{t_1, t_2, \ldots, t_q, t_{q+1}, t_{q+2}, \ldots, t_n\}$ is a basis for $\mathbb{R}^n$, equivalently,

$$T = \begin{bmatrix} t_1 & t_2 & \cdots & t_q \mid t_{q+1} & t_{q+2} & \cdots & t_n \end{bmatrix}$$

is nonsingular. To show first that $\hat{B} = T^{-1}B$, consider the identity $B = T\hat{B}$. Since the jth column of B satisfies

$$b_j \in \text{Im} \begin{bmatrix} B & AB & \cdots & A^{n-1}B \end{bmatrix} = \text{span}\{t_1, t_2, \ldots, t_q\}$$

and the jth column of $\hat{B}, \hat{b}_j$, contains the n coefficients of the unique linear combination

$$b_j = \hat{b}_{1j}t_1 + \hat{b}_{2j}t_2 + \cdots + \hat{b}_{qj}t_q + \hat{b}_{q+1j}t_{q+1} + \hat{b}_{q+2j}t_{q+2} + \cdots + \hat{b}_{nj}t_n$$

it follows that $\hat{b}_{q+1j} = \hat{b}_{q+2j} = \cdots = \hat{b}_{nj} = 0$. That is,

$$\hat{b}_j = \begin{bmatrix} \hat{b}_{1j} \\ \hat{b}_{2j} \\ \vdots \\ \hat{b}_{qj} \\ 0 \\ 0 \\ \vdots \\ 0 \end{bmatrix}$$

Applying this argument to every column of B, it follows that every column of $\hat{B}$ has zeros in the last $n - q$ components and therefore that $\hat{B}$ has the required form.

To show that $\hat{A}$ has the required upper block triangular form, recall that, denoting the characteristic polynomial of A by

$$|\lambda I - A| = \lambda^n + a_{n-1}\lambda^{n-1} + \cdots + a_1\lambda + a_0$$

the Cayley-Hamilton theorem leads to the identity

$$A^n = -a_0 I - a_1 A - \cdots - a_{n-1} A^{n-1}$$

and consequently, for each $j = 1, 2, \ldots, m$,

$$A^n b_j \in \text{span } \{b_j, Ab_j, \ldots, A^{n-1} b_j\} \subset \text{span } \{t_1, t_2, \ldots, t_q\} \qquad (3.9)$$

Now, for each $i = 1, 2, \ldots, q$, $t_i = A^k b_j$ for some $j = 1, 2, \ldots, m$ and $k = 0, 1, 2, \ldots, n - 1$.

There are two cases to consider:

1. $k < n - 1$ so that directly

$$At_i = A^{k+1} b_j \in \text{span } \{b_j, Ab_j, \ldots, A^{n-1} b_j\} \subset \text{span } \{t_1, t_2, \ldots, t_q\}$$

2. $k = n - 1$ so that by Equation (3.9)

$$At_i = A^n b_j \in \text{span } \{b_j, Ab_j, \ldots, A^{n-1} b_j\} \subset \text{span } \{t_1, t_2, \ldots, t_q\}$$

Thus, in either case, $At_i \in \text{span } \{t_1, t_2, \ldots, t_q\}$, for $i = 1, 2, \ldots, q$. Using similar reasoning as above,

$$A \begin{bmatrix} t_1 & t_2 & \cdots & t_q \end{bmatrix} = \begin{bmatrix} At_1 & At_2 & \cdots & At_q \end{bmatrix}$$

$$= \begin{bmatrix} t_1 & t_2 & \cdots & t_q \,|\, t_{q+1} & t_{q+2} & \cdots & t_n \end{bmatrix} \begin{bmatrix} A_{11} \\ 0 \end{bmatrix}$$

where the ith column of A_{11} contains the q coefficients that uniquely characterize At_i as a linear combination of $\{t_1, t_2, \ldots, t_q\}$. Also, since $At_i \in \text{span } \{t_1, t_2, \ldots, t_q, t_{q+1}, t_{q+2}, \ldots, t_n\}$ for $i = q + 1, \ldots, n$, we have

$$A \begin{bmatrix} t_{q+1} & t_{q+2} & \cdots & t_n \end{bmatrix} = \begin{bmatrix} At_{q+1} & At_{q+2} & \cdots & At_n \end{bmatrix}$$

$$= \begin{bmatrix} t_1 & t_2 & \cdots & t_q \,|\, t_{q+1} & t_{q+2} & \cdots & t_n \end{bmatrix} \begin{bmatrix} A_{12} \\ A_{22} \end{bmatrix}$$

Putting everything together,

$$A \begin{bmatrix} t_1 & t_2 & \cdots & t_q \,|\, t_{q+1} & t_{q+2} & \cdots & t_n \end{bmatrix}$$

$$= \begin{bmatrix} t_1 & t_2 & \cdots & t_q \,|\, t_{q+1} & t_{q+2} & \cdots & t_n \end{bmatrix} \begin{bmatrix} A_{11} & A_{12} \\ 0 & A_{22} \end{bmatrix}$$

Thus

$$AT = T \begin{bmatrix} A_{11} & A_{12} \\ 0 & A_{22} \end{bmatrix} \qquad \begin{aligned} \hat{A} &= T^{-1}AT \\ &= \begin{bmatrix} A_{11} & A_{12} \\ 0 & A_{22} \end{bmatrix} \end{aligned}$$

as required. Since

$$T^{-1} \begin{bmatrix} B & AB & \cdots & A^{n-1}B \end{bmatrix} = \begin{bmatrix} \hat{B} & \hat{A}\hat{B} & \cdots & \hat{A}^{n-1}\hat{B} \end{bmatrix}$$

$$= \begin{bmatrix} B_1 & A_{11}B_1 & \cdots & A_{11}^{n-1}B_1 \\ 0 & 0 & \cdots & 0 \end{bmatrix}$$

and multiplication by a nonsingular matrix does not affect matrix rank, we conclude that

$$\text{rank} \begin{bmatrix} B_1 & A_{11}B_1 & \cdots & A_{11}^{q-1}B_1 \,|\, A_{11}^{q}B_1 & \cdots & A_{11}^{n-1}B_1 \end{bmatrix} = q$$

Finally, an argument involving the Cayley-Hamilton theorem applied to the $q \times q$ submatrix A_{11} shows that

$$\text{rank} \begin{bmatrix} B_1 & A_{11}B_1 & \cdots & A_{11}^{q-1}B_1 \end{bmatrix} = q$$

so that the pair (A_{11}, B_1) defines a controllable q-dimensional state equation. $\qquad\square$

Example 3.7 Recall that the uncontrollable state equation from Example 3.2, namely,

$$\begin{bmatrix} \dot{x}_1(t) \\ \dot{x}_2(t) \\ \dot{x}_3(t) \end{bmatrix} = \begin{bmatrix} 0 & 1 & 0 \\ 0 & 0 & 1 \\ -6 & -11 & -6 \end{bmatrix} \begin{bmatrix} x_1(t) \\ x_2(t) \\ x_3(t) \end{bmatrix} + \begin{bmatrix} 0 \\ 1 \\ -3 \end{bmatrix} u(t)$$

has the rank 2 controllability matrix

$$\begin{bmatrix} B & AB & A^2B \end{bmatrix} = \begin{bmatrix} 0 & 1 & -3 \\ 1 & -3 & 7 \\ -3 & 7 & -15 \end{bmatrix}$$

in which the first two columns are linearly independent. Appending to these column vectors the third standard basis vector gives

$$T = \begin{bmatrix} 0 & 1 & 0 \\ 1 & -3 & 0 \\ -3 & 7 & 1 \end{bmatrix}$$

which is nonsingular, as can be verified with a straightforward determinant computation. A direct calculation shows that

$$\hat{A} = T^{-1}AT \qquad\qquad \hat{B} = T^{-1}B$$

$$= \left[\begin{array}{cc|c} 0 & -2 & 1 \\ 1 & -3 & 0 \\ \hline 0 & 0 & -3 \end{array}\right] \qquad\qquad = \left[\begin{array}{c} 1 \\ 0 \\ \hline 0 \end{array}\right]$$

which is in the standard form for an uncontrollable state equation because the $q =$ two-dimensional state equation specified by

$$A_{11} = \begin{bmatrix} 0 & -2 \\ 1 & -3 \end{bmatrix} \qquad B_1 = \begin{bmatrix} 1 \\ 0 \end{bmatrix}$$

is easily seen to be controllable.   □

3.4   POPOV-BELEVITCH-HAUTUS TESTS FOR CONTROLLABILITY

Checking the rank of the controllability matrix provides an algebraic test for controllability. Here we present two others, referred to as the Popov-Belevitch-Hautus eigenvector and rank tests, respectively.

Theorem 3.7   *(Popov-Belevitch-Hautus Eigenvector Test for Controllability). The state equation specified by the pair (A, B) is controllable if and only if there exists no left eigenvector of A orthogonal to the columns of B.*

Proof.   We must prove two implications:

If the state equation specified by the pair (A, B) is controllable, then there exists no left eigenvector of A orthogonal to the columns of B.

If there exists no left eigenvector of A orthogonal to the columns of B, then the state equation specified by the pair (A, B) is controllable.

For necessity, we prove the contrapositive statement: If there *does* exist a left eigenvector of A orthogonal to the columns of B, then the state equation specified by the pair (A, B) is *not* controllable. To proceed, suppose that for $w \in \mathbb{C}^n$ is a left eigenvector of A associated with $\lambda \in \sigma(A)$ that is orthogonal to the columns of B. Then

$$w \neq 0 \quad w^*A = \lambda w^* \quad w^*B = 0.$$

A straightforward induction argument shows that

$$w^*A^k B = \lambda^k (w^*B) = 0 \qquad \text{for all } k \geq 0$$

so that
$$w^* \begin{bmatrix} B & AB & A^2 B & \cdots & A^{n-1}B \end{bmatrix} = \begin{bmatrix} 0 & 0 & 0 & \cdots & 0 \end{bmatrix}$$

Since $w \neq 0$, the preceding identity indicates that there is a nontrivial linear combination of the rows of the controllability matrix that yields the $1 \times (nm)$ zero vector, so, by definition, the controllability matrix has linearly dependent rows. In other words,

$$\text{rank} \begin{bmatrix} B & AB & A^2 B & \cdots & A^{n-1}B \end{bmatrix} < n$$

from which it follows that the state equation specified by the pair (A, B) is not controllable.

For sufficiency, we again prove the contrapositive statement: If the state equation specified by the pair (A, B) is not controllable, then there does exist a left eigenvector of A orthogonal to the columns of B. For this, suppose that the state equation is not controllable, so

$$q = \text{rank} \begin{bmatrix} B & AB & A^2 B & \cdots & A^{n-1}B \end{bmatrix} < n$$

By Theorem 3.6, there exists a state coordinate transformation $x(t) = T\, z(t)$ such that the transformed state equation has

$$\hat{A} = T^{-1}AT \qquad\qquad \hat{B} = T^{-1}B$$
$$= \begin{bmatrix} A_{11} & A_{12} \\ 0 & A_{22} \end{bmatrix} \qquad = \begin{bmatrix} B_1 \\ 0 \end{bmatrix}$$

Let λ be any eigenvalue of the submatrix A_{22} and $w_2 \in \mathbb{C}^{n-q}$ be an associated left eigenvector. Define

$$w = T^{-T} \begin{bmatrix} 0 \\ w_2 \end{bmatrix} \neq 0$$

which satisfies

$$w^*A = \begin{bmatrix} 0 & w_2^* \end{bmatrix} T^{-1} \left(T \begin{bmatrix} A_{11} & A_{12} \\ 0 & A_{22} \end{bmatrix} T^{-1} \right)$$

$$= \begin{bmatrix} 0 & w_2^* A_{22} \end{bmatrix} T^{-1}$$

$$= \begin{bmatrix} 0 & \lambda w_2^* \end{bmatrix} T^{-1}$$

$$= \lambda \begin{bmatrix} 0 & w_2^* \end{bmatrix} T^{-1}$$

$$= \lambda w^*$$

along with

$$w^*B = \begin{bmatrix} 0 & w_2^* \end{bmatrix} T^{-1} \left(T \begin{bmatrix} B_1 \\ 0 \end{bmatrix} \right)$$

$$= \begin{bmatrix} 0 & w_2^* \end{bmatrix} \begin{bmatrix} B_1 \\ 0 \end{bmatrix}$$

$$= 0$$

Thus we have constructed a left eigenvector w of A orthogonal to the columns of B.   □

Example 3.8   We again return to the uncontrollable state equation from Examples 3.2 and 3.7:

$$\begin{bmatrix} \dot{x}_1(t) \\ \dot{x}_2(t) \\ \dot{x}_3(t) \end{bmatrix} = \begin{bmatrix} 0 & 1 & 0 \\ 0 & 0 & 1 \\ -6 & -11 & -6 \end{bmatrix} \begin{bmatrix} x_1(t) \\ x_2(t) \\ x_3(t) \end{bmatrix} + \begin{bmatrix} 0 \\ 1 \\ -3 \end{bmatrix} u(t)$$

The eigenvalues of A are found to be $\lambda_1 = -1$, $\lambda_2 = -2$, and $\lambda_3 = -3$, with associated left eigenvectors

$$w_1 = \begin{bmatrix} 6 \\ 5 \\ 1 \end{bmatrix} \quad w_2 = \begin{bmatrix} 3 \\ 4 \\ 1 \end{bmatrix} \quad w_3 = \begin{bmatrix} 2 \\ 3 \\ 1 \end{bmatrix}$$

Of these, $w_3^T B = 0$, which again confirms that this state equation is not controllable. Furthermore, as expected from the proof of Theorem 3.7,

$$w_3^T \begin{bmatrix} B & AB & A^2B \end{bmatrix} = \begin{bmatrix} 2 & 3 & 1 \end{bmatrix} \begin{bmatrix} 0 & 1 & -3 \\ 1 & -3 & 7 \\ -3 & 7 & -15 \end{bmatrix}$$

$$= \begin{bmatrix} 0 & 0 & 0 \end{bmatrix}$$ □

Theorem 3.8 *(Popov-Belevitch-Hautus Rank Test for Controllability). The state equation specified by the pair (A, B) is controllable if and only if*

$$\text{rank} \begin{bmatrix} \lambda I - A & B \end{bmatrix} = n \quad \text{for all } \lambda \in \mathbb{C}$$

Proof. First, observe that by definition

$$\text{rank}(\lambda I - A) < n \quad \text{equivalently } |\lambda I - A| = 0$$

when and only when $\lambda \in \mathbb{C}$ is an eigenvalue of A. Thus,

$$\text{rank}(\lambda I - A) = n$$

for all $\lambda \in \mathbb{C}$ except at the eigenvalues of A, and consequently,

$$\text{rank} \begin{bmatrix} \lambda I - A & B \end{bmatrix} = n \quad \text{for all } \lambda \in \mathbb{C} - \sigma(A)$$

Thus it remains to show that this rank condition also holds for $\lambda \in \sigma(A)$ when and only when the state equation is controllable.

First, suppose that

$$\text{rank} \begin{bmatrix} \lambda I - A & B \end{bmatrix} < n$$

for some $\lambda \in \sigma(A)$ so that the $n \times (n + m)$ matrix

$$\begin{bmatrix} \lambda I - A & B \end{bmatrix}$$

has linearly dependent rows. Consequently, there is a nonzero vector $w \in \mathbb{C}^n$ such that

$$w^* \begin{bmatrix} \lambda I - A & B \end{bmatrix} = \begin{bmatrix} 0 & 0 \end{bmatrix}$$

In other words,

$$w^*(\lambda I - A) = 0 \quad w^* B = 0$$

so that w is necessarily a left eigenvector of A orthogonal to the columns of B. By the Popov-Belevitch-Hautus eigenvector test, the state equation is not controllable.

Conversely, suppose that the state equation is not controllable so that again by the Popov-Belevitch-Hautus eigenvector test corresponding to an eigenvalue λ of A there is a left eigenvector w that is orthogonal to the columns of B. Reversing the preceding steps, we find

$$w^* \begin{bmatrix} \lambda I - A & B \end{bmatrix} = \begin{bmatrix} 0 & 0 \end{bmatrix}$$

Thus we have identified a $\lambda \in \sigma(A) \subset \mathbb{C}$ for which

$$\begin{bmatrix} \lambda I - A & B \end{bmatrix}$$

has linearly dependent rows so that

$$\text{rank} \begin{bmatrix} \lambda I - A & B \end{bmatrix} < n \qquad \square$$

As an application of the Popov-Belevitch-Hautus tests, consider the linear time-invariant state equation (3.1) together with the state coordinate transformation $x(t) = T z(t)$ and the state feedback law

$$u(t) = -Kx(t) + Gr(t)$$

where $r(t)$ is a new external input signal, and G is an input gain matrix. Note that this can be viewed as a combined state and input transformation, that is,

$$\begin{bmatrix} x(t) \\ u(t) \end{bmatrix} = \begin{bmatrix} T & 0 \\ -KT & G \end{bmatrix} \begin{bmatrix} z(t) \\ r(t) \end{bmatrix} \tag{3.10}$$

This relationship can be inverted provided that both T and G are nonsingular because

$$\begin{bmatrix} z(t) \\ r(t) \end{bmatrix} = \begin{bmatrix} T & 0 \\ -KT & G \end{bmatrix}^{-1} \begin{bmatrix} x(t) \\ u(t) \end{bmatrix}$$

$$= \begin{bmatrix} T^{-1} & 0 \\ G^{-1}K & G^{-1} \end{bmatrix} \begin{bmatrix} x(t) \\ u(t) \end{bmatrix}$$

In this case necessarily, G is $m \times m$, and $r(t)$ is $m \times 1$. The transformed state equation is easily found to be

$$\dot{z}(t) = T^{-1}(A - BK)T \, z(t) + T^{-1}BG \, r(t) \tag{3.11}$$

We have already seen that controllability is invariant with respect to state coordinate transformations. That the same is true for this larger class of state and input transformations is a direct consequence of the Popov-Belevitch-Hautus rank test.

Theorem 3.9   *For any invertible state and input transformation* (3.10), *the state equation* (3.11) *is controllable if and only if the state equation* (3.1) *is controllable.*

Proof.   We first observe that

$$\left[\lambda I - T^{-1}(A - BK)T \quad T^{-1}BG\right] = \left[T^{-1}(\lambda I - (A - BK))T | T^{-1}BG\right]$$

$$= T^{-1}\left[(\lambda I - (A - BK))T | BG\right]$$

$$= T^{-1}\left[\lambda I - A \quad B\right]\begin{bmatrix} T & 0 \\ KT & G \end{bmatrix}$$

Since the rightmost factor is nonsingular and matrix rank is unaffected by pre- and postmultiplication by nonsingular matrices, we conclude that

$$\text{rank}\left[\lambda I - T^{-1}(A - BK)T \quad T^{-1}BG\right] = \text{rank}\left[\lambda I - A \quad B\right]$$

$$\text{for all } \lambda \in \mathbb{C}$$

The desired conclusion follows immediately from the Popov-Belevitch-Hautus rank test.   □

3.5   MATLAB **FOR CONTROLLABILITY AND CONTROLLER CANONICAL FORM**

MATLAB **for Controllability**

Some MATLAB functions that are useful for controllability analysis and decomposition are

`P = ctrb(JbkR)`   Calculate the controllability matrix associated with the linear time-invariant system **JbkR**; only matrices **A** and **B** are used in the calculation.

`rank(P)`   Calculate the rank of matrix **P**.

`det(P)`   Calculate the determinant of square matrix **P**.

`size(A,1)`   Determine the system order **n**.

`ctrbf`   Decomposition into controllable/uncontrollable subsystems (if not controllable).

MATLAB **for Controller Canonical Form**

The MATLAB functions that are useful for coordinate transformations and the controller canonical form realization have been given in previous MATLAB sections in Chapters 1–2. There is no controller canonical form switch for the **canon** function.

Continuing MATLAB Example

Controllability   We now assess the controllability of the open-loop system for the Continuing MATLAB Example (rotational mechanical system). The following MATLAB code performs this determination for the Continuing MATLAB Example.

```
%------------------------------------------------------------
%   Chapter 3.  Controllability
%------------------------------------------------------------

P = ctrb(JbkR);                    % Calculate
                                   % controllability
                                   % matrix P
if (rank(P) == size(A,1))          % Logic to assess
                                   % controllability
    disp('System is controllable.');
else
    disp('System is NOT controllable.');
end

P1 = [B A*B];                      % Check P via the
                                   % formula
```

This m-file, combined with the m-files from previous chapters, performs the controllability check for the Continuing MATLAB Example:

```
P =
     0    1
     1   -4


System is controllable.
```

Coordinate Transformations and Controller Canonical Form   For the Continuing MATLAB Example, we now calculate the controller canonical form state-space realization. The following MATLAB code, along with the code from previous chapters, performs this computation:

```
%------------------------------------------------------------
%   Chapter 3.  Coordinate Transformations and
%  Controller Canonical Form
%------------------------------------------------------------
```

```
CharPoly = poly(A);         % Determine the system
                            % characteristic polynomial
a1 = CharPoly(2);           % Extract a1

Pccfi = [a1 1;1 0];         % Calculate the inverse of
                            % matrix Pccf
Tccf = P*Pccfi;             % Calculate the CCF
                            % transformation matrix

Accf = inv(Tccf)*A*Tccf;    % Transform to CCF via
                            % formula
Bccf = inv(Tccf)*B;
Cccf = C*Tccf;
Dccf = D;
```

The following output is produced:

```
CharPoly =
    1.0000     4.0000    40.0000

Tccf =
    1.0000          0
   -0.0000     1.0000

Accf =
   -0.0000     1.0000
  -40.0000    -4.0000

Bccf =
    0
    1

Cccf =
    1     0

Dccf =
    0
```

Note that the coordinate transformation matrix **Tccf** in this example is a 2 × 2 identity matrix, which means that our original state-space realization was already in controller canonical form. The resulting state-space coefficient matrices are identical to those originally derived and entered into the linear time-invariant system data structure **JbkR** in Chapter 1.

3.6 CONTINUING EXAMPLES FOR CONTROLLABILITY AND CONTROLLER CANONICAL FORM

Continuing Example 1: Two-Mass Translational Mechanical System

Controllability   We now assess controllability of Continuing Example 1 (two-mass translational mechanical system), for both case a (multiple-input, multiple-output) and case b [input $u_2(t)$ and output $y_1(t)$].

Case a.   The 4×8 controllability matrix P is

$$P = \begin{bmatrix} B & AB & A^2B & A^3B \end{bmatrix}$$

$$= \begin{bmatrix} 0 & 0 & 0.03 & 0 & -0.02 & 0.01 & -0.36 & 0.23 \\ 0.03 & 0 & -0.02 & 0.01 & -0.36 & 0.23 & 0.67 & -0.61 \\ 0 & 0 & 0 & 0.05 & 0.01 & -0.03 & 0.23 & -0.48 \\ 0 & 0.05 & 0.01 & -0.03 & 0.23 & -0.48 & -0.61 & 0.73 \end{bmatrix}$$

This controllability matrix is of full rank, i.e. rank$(P) = 4$, which matches the system order $n = 4$. Therefore, the state equation is controllable.

Case b.   In case b, the system dynamics matrix A is identical to case a; however, since B is different owing to the single input $u_2(t)$, we must again check for controllability. The 4×4 controllability matrix P is

$$P = \begin{bmatrix} B & AB & A^2B & A^3B \end{bmatrix}$$

$$= \begin{bmatrix} 0 & 0 & 0.01 & 0.23 \\ 0 & 0.01 & 0.23 & -0.61 \\ 0 & 0.05 & -0.03 & -0.48 \\ 0.05 & -0.03 & -0.48 & 0.73 \end{bmatrix}$$

This controllability matrix is of full rank, i.e. rank$(P) = 4$, which matches the system order $n = 4$. Also, the determinant of this square matrix P is nonzero, $|P| = 1.5625e - 004$, confirming that P is nonsingular. Therefore, the state equation is controllable.

Now let us return to the diagonal canonical form of Section 2.7. Since the system eigenvalues are distinct and B_{DCF} is fully populated (no zero elements), this means that the state equation is controllable, which agrees with our conclusion above.

Coordinate Transformations and Controller Canonical Form   We now construct the controller canonical form for Continuing Example 1, for case b (single-input $u_2(t)$ and single-output $y_1(t)$). The coordinate transformation matrix for controller canonical form is $T_{CCF} = PP_{CCF}^{-1}$, where the controllability matrix P was given earlier and the inverse of the controller canonical form controllability matrix P_{CCF}, is given below.

$$P_{CCF}^{-1} = \begin{bmatrix} 10 & 25.25 & 1.25 & 1 \\ 25.25 & 1.25 & 1 & 0 \\ 1.25 & 1 & 0 & 0 \\ 1 & 0 & 0 & 0 \end{bmatrix}$$

This produces

$$T_{CCF} = PP_{CCF}^{-1}$$

$$= \begin{bmatrix} 0.25 & 0.0125 & 0 & 0 \\ 0 & 0.25 & 0.0125 & 0 \\ 0.75 & 0.0375 & 0.05 & 0 \\ 0 & 0.75 & 0.0375 & 0.05 \end{bmatrix}$$

Using this coordinate transformation, we find the controller canonical form:

$$A_{CCF} = T_{CCF}^{-1}AT_{CCF} \qquad\qquad B_{CCF} = T_{CCF}^{-1}B$$

$$= \begin{bmatrix} 0 & 1 & 0 & 0 \\ 0 & 0 & 1 & 0 \\ 0 & 0 & 0 & 1 \\ -100 & -10 & -25.25 & -1.25 \end{bmatrix} \qquad = \begin{bmatrix} 0 \\ 0 \\ 0 \\ 1 \end{bmatrix}$$

$$C_{CCF} = CT_{CCF} \qquad\qquad D_{CCF} = D$$

$$= \begin{bmatrix} 0.25 & 0.0125 & 0 & 0 \end{bmatrix} \qquad\qquad = 0$$

Note that controller canonical form is as expected; i.e., the placement of ones and zeros is correct in the first three rows of A_{CCF}, and the fourth row displays the coefficients of the system characteristic polynomial (except for the coefficient 1 of s^4) in ascending order left to right, with negative signs. Furthermore the form of B_{CCF} is correct, with zeros in rows 1 through 3 and a one in the fourth row. Also note that C_{CCF} is composed of coefficients of the numerator of the single-input, single-output transfer function presented for case b in Chapter 2, $0.0125s + 0.25$, again in ascending powers of s.

Continuing Example 2: Rotational Electromechanical System

Controllability Here we assess controllability for Continuing Example 2 (rotational electromechanical system). The 3×3 controllability matrix P is

$$P = \begin{bmatrix} B & AB & A^2B \end{bmatrix}$$

$$= \begin{bmatrix} 0 & 0 & 2 \\ 0 & 2 & -6 \\ 2 & -6 & 14 \end{bmatrix}$$

This controllability matrix is of full rank, i.e., rank$(P) = 3$, which matches the system order $n = 3$. Therefore, the state equation is controllable. Also $|P| = -8 \neq 0$, leading to the same conclusion.

Again, since the system eigenvalues are distinct and B_{DCF} is fully populated (no zero elements) in the diagonal canonical form from Chapter 2, this means that the state equation is controllable, which agrees with our conclusion above.

Coordinate Transformations and Controller Canonical Form
Now we calculate the controller canonical form for Continuing Example 2. The original realization is nearly in controller canonical form already; the 2 in B need only be scaled to a 1. Therefore, the coordinate transformation matrix for controller canonical form is simple (given below). The controllability matrix P was given earlier, and the inverse of the controller canonical form controllability matrix P_{CCF}, is given below:

$$P_{\text{CCF}}^{-1} = \begin{bmatrix} 2 & 3 & 1 \\ 3 & 1 & 0 \\ 1 & 0 & 0 \end{bmatrix}$$

This gives

$$T_{\text{CCF}} = P P_{\text{CCF}}^{-1}$$

$$= \begin{bmatrix} 2 & 0 & 0 \\ 0 & 2 & 0 \\ 0 & 0 & 2 \end{bmatrix}$$

Using this coordinate transformation, we find the controller canonical form:

$$A_{\text{CCF}} = T_{\text{CCF}}^{-1} A T_{\text{CCF}} \qquad B_{\text{CCF}} = T_{\text{CCF}}^{-1} B$$

$$= \begin{bmatrix} 0 & 1 & 0 \\ 0 & 0 & 1 \\ 0 & -2 & -3 \end{bmatrix} \qquad = \begin{bmatrix} 0 \\ 0 \\ 1 \end{bmatrix}$$

$$C_{CCF} = CT_{CCF} \qquad\qquad D_{CCF} = D$$

$$= \begin{bmatrix} 2 & 0 & 0 \end{bmatrix} \qquad\qquad = 0$$

Note that controller canonical form is as expected; i.e., the placement of ones and zeros is correct in the first two rows of A_{CCF}, and the third row displays the coefficients of the system characteristic polynomial (except for the unity coefficient 1 of s^3) in ascending order left to right, with negative signs. Furthermore, the form of B_{CCF} is correct, with zeros in rows 1 and 2 and a one in the third row. Again, C_{CCF} is composed of the coefficients of the numerator polynomial in ascending powers of s of the single-input, single-output transfer function presented for this example in Chapter (simply a constant 2).

3.7 HOMEWORK EXERCISES

Numerical Exercises

NE3.1 Assess the controllability of the following systems, represented by the matrix pair (A, B).

a. $A = \begin{bmatrix} -4 & 0 \\ 0 & -5 \end{bmatrix}$ $B = \begin{bmatrix} 1 \\ 1 \end{bmatrix}$

b. $A = \begin{bmatrix} -4 & 0 \\ 0 & -5 \end{bmatrix}$ $B = \begin{bmatrix} 1 \\ 0 \end{bmatrix}$

c. $A = \begin{bmatrix} 0 & -10 \\ 1 & -2 \end{bmatrix}$ $B = \begin{bmatrix} 1 \\ 2 \end{bmatrix}$

d. $A = \begin{bmatrix} 0 & 1 \\ -10 & -2 \end{bmatrix}$ $B = \begin{bmatrix} 0 \\ 1 \end{bmatrix}$

e. $A = \begin{bmatrix} 2 & 0 \\ -1 & 1 \end{bmatrix}$ $B = \begin{bmatrix} 1 \\ -1 \end{bmatrix}$

NE3.2 Compute the controller canonical form of the following systems ($D = 0$ for all cases).

a. $A = \begin{bmatrix} -4 & 0 \\ 0 & -5 \end{bmatrix}$ $B = \begin{bmatrix} 1 \\ 1 \end{bmatrix}$ $C = \begin{bmatrix} 1 & 1 \end{bmatrix}$

b. $A = \begin{bmatrix} -4 & 0 \\ 0 & -5 \end{bmatrix}$ $B = \begin{bmatrix} 1 \\ 0 \end{bmatrix}$ $C = \begin{bmatrix} 1 & 0 \end{bmatrix}$

c. $A = \begin{bmatrix} 0 & -10 \\ 1 & -2 \end{bmatrix}$ $B = \begin{bmatrix} 1 \\ 2 \end{bmatrix}$ $C = \begin{bmatrix} 0 & 1 \end{bmatrix}$

d. $A = \begin{bmatrix} 0 & 1 \\ -10 & -2 \end{bmatrix}$ $B = \begin{bmatrix} 0 \\ 1 \end{bmatrix}$ $C = \begin{bmatrix} 1 & 2 \end{bmatrix}$

e. $A = \begin{bmatrix} 2 & 0 \\ -1 & 1 \end{bmatrix}$ $B = \begin{bmatrix} 1 \\ -1 \end{bmatrix}$ $C = \begin{bmatrix} 1 & 1 \end{bmatrix}$

NE3.3 Repeat **NE 3.1** using the Popov-Belevitch-Hautus tests for controllability.

Analytical Exercises

AE 3.1 For the $n \times n$ matrix

$$A = \begin{bmatrix} 0 & 1 & 0 & \cdots & 0 & 0 \\ 0 & 0 & 1 & \cdots & 0 & 0 \\ \vdots & \vdots & \vdots & \ddots & \vdots & \vdots \\ 0 & 0 & 0 & \cdots & 1 & 0 \\ 0 & 0 & 0 & \cdots & 0 & 1 \\ -a_0 & -a_1 & -a_2 & \cdots & -a_{n-2} & -a_{n-1} \end{bmatrix}$$

a. Use mathematical induction on n to show that the characteristic polynomial of A is

$$\lambda^n + a_{n-1}\lambda^{n-1} + a_{n-2}\lambda^{n-2} + \cdots + a_2\lambda^2 + a_1\lambda + a_0$$

b. If λ_i is an eigenvalue of A, show that a corresponding eigenvector is given by

$$v_i = \begin{bmatrix} 1 \\ \lambda_i \\ \lambda_i^2 \\ \vdots \\ \lambda_i^{n-1} \end{bmatrix}$$

c. Show that the geometric multiplicity of each distinct eigenvalue is one.

AE3.2 Show that the controllability Gramian satisfies the matrix differential equation

$$\frac{d}{dt}W(t, t_f) - AW(t, t_f) - W(t, t_f)A^T + BB^T = 0$$

$$W(t_f, t_f) = 0$$

AE3.3 Show that the matrix exponential for

$$\begin{bmatrix} A & BB^T \\ 0 & -A^T \end{bmatrix}$$

is given by

$$\begin{bmatrix} e^{At} & e^{At}W(0,t) \\ 0 & e^{-A^Tt} \end{bmatrix}$$

AE 3.4 Consider the reachability Gramian, defined as follows for any initial time t_0 and any finite final time $t_f > t_0$:

$$W_R(t_0, t_f) = \int_{t_0}^{t_f} e^{A(t_f - \tau)} BB^T e^{A^T(t_f - \tau)} d\tau$$

Show that $W_R(t_0, t_f)$ is nonsingular for any t_0 and $t_f > t_0$ if and only if the pair (A, B) is controllable. Assuming that (A, B) is controllable, use the reachability Gramian to construct, for any $x_f \in \mathbb{R}^n$, an input signal on $[t_0, t_f]$ such that $x(t_0) = 0$ and $x(t_f) = x_f$.

AE 3.5 Suppose that a single-input n-dimensional state equation has

$$q \triangleq \operatorname{rank}\begin{bmatrix} B & AB & A^2B & \cdots & A^{n-1}B \end{bmatrix} \le n$$

Show that the first q columns $\{B, AB, A^2B, \ldots, A^{q-1}B\}$ are linearly independent.

AE 3.6 Show that the single-input state equation characterized by

$$A = \begin{bmatrix} \mu & 1 & 0 & \cdots & 0 \\ 0 & \mu & 1 & \cdots & 0 \\ \vdots & \vdots & \vdots & \ddots & \vdots \\ 0 & 0 & 0 & \cdots & 1 \\ 0 & 0 & 0 & \cdots & \mu \end{bmatrix} \qquad B = \begin{bmatrix} b_1 \\ b_2 \\ \vdots \\ b_{n-1} \\ b_n \end{bmatrix}$$

is controllable if and only if $b_n \neq 0$.

AE 3.7 For the controllable system with two separate scalar outputs

$$\dot{x}(t) = Ax(t) + Bu(t)$$
$$y_1(t) = C_1 x(t)$$
$$y_2(t) = C_2 x(t)$$

show that the related impulse responses $h_1(t)$ and $h_2(t)$ are identical if and only if $C_1 = C_2$.

AE3.8 Suppose that $H_1(s)$ and $H_2(s)$ are two strictly proper single-input, single-output transfer functions with controllable state-space realizations (A_1, B_1, C_1) and (A_2, B_2, C_2), respectively. Construct a state-space realization for the parallel interconnection $H_1(s) + H_2(s)$, and use the Popov-Belevitch-Hautus eigenvector test to show that the parallel realization is controllable if and only if A_1 and A_2 have no common eigenvalues.

AE3.9 Suppose that $H_1(s)$ and $H_2(s)$ are two strictly proper single-input, single-output transfer functions with controllable state-space realizations (A_1, B_1, C_1) and (A_2, B_2, C_2), respectively. Construct a state-space realization for the series interconnection $H_1(s)H_2(s)$, and show that this realization is controllable if and only if no eigenvalue of A_2 is a zero of $H_1(s)$.

AE3.10 Show that the pair (A, B) is controllable if and only if the only square matrix X that satisfies

$$AX = XA \quad XB = 0$$

is $X = 0$.

Continuing MATLAB Exercises

CME3.1 For the CME1.1 system:
   a. Assess the system controllability.
   b. Compute the controller canonical form.

CME3.2 For the CME1.2 system:
   a. Assess the system controllability.
   b. Compute the controller canonical form.

CME3.3 For the CME1.3 system:
   a. Assess the system controllability.
   b. Compute the controller canonical form.

CME3.4 For the CME1.4 system:
   a. Assess the system controllability.
   b. Compute the controller canonical form.

Continuing Exercises

CE3.1a Determine if the CE1 system is controllable for all three cases (cases from CE1.1b and numeric parameters from CE2.1a). Give the mathematical details to justify your answers; explain your results in all cases by looking at the physical problem.

CE3.1b Compute the controller canonical form for the CE1 system, case iii only. Comment on the structure of your results. Determine the system controllability by looking at the diagonal canonical form realization of CE 2.1b; compare to your controllability results from CE3.1a.

CE3.2a Determine if the CE2 system is controllable for all three cases (cases from CE1.2b and numeric parameters from CE2.2a). Give the mathematical details to justify your answers; explain your results in all cases by looking at the physical problem.

CE3.2b Compute the controller canonical form for the CE2 system, Case i only. Comment on the structure of your results.

CE3.3a Determine if the CE3 system is controllable for both cases (cases from CE1.3b and numeric parameters from CE2.3a). Give the mathematical details to justify your answers; explain your results in all cases by looking at the physical problem.

CE3.3b Compute the controller canonical form for the CE3 system for both cases. Comment on the structure of your results. Determine the system controllability by looking at the diagonal canonical form realization from CE 2.3b for case i; compare to your results from CE3.3a.

CE3.4a Determine if the CE4 system is controllable (for the CE1.4b single-input, single-output case with the numeric parameters from CE2.4a). Give the mathematical details to justify your answers; explain your results by looking at the physical problem.

CE3.4b Compute the controller canonical form for the CE4 system. Comment on the structure of your results. Determine the system controllability by looking at the diagonal canonical form realization from CE 2.4b; compare with your results from CE3.4a.

CE3.5a Determine if the CE5 system is controllable (for the CE1.5b single-input, single-output case with the numeric parameters from CE2.5a). Give the mathematical details to justify your answers; explain your results by looking at the physical problem.

CE3.5b Compute the controller canonical form for the CE5 system. Comment on the structure of your results. Determine the system controllability by looking at the diagonal canonical form realization from CE 2.5b; compare with your results from CE3.5a.

4

OBSERVABILITY

In our state-space description of linear time-invariant systems, the state vector constitutes an internal quantity that is influenced by the input signal and, in turn, affects the output signal. We have seen in our examples and it is generally true in practice that the dimension of the state vector, equivalently the dimension of the system modeled by the state equation, is greater than the number of input signals or output signals. This reflects the fact that the complexity of real-world systems precludes the ability to directly actuate or sense each state variable. Nevertheless, we are often interested in somehow estimating the state vector because it characterizes the system's complex inner workings and, as we shall see in Chapters 7 and 8, figures prominently in state-space methods of control system design.

The fundamental question we address in this chapter is whether or not measurements of the input and output signals of our linear state equation over a finite time interval can be processed in order to uniquely determine the initial state. If so, knowledge of the initial state and the input signal allows the entire state trajectory to be reconstructed according to the state equation solution formula. This, in essence, characterizes the system property of *observability*. As with our treatment of controllability in Chapter 3, our aim is to establish algebraic criteria for observability expressed in terms of the state-equation coefficient matrices.

Linear State-Space Control Systems, by Robert L. Williams II and Douglas A. Lawrence
Copyright © 2007 John Wiley & Sons, Inc.

We begin the chapter with an analysis of observability patterned after our introduction to controllability. This suggests a duality that exists between controllability and observability that we develop in detail. This pays immediate dividends in that various observability-related results can be established with modest effort by referring to the corresponding result for controllability and invoking duality. In particular, we investigate relationships between observability and state coordinate transformations as well as formulate Popov-Belevich-Hautus tests for observability. We conclude the chapter by illustrating the use of MATLAB for observability analysis and revisit the MATLAB Continuing Example as well as Continuing Examples 1 and 2.

4.1 FUNDAMENTAL RESULTS

For the n–dimensional linear time-invariant state equation

$$\begin{aligned} \dot{x}(t) &= Ax(t) + Bu(t) \\ y(t) &= Cx(t) + Du(t) \end{aligned} \qquad x(t_0) = x_0 \qquad (4.1)$$

we assume that the input signal $u(t)$ and the output signal $y(t)$ can be measured over a finite time interval and seek to deduce the initial state $x(t_0) = x_0$ by processing this information in some way. As noted earlier, if the initial state can be uniquely determined, then this, along with knowledge of the input signal, yields the entire state trajectory via

$$x(t) = e^{A(t-t_0)}x_0 + \int_{t_0}^{t} e^{A(t-\tau)} Bu(\tau)\, d\tau \quad \text{for} \quad t \geq t_0$$

Since $u(t)$ is assumed to be known, the zero-state response can be extracted from the complete response $y(t)$, also known, in order to isolate the zero-input response component via

$$y(t) - \left[\int_{t_0}^{t} Ce^{A(t-\tau)} Bu(\tau)d\tau + Du(t) \right] = Ce^{A(t-t_0)}x_0$$

which depends directly on the unknown initial state. Consequently, we can assume without loss of generality that $u(t) \equiv 0$ for all $t \geq t_0$ and instead consider the homogeneous state equation

$$\begin{aligned} \dot{x}(t) &= Ax(t) \\ y(t) &= Cx(t) \end{aligned} \qquad x(t_0) = x_0 \qquad (4.2)$$

which directly produces the zero-input response component of Equation (4.1).

Definition 4.1   *A state $x_0 \in \mathbb{R}^n$ is* **unobservable** *if the zero-input response of the linear state equation (4.1) with initial state $x(t_0) = x_0$ is $y(t) \equiv 0$ for all $t \geq t_0$. The state equation (4.1) is* **observable** *if the zero vector $0 \in \mathbb{R}^n$ is the only unobservable state.*   □

Note that, by definition, $0 \in \mathbb{R}^n$ is an unobservable state because $x(t_0) = 0$ yields $y(t) \equiv 0$ for all $t \geq t_0$ for the zero-input response of Equation (4.1) and, equivalently, the complete response of the homogeneous state equation (4.2). Therefore, a nonzero unobservable state is sometimes called *indistinguishable from* $0 \in \mathbb{R}^n$. The existence of nonzero unobservable states clearly hinders our ability to uniquely ascertain the initial state from measurements of the input and output, so we are interested in characterizing observable state equations in the sense of Definition 4.1. As in Chapter 3, we first seek an equivalent algebraic characterization for observability. Again noting that the underlying definition involves the response of the homogeneous state equation (4.2) characterized by the A and C coefficient matrices, we should not be surprised that our first algebraic characterization is cast in terms of these matrices.

Theorem 4.2   *The linear state equation* (4.1) *is observable if and only if*

$$\text{rank} \begin{bmatrix} C \\ CA \\ CA^2 \\ \vdots \\ CA^{n-1} \end{bmatrix} = n$$

We refer to this matrix as the *observability matrix* and henceforth denote it as Q. Since the algebraic test for observability established in the theorem involves only the coefficient matrices A and C, we will have occasion for reasons of brevity to refer to observability of the matrix pair (A, C) with an obvious connection to either the state equation (4.1) or the homogeneous state equation (4.2).

For the general multiple-output case, we see that Q consists of n matrix blocks, $C, CA, CA^2, \ldots, CA^{n-1}$, each with dimension $p \times n$, stacked one on top of another. Hence Q has dimension $(np) \times n$ and therefore has more rows than columns in the multiple-output case. The rank condition in the theorem requires that the n columns of Q are linearly independent when viewed as column vectors of dimension $(np) \times 1$. Consequently, Q satisfying the preceding rank condition is said to have *full-column rank*.

Alternatively, this rank condition means that out of the np rows of Q written as

$$
\begin{bmatrix}
c_1 \\
c_1 \\
\vdots \\
c_p \\
\hline
c_1 A \\
c_2 A \\
\vdots \\
c_p A \\
\hline
\vdots \\
\hline
c_1 A^{n-1} \\
c_2 A^{n-1} \\
\vdots \\
c_p A^{n-1}
\end{bmatrix}
$$

there must be at least one way to select n linearly independent rows.

For the single-output case, C consists of a single row, and Q is a square $n \times n$ matrix. Hence a single-output linear state equation is observable if and only if the observability matrix is nonsingular. We can check that Q is nonsingular by verifying that it has a nonzero determinant.

Proof of Theorem 4.2. The theorem involves two implications:

If the state equation (4.1) is observable, then $\text{rank} Q = n$.

If $\text{rank} Q = n$, then the state equation (4.1) is observable.

We begin by proving the contrapositive of the first implication: If rank $Q < n$, then the state equation (4.1) is *not* observable. Assuming rank $Q < n$, Q has linearly dependent columns, so there exists a nontrivial linear combination of the columns of Q that yields an $(np) \times 1$ zero vector. Stacking the coefficients of this linear combination into a nonzero $n \times 1$ vector x_0, we have

$$
\begin{bmatrix}
C \\
CA \\
CA^2 \\
\vdots \\
CA^{n-1}
\end{bmatrix}
x_0 =
\begin{bmatrix}
0 \\
0 \\
0 \\
\vdots \\
0
\end{bmatrix}
$$

Another way to arrive at this is to observe that $\text{rank}\,Q < n$ implies, by Sylvester's law of nullity, that $\text{nullity}\,Q \geq 1$. The vector x_0 just introduced is a nonzero $n \times 1$ vector lying in the null space of Q. In any case, the preceding identity can be divided up into

$$Cx_0 = CAx_0 = CA^2x_0 = \cdots = CA^{n-1}x_0 = 0 \in \mathbb{R}^p$$

Using the finite series expansion

$$e^{At} = \sum_{k=0}^{n-1} \alpha_k(t)A^k$$

for scalar functions $\alpha_0(t), \alpha_1(t), \ldots, \alpha_{n-1}(t)$, we see that the response of the homogeneous system (4.2) with initial state $x(t_0) = x_0$ satisfies

$$\begin{aligned}
y(t) &= Ce^{A(t-t_0)}x_0 \\
&= C\left[\sum_{k=0}^{n-1} \alpha_k(t-t_0)A^k\right]x_0 \\
&= \sum_{k=0}^{n-1} \alpha_k(t-t_0)(CA^kx_0) \\
&\equiv 0 \quad \text{for all} \quad t \geq t_0
\end{aligned}$$

Thus x_0 is an unobservable state. Since x_0 was taken initially to be a nonzero vector, we have, as a consequence of Definition 4.1, that the linear state equation (4.1) is *not* observable.

We next prove the contrapositive of the second implication: If the linear state equation (4.1) is *not* observable, then $\text{rank}\,Q < n$. Assuming that the linear state equation (4.1) is not observable, there exists by Definition 4.1 a nonzero unobservable state. That is, there exists a nonzero $n \times 1$ vector x_0 for which $Ce^{A(t-t_0)}x_0 \equiv 0$ for all $t \geq t_0$. We can repeatedly differentiate this identity with respect to t and evaluate at $t = t_0$ to obtain

$$0 = \frac{d^k}{dt^k}(Ce^{A(t-t_0)}x_0)\Big|_{t=t_0} = CA^k e^{A(t-t_0)}x_0\Big|_{t=t_0}$$

$$= CA^kx_0 \quad \text{for} \quad k = 0, 1, \ldots, n-1$$

These identities can be repackaged into

$$
\begin{bmatrix} 0 \\ 0 \\ 0 \\ \vdots \\ 0 \end{bmatrix} = \begin{bmatrix} Cx_0 \\ CAx_0 \\ CA^2x_0 \\ \vdots \\ CA^{n-1}x_0 \end{bmatrix} = \begin{bmatrix} C \\ CA \\ CA^2 \\ \vdots \\ CA^{n-1} \end{bmatrix} x_0
$$

which indicates that the unobservable state x_0 lies in the null space of Q. In terms of the components of x_0, the preceding identity specifies a linear combination of the columns of Q that yields the zero vector. Since x_0 was taken initially to be a nonzero vector, we conclude that Q has linearly dependent columns and hence less than full column rank n. Alternatively, we have that nullity $Q \geq 1$, so an appeal to Sylvester's law of nullity yields rank $Q < n$. $\qquad\square$

We conclude this section by answering the fundamental question posed at the outset: It is possible to uniquely determine the initial state by processing measurements of the input and output signals over a finite time interval if and only if the linear state equation (4.1) is observable. To argue the necessity of observability, suppose that $x(t_0) = x_0$ is a nonzero unobservable initial state that yields, by definition, $y_{zi}(t) = Ce^{A(t-t_0)}x_0 \equiv 0$ for all $t \geq t_0$. Thus $x(t_0) = x_0$ is indistinguishable from the zero initial state, and it is not possible to resolve this ambiguity by processing the zero-input response. To show that observability is sufficient to uniquely recover the initial state from the input and output signals as above, we first show that observability also be can characterized by an *observability Gramian* defined as follows for any initial time t_0 and any finite final time $t_f > t_0$:

$$
M(t_0, t_f) = \int_{t_0}^{t_f} e^{A^{\mathrm{T}}(\tau - t_0)} C^{\mathrm{T}} C e^{A(\tau - t_0)} \, d\tau
$$

As with the controllability Gramian introduced in Chapter 3, the observability Gramian is a square $n \times n$ symmetric matrix that is also positive semidefinite because the associated real-valued quadratic form $x^T M(t_0, t_f)x$ satisfies the following for all vectors $x \in \mathbb{R}^n$

$$
x^T M(t_0, t_f)x = x^T \int_{t_0}^{t_f} e^{A^{\mathrm{T}}(\tau - t_0)} C^{\mathrm{T}} C e^{A(\tau - t_0)} \, d\tau \; x
$$

$$= \int_{t_0}^{t_f} \|Ce^{A(\tau - t_0)}x\|^2 \, d\tau$$

$$\geq 0$$

Lemma 4.3

$$\text{rank} \begin{bmatrix} C \\ CA \\ CA^2 \\ \vdots \\ CA^{n-1} \end{bmatrix} = n$$

if and only if for any initial time t_0 and any finite final time $t_f > t_0$ the observability Gramian $M(t_0, t_f)$ is nonsingular.

Proof. This proof is very similar to the proof of Lemma 3.3 from the previous chapter, and so the reader is encouraged to independently adapt that proof to the case at hand and then compare the result with what follows.

Lemma 4.3 involves two implications:

If rank $Q = n$, then $M(t_0, t_f)$ is nonsingular for any t_0 and any finite $t_f > t_0$.
If $M(t_0, t_f)$ is nonsingular for any t_0 and any finite $t_f > t_0$, then rank $Q = n$.

To begin, we prove the contrapositive of the first implication: If $M(t_0, t_f)$ is singular for some t_0 and finite $t_f > t_0$, then rank $Q < n$. If we assume that $M(t_0, t_f)$ is singular for some t_0 and finite $t_f > t_0$, then there exists a nonzero vector $x_0 \in \mathbb{R}^n$ for which $M(t_0, t_f)x_0 = 0 \in \mathbb{R}^n$ and consequently $x_0^T M(t_0, t_f)x_0 = 0 \in \mathbb{R}$. Using the observability Gramian definition, we have

$$0 = x_0^T M(t_0, t_f)x_0$$

$$= x_0^T \int_{t_0}^{t_f} e^{A^T(\tau - t_0)} C^T C e^{A(\tau - t_0)} d\tau \, x_0$$

$$= \int_{t_0}^{t_f} \|Ce^{A(\tau - t_0)}x_0\|^2 \, d\tau$$

The integrand in this expression involves the Euclidean norm of the $p \times 1$ vector quantity $C e^{A(\tau - t_0)} x_0$, which we view as an analytic function of the integration variable $\tau \in [t_0, t_f]$. Since this integrand never can be negative, the only way the integral over a positive time interval can evaluate to zero is if the integrand is identically zero for all $\tau \in [t_0, t_f]$. Because only the zero vector has zero Euclidean norm, we must have

$$C e^{A(\tau - t_0)} x_0 = 0 \in \mathbb{R}^p \quad \text{for all} \quad \tau \in [t_0, t_f]$$

This means that the derivative of this expression of any order with respect to τ also must be the zero vector for all $\tau \in [t_0, t_f]$ and in particular at $\tau = t_0$. That is,

$$0 = \left. \frac{d^k}{dt^k} C e^{A(\tau - t_0)} x_0 \right|_{\tau = t_0} = \left. C A^k e^{A(\tau - t_0)} x_0 \right|_{\tau = t_0} = C A^k x_0 \quad \text{for all} \quad k \geq 0$$

We conclude that

$$\begin{bmatrix} C \\ CA \\ CA^2 \\ \vdots \\ CA^{n-1} \end{bmatrix} x_0 = \begin{bmatrix} C x_0 \\ C A x_0 \\ C A^2 x_0 \\ \vdots \\ C A^{n-1} x_0 \end{bmatrix} = \begin{bmatrix} 0 \\ 0 \\ 0 \\ \vdots \\ 0 \end{bmatrix}$$

which implies, since $x_0 \in \mathbb{R}^n$ is a nonzero vector, that Q has less than full-column rank, or rank $Q < n$.

We next prove the contrapositive of the second implication: If rank $Q < n$, then $M(t_0, t_f)$ is singular for some t_0 and finite $t_f > t_0$. Assuming that Q has less than full column rank, there exists a nonzero vector $x_0 \in \mathbb{R}^n$ whose components specify a linear combination of the columns of Q that yield an $(np) \times 1$ zero vector. That is,

$$\begin{bmatrix} C \\ CA \\ CA^2 \\ \vdots \\ CA^{n-1} \end{bmatrix} x_0 = \begin{bmatrix} C x_0 \\ C A x_0 \\ C A^2 x_0 \\ \vdots \\ C A^{n-1} x_0 \end{bmatrix} = \begin{bmatrix} 0 \\ 0 \\ 0 \\ \vdots \\ 0 \end{bmatrix}$$

which shows that $CA^k x_0 = 0$ for $k = 0, 1, \ldots, n - 1$. The finite series expansion

$$e^{At} = \sum_{k=0}^{n-1} \alpha_k(t) A^k$$

for scalar functions $\alpha_0(t), \alpha_1(t), \ldots, \alpha_{n-1}(t)$ now gives

$$Ce^{At} x_0 = C \left[\sum_{k=0}^{n-1} \alpha_k(t) A^k \right] x_0 = \sum_{k=0}^{n-1} \alpha_k(t)(CA^k x_0) = 0 \quad \text{for all} \quad t$$

It follows from this that for any t_0 and finite $t_f > t_0$,

$$M(t_0, t_f) x_0 = \int_{t_0}^{t_f} e^{A^T(\tau - t_0)} C^T C e^{A(\tau - t_0)} \, d\tau \, x_0$$

$$= \int_{t_0}^{t_f} e^{A^T(\tau - t_0)} C^T (C e^{A(\tau - t_0)} x_0) \, d\tau$$

$$= 0$$

which implies, since $x_0 \in \mathbb{R}^n$ is a nonzero vector, that $M(t_0, t_f)$ is singular. $\square$

As a consequence of Lemma 4.3, an observable state equation has, for any initial time t_0 and finite final time $t_f > t_0$, a nonsingular observability Gramian. This can be used to process input and output measurements $u(t), y(t)$ on the interval $[t_0, t_f]$ to yield the initial state $x(t_0) = x_0$ as follows: We first form the zero-input response component

$$y_{zi}(t) = y(t) - \left[\int_{t_0}^{t} C e^{A(t - \tau)} B u(\tau) \, d\tau + D u(t) \right]$$

$$= C e^{A(t - t_0)} x_0$$

Then we process the zero-input response according to

$$M^{-1}(t_0, t_f) \int_{t_0}^{t_f} e^{A^T(\tau - t_0)} C^T y_{zi}(\tau) \, d\tau$$

$$= M^{-1}(t_0, t_f) \int_{0}^{t_f} e^{A^T(\tau - t_0)} C^T C e^{A(\tau - t_0)} x_0 \, d\sigma$$

$$= M^{-1}(t_0, t_f) \int_{0}^{t_f} e^{A^T(\tau - t_0)} C^T C e^{A(\tau - t_0)} \, d\tau \, x_0$$

$$= M^{-1}(t_0, t_f)M(t_0, t_f)\, x_0$$

$$= x_0$$

which uniquely recovers the initial state.

4.2 OBSERVABILITY EXAMPLES

Example 4.1   Consider the two–dimensional single-output state equation

$$\begin{bmatrix} \dot{x}_1(t) \\ \dot{x}_2(t) \end{bmatrix} = \begin{bmatrix} 1 & 5 \\ 8 & 4 \end{bmatrix} \begin{bmatrix} x_1(t) \\ x_2(t) \end{bmatrix} + \begin{bmatrix} -2 \\ 2 \end{bmatrix} u(t)$$

$$y(t) = [2 \quad 2] \begin{bmatrix} x_1(t) \\ x_2(t) \end{bmatrix} + [0]u(t)$$

for which the associated (A, B) pair is the same as in Example 3.1. The observability matrix Q is found as follows:

$$C = [2 \quad 2]$$

$$CA = [18 \quad 18]$$

so

$$Q = \begin{bmatrix} 2 & 2 \\ 18 & 18 \end{bmatrix}$$

Clearly $|Q| = 0$ so the state equation is not observable. Because rank $Q < 2$ but Q is not the 2×2 zero matrix, we have rank $Q = 1$ and nullity $Q = 1$.

To see why this state equation is not observable, we again use the state coordinate transformation given by:

$$\begin{bmatrix} z_1(t) \\ z_2(t) \end{bmatrix} = \begin{bmatrix} x_1(t) \\ x_1(t) + x_2(t) \end{bmatrix} = \begin{bmatrix} 1 & 0 \\ 1 & 1 \end{bmatrix} \begin{bmatrix} x_1(t) \\ x_2(t) \end{bmatrix}$$

which yields the transformed state equation

$$\begin{bmatrix} \dot{z}_1(t) \\ \dot{z}_2(t) \end{bmatrix} = \begin{bmatrix} -4 & 5 \\ 0 & 9 \end{bmatrix} \begin{bmatrix} z_1(t) \\ z_2(t) \end{bmatrix} + \begin{bmatrix} -2 \\ 0 \end{bmatrix} u(t)$$

$$y(t) = [0 \quad 2] \begin{bmatrix} z_1(t) \\ z_2(t) \end{bmatrix} + [0]u(t)$$

Here both the state variable $z_2(t)$ and the output $y(t)$ are decoupled from $z_1(t)$. Thus, $z_1(0)$ cannot be determined from measurements of the zero-input response $y_{zi}(t) = 2e^{9t}z_2(0)$. This is why the given state equation is not observable.

Also, note that $x_0 = [1, -1]^T$ satisfies $Qx_0 = [0, 0]^T$ and we conclude from the proof of Theorem 4.2 that x_0 is a nonzero unobservable state. $\square$

Example 4.2 Given the following three-dimensional single-output homogeneous state equation, that is,

$$\begin{bmatrix} \dot{x}_1(t) \\ \dot{x}_2(t) \\ \dot{x}_3(t) \end{bmatrix} = \begin{bmatrix} 0 & 0 & -6 \\ 1 & 0 & -11 \\ 0 & 1 & -6 \end{bmatrix} \begin{bmatrix} x_1(t) \\ x_2(t) \\ x_3(t) \end{bmatrix}$$

$$y(t) = [0 \quad 1 \quad -3] \begin{bmatrix} x_1(t) \\ x_2(t) \\ x_3(t) \end{bmatrix}$$

we construct the observability matrix as follows:

$$C = [0 \quad 1 \quad -3]$$

$$CA = [0 \quad 1 \quad -3] \begin{bmatrix} 0 & 0 & -6 \\ 1 & 0 & -11 \\ 0 & 1 & -6 \end{bmatrix} = [1 \quad -3 \quad 7]$$

$$CA^2 = (CA)A = [1 \quad -3 \quad 7] \begin{bmatrix} 0 & 0 & -6 \\ 1 & 0 & -11 \\ 0 & 1 & -6 \end{bmatrix} = [-3 \quad 7 \quad -15]$$

yielding

$$Q = \begin{bmatrix} C \\ CA \\ CA^2 \end{bmatrix} = \begin{bmatrix} 0 & 0 & -6 \\ 1 & -3 & 7 \\ -3 & 7 & -15 \end{bmatrix}$$

To check observability, we calculate

$$|Q| = \begin{vmatrix} C \\ CA \\ CA^2 \end{vmatrix}$$

$$= \begin{vmatrix} 0 & 1 & -3 \\ 1 & -3 & 7 \\ -3 & 7 & -15 \end{vmatrix}$$

$$= [0 + (-21) + (-21)] - [(-27) + 0 + (-15)]$$

$$= -42 - (-42)$$

$$= 0$$

and thus rank $Q < 3$. This indicates that the state equation is not observable, so there exist nonzero unobservable states for this state equation. The upper left 2×2 submatrix

$$\begin{bmatrix} 0 & 1 \\ 1 & -3 \end{bmatrix}$$

has nonzero determinant, indicating that rank $Q = 2$ and nullity $Q = 3 - 2 = 1$ (by Sylvester's law of nullity). Consequently, any nonzero solution to the homogeneous equation

$$Qx_0 = 0$$

will yield a nonzero unobservable state. Applying elementary row operations to the observability matrix Q yields the row-reduced echelon form

$$Q_R = \begin{bmatrix} 1 & 0 & -2 \\ 0 & 1 & -3 \\ 0 & 0 & 0 \end{bmatrix}$$

from which an easily identified solution to

$$Q_R x_0 = 0$$

is

$$x_0 = \begin{bmatrix} 2 \\ 3 \\ 1 \end{bmatrix}$$

Moreover, any nonzero scalar multiple of this solution also yields a nonzero unobservable state.   □

Example 4.3   We investigate the observability of the three-dimensional state equation

$$
\begin{bmatrix} \dot{x}_1(t) \\ \dot{x}_2(t) \\ \dot{x}_2(t) \end{bmatrix} = \begin{bmatrix} 0 & 0 & -a_0 \\ 1 & 0 & -a_1 \\ 0 & 1 & -a_2 \end{bmatrix} \begin{bmatrix} x_1(t) \\ x_2(t) \\ x_2(t) \end{bmatrix} + \begin{bmatrix} b_0 \\ b_1 \\ b_2 \end{bmatrix} u(t)
$$

$$
y(t) = \begin{bmatrix} 0 & 0 & 1 \end{bmatrix} \begin{bmatrix} x_1(t) \\ x_2(t) \\ x_2(t) \end{bmatrix}
$$

which the reader will recall is a state-space realization of the transfer function

$$
H(s) = \frac{b_2 s^2 + b_1 s + b_0}{s^3 + a_2 s^2 + a_1 s + a_0}
$$

The observability matrix Q is found as follows:

$$
Q = \begin{bmatrix} C \\ CA \\ CA^2 \end{bmatrix}
$$

$$
= \begin{bmatrix} 0 & 0 & 1 \\ 0 & 1 & -a_2 \\ 1 & -a_2 & a_2^2 - a_1 \end{bmatrix}
$$

This observability matrix is identical to the controllability matrix P from Example 3.3. The observability matrix Q is independent of the transfer function-numerator coefficients b_0, b_1, and b_2. The determinant of the observability matrix is $|Q| = -1 \neq 0$, so the state equation is observable. Note that this outcome is independent of the characteristic polynomial coefficients a_0, a_1, and a_2, so a state-space realization in this form is always observable. This is also true for any system order n, as we will demonstrate shortly. ☐

Example 4.4   Consider the five-dimensional, two-output homogeneous state equation

$$
\begin{bmatrix} \dot{x}_1(t) \\ \dot{x}_2(t) \\ \dot{x}_3(t) \\ \dot{x}_4(t) \\ \dot{x}_5(t) \end{bmatrix} = \begin{bmatrix} 0 & 0 & 0 & 0 & 0 \\ 1 & 0 & 0 & 0 & 0 \\ 0 & 0 & 0 & 0 & 0 \\ 0 & 0 & 1 & 0 & 0 \\ 0 & 0 & 0 & 1 & 0 \end{bmatrix} \begin{bmatrix} x_1(t) \\ x_2(t) \\ x_3(t) \\ x_4(t) \\ x_5(t) \end{bmatrix}
$$

$$\begin{bmatrix} y_1(t) \\ y_2(t) \end{bmatrix} = \begin{bmatrix} 0 & 1 & 0 & 0 & 0 \\ 0 & 0 & 0 & 0 & 1 \end{bmatrix} \begin{bmatrix} x_1(t) \\ x_2(t) \\ x_3(t) \\ x_4(t) \\ x_5(t) \end{bmatrix}$$

The observability matrix is constructed as follows

$$Q = \begin{bmatrix} C \\ \hline CA \\ \hline CA^2 \\ \hline CA^3 \\ \hline CA^4 \end{bmatrix} = \begin{bmatrix} 0 & 1 & 0 & 0 & 0 \\ 0 & 0 & 0 & 0 & 1 \\ \hline 1 & 0 & 0 & 0 & 0 \\ 0 & 0 & 0 & 1 & 0 \\ \hline 0 & 0 & 0 & 0 & 0 \\ 0 & 0 & 1 & 0 & 0 \\ \hline 0 & 0 & 0 & 0 & 0 \\ 0 & 0 & 0 & 0 & 0 \\ \hline 0 & 0 & 0 & 0 & 0 \\ 0 & 0 & 0 & 0 & 0 \end{bmatrix}$$

Q has full-column rank 5 because of the pattern of ones and zeros. Therefore, the state equation is observable. Furthermore, rows 1, 2, 3, 4, and 6 form a linearly independent set of 1×5 row vectors. Since the remaining rows are each 1×5 zero vectors, there is only one way to select five linearly independent rows from Q. □

Example 4.5 Consider now the five-dimensional, two-output homogeneous state equation

$$\begin{bmatrix} \dot{x}_1(t) \\ \dot{x}_2(t) \\ \dot{x}_3(t) \\ \dot{x}_4(t) \\ \dot{x}_5(t) \end{bmatrix} = \begin{bmatrix} 0 & 1 & 0 & 0 & 0 \\ 1 & 0 & 0 & 0 & -1 \\ 0 & -1 & 0 & 0 & 0 \\ 0 & 0 & 1 & 0 & -1 \\ 0 & 1 & 0 & 1 & 0 \end{bmatrix} \begin{bmatrix} x_1(t) \\ x_2(t) \\ x_3(t) \\ x_4(t) \\ x_5(t) \end{bmatrix}$$

$$\begin{bmatrix} y_1(t) \\ y_2(t) \end{bmatrix} = \begin{bmatrix} 0 & 1 & 0 & 0 & 1 \\ 0 & -1 & 0 & 0 & 1 \end{bmatrix} \begin{bmatrix} x_1(t) \\ x_2(t) \\ x_3(t) \\ x_4(t) \\ x_5(t) \end{bmatrix}$$

differing from the preceding example only in the second and fifth columns of both A and C. These modifications lead to the observability matrix

$$
Q = \begin{bmatrix} C \\ CA \\ CA^2 \\ CA^3 \\ CA^4 \end{bmatrix} = \left[\begin{array}{ccccc}
0 & 1 & 0 & 0 & 1 \\
0 & -1 & 0 & 0 & 1 \\
\hline
1 & 1 & 0 & 1 & -1 \\
-1 & 1 & 0 & 1 & 1 \\
\hline
1 & 0 & 1 & -1 & -2 \\
1 & 0 & 1 & 1 & -2 \\
\hline
0 & -2 & -1 & -2 & 1 \\
0 & -2 & 1 & -2 & -1 \\
\hline
-2 & 2 & -2 & 1 & 4 \\
-2 & -2 & -2 & -1 & 4
\end{array} \right]
$$

This observability matrix also has rank equal to 5, indicating that the state equation is observable. In contrast to the preceding example, however, there are many ways to select five linearly independent rows from this observability matrix, as the reader may verify with the aid of MATLAB.

□

4.3 DUALITY

In this section we establish an interesting connection between controllability and observability known as *duality*. To begin, consider the following state equation related to Equation (4.1)

$$
\begin{aligned}
\dot{z}(t) &= A^{\mathrm{T}}z(t) + C^{\mathrm{T}}v(t) \\
w(t) &= B^{\mathrm{T}}z(t) + D^{\mathrm{T}}v(t)
\end{aligned}
\qquad z(0) = z_0 \qquad (4.3)
$$

having n-dimensional state vector $z(t)$, p-dimensional input vector $v(t)$, and m-dimensional output vector $w(t)$ (note that input and output dimensions have swapped roles here; in the original state equation (4.1) p is the dimension of the output vector and m is the dimension of the input vector). Although differing slightly from standard convention, we will refer to Equation (4.3) as the *dual* state equation for Equation (4.1). An immediate relationship exists between the transfer-function matrix of Equation (4.1) and that of its dual, that is,

$$
[C(sI - A)^{-1}B + D]^{\mathrm{T}} = B^{\mathrm{T}}(sI - A^{\mathrm{T}})^{-1}C^{\mathrm{T}} + D^{\mathrm{T}}
$$

further reinforcing the fact that the input and output dimensions for Equation (4.3) are reversed in comparison with those of Equation (4.1).

In the single-input, single output case, we have

$$C(sI - A)^{-1}B + D = [C(sI - A)^{-1}B + D]^{\mathrm{T}}$$
$$= B^{\mathrm{T}}(sI - A^{\mathrm{T}})^{-1}C^{\mathrm{T}} + D^{\mathrm{T}}$$

indicating that the original state equation and its dual are both realizations of the same scalar transfer function.

For the original state equation (4.1), we have previously introduced the following matrices associated with controllability and observability:

$$P_{(A,B)} = [B \quad AB \quad A^2B \quad \cdots \quad A^{n-1}B] \qquad Q_{(A,C)} = \begin{bmatrix} C \\ CA \\ CA^2 \\ \vdots \\ CA^{n-1} \end{bmatrix}$$

$$\tag{4.4}$$

and for the dual state equation (4.3), we analogously have

$$P_{(A^T,C^T)} = [C^{\mathrm{T}} \quad A^{\mathrm{T}}C^{\mathrm{T}} \quad (A^{\mathrm{T}})^2C^{\mathrm{T}} \quad \cdots \quad (A^{\mathrm{T}})^{n-1}C^{\mathrm{T}}]$$

$$Q_{(A^T,B^T)} = \begin{bmatrix} B^{\mathrm{T}} \\ B^{\mathrm{T}}A^{\mathrm{T}} \\ B^{\mathrm{T}}(A^{\mathrm{T}})^2 \\ \vdots \\ B^{\mathrm{T}}(A^{\mathrm{T}})^{n-1} \end{bmatrix} \tag{4.5}$$

Since

$$[C^{\mathrm{T}} \quad A^{\mathrm{T}}C^{\mathrm{T}} \quad (A^{\mathrm{T}})^2C^{\mathrm{T}} \quad \cdots \quad (A^{\mathrm{T}})^{n-1}C^{\mathrm{T}}] = \begin{bmatrix} C \\ CA \\ CA^2 \\ \vdots \\ CA^{n-1} \end{bmatrix}^{\mathrm{T}}$$

and

$$\begin{bmatrix} B^{\mathrm{T}} \\ B^{\mathrm{T}}A^{\mathrm{T}} \\ B^{\mathrm{T}}(A^{\mathrm{T}})^2 \\ \vdots \\ B^{\mathrm{T}}(A^{\mathrm{T}})^{n-1} \end{bmatrix} = [B \quad AB \quad A^2B \quad \cdots \quad A^{n-1}B]^{\mathrm{T}}$$

it follows from the fact that matrix rank is unaffected by the matrix transpose operation that

$$\text{rank } P_{(A^T, C^T)} = \text{rank } Q_{(A,C)} \quad \text{and} \quad \text{rank } Q_{(A^T, B^T)} = \text{rank } P_{(A,B)}$$
(4.6)

These relationships have the following implications:

- The dual state equation (4.3) is controllable if and only if the original state equation (4.1) is observable.
- The dual state equation (4.3) is observable if and only if the original state equation (4.1) is controllable.

The reader is invited to check that Examples 4.3, 4.4, and 4.5 are linked via duality to Examples 3.3, 3.4, and 3.5, respectively.

4.4 COORDINATE TRANSFORMATIONS AND OBSERVABILITY

The linear time-invariant state equation (4.1), together with the state coordinate transformation $x(t) = Tz(t)$ (see Section 2.5), yields the transformed state equation

$$\begin{aligned} \dot{z}(t) &= \hat{A}z(t) + \hat{B}u(t) \\ y(t) &= \hat{C}z(t) + \hat{D}u(t) \end{aligned} \quad z(t_0) = z_0$$
(4.7)

in which

$$\hat{A} = T^{-1}AT \qquad \hat{B} = T^{-1}B \qquad \hat{C} = CT \qquad \hat{D} = D$$

$$z_0 = T^{-1}x_0$$

We see directly from Definition 4.1 that if is x_0 is an unobservable state for the state equation (4.1) so that $Ce^{A(t-t_0)}x_0 \equiv 0$ for all $t \geq t_0$, then $z_0 = T^{-1}x_0$ satisfies

$$\begin{aligned} \hat{C}e^{\hat{A}(t-t_0)}z_0 &= (CT)(T^{-1}e^{A(t-t_0)}T)(T^{-1}x_0) \\ &= Ce^{A(t-t_0)}x_0 \\ &\equiv 0 \quad \text{for all} \quad t \geq t_0 \end{aligned}$$

from which we conclude that z_0 is an unobservable state for the transformed state equation (4.7). Since z_0 is a nonzero vector if and only if x_0 is a nonzero vector, we conclude that the transformed state equation (4.7) is observable if and only if the original state equation (4.1) is observable.

We therefore say that observability is invariant with respect to coordinate transformations, as is controllability.

Any equivalent characterization of observability must preserve this invariance with respect to coordinate transformations. For example, we expect that the rank test for observability established in Theorem 4.2 should yield consistent results across all state equations related by a coordinate transformation. To confirm this, a derivation analogous to that in Section 3.4 shows that the observability matrix for Equation (4.7) is related to the observability matrix for Equation (4.1) via

$$\hat{Q} = \begin{bmatrix} \hat{C} \\ \hat{C}\hat{A} \\ \hat{C}\hat{A}^2 \\ \vdots \\ \hat{C}\hat{A}^{n-1} \end{bmatrix} = \begin{bmatrix} C \\ CA \\ CA^2 \\ \vdots \\ CA^{n-1} \end{bmatrix} T = QT.$$

Again, since either pre- or postmultiplication by a square nonsingular matrix does not affect matrix rank, we see that

$$\text{rank } \hat{Q} = \text{rank } Q$$

In addition, the observability Gramians for Equations (4.7) and (4.1) are related according to

$$\hat{M}(t_0, t_f) = \int_{t_0}^{t_f} e^{\hat{A}^{\mathrm{T}}(\tau - t_0)} \hat{C}^{\mathrm{T}} \hat{C} e^{\hat{A}(\tau - t_0)} \, d\tau$$

$$= T^T M(t_0, t_f) T$$

from which we conclude that that $\hat{M}(t_0, t_f)$ is nonsingular for some initial time t_0 and finite final time $t_f > t_0$ if and only if $M(t_0, t_f)$ is nonsingular for the same t_0 and t_f.

This section concludes with three results relating coordinate transformations and observability that are the counterparts of the controllability results presented in Chapter 3. Various technical claims are easily proven by referring to the corresponding fact for controllability and invoking duality.

Observable Realizations of the Same Transfer Function

Consider two single-input, single-output state equations

$$\dot{x}_1(t) = A_1 x_1(t) + B_1 u(t) \qquad\qquad \dot{x}_2(t) = A_2 x_2(t) + B_2 u(t)$$
$$y(t) = C_1 x_1(t) + D_1 u(t) \qquad\qquad y(t) = C_2 x_2(t) + D_2 u(t)$$

$$(4.8)$$

that are both n-dimensional realizations of the same transfer function. When both realizations are observable, they are related by a uniquely and explicitly defined coordinate transformation.

Lemma 4.4   *Two observable n-dimensional single-input, single-output realizations (4.8) of the same transfer function are related by the unique state coordinate transformation $x_1(t) = T x_2(t)$, where*

$$
T = \begin{bmatrix} C_1 \\ C_1 A_1 \\ \vdots \\ C_1 A_1^{n-1} \end{bmatrix}^{-1} \begin{bmatrix} C_2 \\ C_2 A_2 \\ \vdots \\ C_2 A_2^{n-1} \end{bmatrix}
$$
$$
= Q_1^{-1} Q_2
$$

Proof.   With this coordinate transformation matrix T, we must verify the identities

$$A_2 = T^{-1} A_1 T \quad B_2 = T^{-1} B_1 \quad C_2 = C_1 T$$

To do so, we associate with the pair of state equations in Equation (4.8) the pair of dual state equations

$$\dot{z}_1(t) = A_1^T z_1(t) + C_1^T v(t) \qquad \dot{z}_2(t) = A_2^T z_2(t) + C_2^T v(t)$$
$$w(t) = B_1^T z_1(t) + D_1^T v(t) \qquad w(t) = B_2^T z_2(t) + D_2^T v(t)$$

Since each state equation in Equation (4.1) is a realization of the same transfer function $H(s)$, each dual state equation is a realization of $H^T(s)$ and in the single-input, single-output case, $H^T(s) = H(s)$. By duality, each dual state equation is a controllable n-dimensional realization of the same transfer function, so we may apply the results of Lemma 3.4 to obtain

$$A_2^T = \left(P_{(A_1^T, C_1^T)} P_{(A_2^T, C_2^T)}^{-1} \right)^{-1} A_1^T \left(P_{(A_1^T, C_1^T)} P_{(A_2^T, C_2^T)}^{-1} \right)$$

Taking transposes through this identity yields

$$
\begin{aligned}
A_2 &= \left(P_{(A_1^T, C_1^T)} P_{(A_2^T, C_2^T)}^{-1} \right)^T A_1 \left(P_{(A_1^T, C_1^T)} P_{(A_2^T, C_2^T)}^{-1} \right)^{-T} \\
&= \left(P_{(A_1^T, C_1^T)}^{-T} P_{(A_2^T, C_2^T)}^{T} \right)^{-1} A_1 \left(P_{(A_1^T, C_1^T)}^{-T} P_{(A_2^T, C_2^T)}^{T} \right) \\
&= \left(Q_{(A_1, C_1)}^{-1} Q_{(A_2, C_2)} \right)^{-1} A_1 \left(Q_{(A_1, C_1)}^{-1} Q_{(A_2, C_2)} \right) \\
&= T^{-1} A_1 T
\end{aligned}
$$

in which we used the alternate notation $Q_{(A_1, C_1)} = Q_1$ and $Q_{(A_2, C_2)} = Q_2$ to emphasize the duality relationships that we exploited. The remaining two identities are verified in a similar fashion, and the details are left to the reader. $\qquad\square$

Observer Canonical Form

Given the scalar transfer function

$$
H(s) = \frac{b(s)}{a(s)} = \frac{b_{n-1} s^{n-1} + \cdots + b_1 s + b_0}{s^n + a_{n-1} s^{n-1} + \cdots + a_1 s + a_0}
$$

we define the *observer canonical form* (OCF) realization to be the dual of the controller canonical form (CCF) realization given by

$$
\dot{x}_{\text{OCF}} = A_{\text{OCF}} \, x_{\text{OCF}}(t) + B_{\text{OCF}} \, u(t)
$$

$$
y(t) = C_{\text{OCF}} \, x_{\text{OCF}}(t)
$$

in which

$$
A_{\text{OCF}} = A_{\text{CCF}}^T \qquad\qquad B_{\text{OCF}} = C_{\text{CCF}}^T
$$

$$
= \begin{bmatrix}
0 & 0 & \cdots & 0 & 0 & -a_0 \\
1 & 0 & \cdots & 0 & 0 & -a_1 \\
0 & 1 & \cdots & 0 & 0 & -a_2 \\
\vdots & \vdots & \ddots & \vdots & \vdots & \vdots \\
0 & 0 & \cdots & 1 & 0 & -a_{n-2} \\
0 & 0 & \cdots & 0 & 1 & -a_{n-1}
\end{bmatrix}
\qquad
= \begin{bmatrix}
b_0 \\
b_1 \\
b_2 \\
\vdots \\
b_{n-2} \\
b_{n-1}
\end{bmatrix}
$$

$$
C_{\text{OCF}} = B_{\text{CCF}}^T
$$

$$
= \begin{bmatrix} 0 & 0 & \cdots & 0 & 0 & 1 \end{bmatrix}
$$

Again, since a scalar transfer function satisfies $H(s) = H^T(s)$, it follows that the observer canonical form is also a realization of $H(s)$. Having previously established controllability of the controller canonical form (in Section 3.4), duality further ensures that the observer canonical form defines an observable state equation with an explicitly defined observability matrix.

The system of Example 4.3 was given in observer canonical form.

Lemma 4.5 *The n-dimensional observer canonical form has the observability matrix*

$$
Q_{OCF} = \begin{bmatrix} C_{OCF} \\ C_{OCF}A_{OCF} \\ C_{OCF}A_{OCF}^2 \\ \vdots \\ C_{OCF}A_{OCF}^{n-1} \end{bmatrix}
$$

$$
= \begin{bmatrix} a_1 & a_2 & \cdots & a_{n-1} & 1 \\ a_2 & a_3 & \cdots & 1 & 0 \\ \vdots & \vdots & \ddots & \vdots & \vdots \\ a_{n-1} & 1 & \cdots & 0 & 0 \\ 1 & 0 & \cdots & 0 & 0 \end{bmatrix}^{-1}
$$

(4.9)

Proof. By duality, $Q_{OCF} = P_{CCF}^T$. However, we observed in Chapter 3 that P_{CCF} is symmetric and by Lemma 3.5 is also given by Equation (4.9). Thus $Q_{OCF} = P_{CCF}^T = P_{CCF}$ and is given by Equation (4.9). □

It follows from Lemma 4.5 that beginning with an arbitrary observable realization of a given transfer function with state $x(t)$, the associated observer canonical form is obtained by applying the coordinate transformation $x(t) = T_{OCF}z(t)$, where $T_{OCF} = Q^{-1}Q_{OCF} = (Q_{OCF}^{-1}Q)^{-1}$, or, more explicitly,

$$
T_{OCF} = \begin{bmatrix} C \\ CA \\ CA^2 \\ \vdots \\ CA^{n-1} \end{bmatrix}^{-1} \begin{bmatrix} a_1 & a_2 & \cdots & a_{n-1} & 1 \\ a_2 & a_3 & \cdots & 1 & 0 \\ \vdots & \vdots & \ddots & \vdots & \vdots \\ a_{n-1} & 1 & \cdots & 0 & 0 \\ 1 & 0 & \cdots & 0 & 0 \end{bmatrix}^{-1}
$$

$$
= \begin{bmatrix} \begin{bmatrix} a_1 & a_2 & \cdots & a_{n-1} & 1 \\ a_2 & a_3 & \cdots & 1 & 0 \\ \vdots & \vdots & \ddots & \vdots & \vdots \\ a_{n-1} & 1 & \cdots & 0 & 0 \\ 1 & 0 & \cdots & 0 & 0 \end{bmatrix} \begin{bmatrix} C \\ CA \\ CA^2 \\ \vdots \\ CA^{n-1} \end{bmatrix} \end{bmatrix}^{-1}
$$

(4.10)

Example 4.6   We return to the three-dimensional single-input, single-output state equation considered previously in Section 2.5 in connection with the diagonal canonical form and in Section 3.4 in connection with the controller canonical form with coefficient matrices given below. We now transform this state equation to observer canonical form.

$$
A = \begin{bmatrix} 8 & -5 & 10 \\ 0 & -1 & 1 \\ -8 & 5 & -9 \end{bmatrix} \quad B = \begin{bmatrix} -1 \\ 0 \\ 1 \end{bmatrix} \quad C = [1 \quad -2 \quad 4] \quad D = 0
$$

As before, the system characteristic polynomial is $|sI - A| = s^3 + 2s^2 + 4s + 8$, and the eigenvalues are $\pm 2i, -2$. The observer canonical form transformation matrix $T_{OCF} = Q^{-1}Q_{OCF} = (Q_{OCF}^{-1}Q)^{-1}$ is computed as follows.

$$
T_{OCF} = \left(\begin{bmatrix} a_1 & a_2 & 1 \\ a_2 & 1 & 0 \\ 1 & 0 & 0 \end{bmatrix} \begin{bmatrix} C \\ CA \\ CA^2 \end{bmatrix} \right)^{-1}
$$

$$
= \left(\begin{bmatrix} 4 & 2 & 1 \\ 2 & 1 & 0 \\ 1 & 0 & 0 \end{bmatrix} \begin{bmatrix} 1 & -2 & 4 \\ -24 & 17 & -28 \\ 32 & -37 & 29 \end{bmatrix} \right)^{-1}
$$

$$
= \begin{bmatrix} -12 & -11 & -11 \\ -22 & 13 & -20 \\ 1 & -2 & 4 \end{bmatrix}^{-1}
$$

$$
= \begin{bmatrix} -0.0097 & -0.0535 & -0.2944 \\ -0.0552 & 0.0300 & -0.0016 \\ -0.0251 & 0.0284 & 0.3228 \end{bmatrix}
$$

The resulting observer canonical form realization is

$$
A_{OCF} = T_{OCF}^{-1} A T_{OCF} \qquad\qquad B_{OCF} = T_{OCF}^{-1} B
$$

$$
= \begin{bmatrix} 0 & 0 & -8 \\ 1 & 0 & -4 \\ 0 & 1 & -2 \end{bmatrix} \qquad\qquad = \begin{bmatrix} 1 \\ 2 \\ 3 \end{bmatrix}
$$

$$
C_{OCF} = C T_{OCF} \qquad\qquad D_{OCF} = D
$$

$$
= [0 \quad 0 \quad 1] \qquad\qquad = 0
$$

As we should expect, the observer canonical form realization is the dual of the controller canonical form realization computed in Example 3.6 for the same state equation.

Conclusions regarding the system's observability can be drawn directly by inspection of the diagonal canonical form (DCF) realization presented in Section 2.5. In particular, for a single-input, single-output system expressed in diagonal canonical form with no repeated eigenvalues, if all C_{DCF} elements are nonzero, then the state equation is observable. Further, any zero elements of C_{DCF} correspond to the nonobservable state variables. The diagonal canonical form realization presented earlier for this example (Sections 2.5 and 3.4) has distinct eigenvalues and nonzero components in C_{DCF}. Therefore, this state equation is observable. This fact is further verified by noting that in this example $|Q| = 1233 \neq 0$, and the rank of Q is 3. □

Unobservable State Equations

It is possible to transform an unobservable state equation into a so-called standard form for unobservable state equations that displays an observable subsystem.

Theorem 4.6 *Suppose that*

$$\text{rank} \begin{bmatrix} C \\ CA \\ CA^2 \\ \vdots \\ CA^{n-1} \end{bmatrix} = q < n$$

Then there exists a state coordinate transformation $x(t) = Tz(t)$ such that the transformed state equation has

$$\hat{A} = T^{-1}AT \qquad \hat{C} = CT$$

$$= \begin{bmatrix} A_{11} & 0 \\ A_{21} & A_{22} \end{bmatrix} \qquad = [\, C_1 \quad 0 \,]$$

where the pair (A_{11}, C_1) defines an observable homogeneous q-dimensional state equation.

Proof. By duality, the pair (A^T, C^T) is not controllable, and rank $P_{(A^T, C^T)} = q$. We let $\tilde{T}$ denote the nonsingular matrix that, by Theorem 3.6

with a slight change in notation to avoid confusion, satisfies

$$\tilde{T}^{-1} A^T \tilde{T} = \begin{bmatrix} \tilde{A}_{11} & \tilde{A}_{12} \\ 0 & \tilde{A}_{22} \end{bmatrix} \qquad \tilde{T}^{-1} C^T = \begin{bmatrix} \tilde{B}_1 \\ 0 \end{bmatrix}$$

where the pair $(\tilde{A}_{11}, \tilde{B}_1)$ defines a controllable q-dimensional state equation. By taking transposes through these identities, we obtain

$$\tilde{T}^T A \tilde{T}^{-T} = \begin{bmatrix} \tilde{A}_{11}^T & 0 \\ \tilde{A}_{12}^T & \tilde{A}_{22}^T \end{bmatrix} \qquad C \tilde{T}^{-T} = [\tilde{B}_1^T \quad 0]$$

We complete the proof by making the following associations $T = \tilde{T}^{-T}$, $A_{11} = \tilde{A}_{11}^T$, $A_{21} = \tilde{A}_{12}^T$, $A_{22} = \tilde{A}_{22}^T$, $C_1 = \tilde{B}_1^T$ and noting that, again by duality, $(A_{11}, C_1) = (\tilde{A}_{11}^T, \tilde{B}_1^T)$ defines an observable homogeneous q-dimensional state equation. $\qquad \square$

Example 4.7 We return to the unobservable homogeneous state equation that was introduced in Example 4.2, that is,

$$\begin{bmatrix} \dot{x}_1(t) \\ \dot{x}_2(t) \\ \dot{x}_3(t) \end{bmatrix} = \begin{bmatrix} 0 & 0 & -6 \\ 1 & 0 & -11 \\ 0 & 1 & -6 \end{bmatrix} \begin{bmatrix} x_1(t) \\ x_2(t) \\ x_3(t) \end{bmatrix}$$

$$y(t) = [0 \quad 1 \quad -3] \begin{bmatrix} x_1(t) \\ x_2(t) \\ x_3(t) \end{bmatrix}$$

We observe that this state equation is related by duality to the uncontrollable state equation of Example 3.7, for which

$$\tilde{T} = \begin{bmatrix} 0 & 1 & 0 \\ 1 & -3 & 0 \\ -3 & 7 & 1 \end{bmatrix}$$

specifies a coordinate transformation yielding the standard form for an uncontrollable state equation. We take

$$T = \tilde{T}^{-T} = \begin{bmatrix} 0 & 1 & -3 \\ 1 & -3 & 7 \\ 0 & 0 & 1 \end{bmatrix}^{-1} = \begin{bmatrix} 3 & 1 & 2 \\ 1 & 0 & 3 \\ 0 & 0 & 1 \end{bmatrix}$$

which gives

$$\hat{A} = T^{-1}AT$$

$$= \left[\begin{array}{cc|c} 0 & 1 & 0 \\ -2 & -3 & 0 \\ \hline 1 & 0 & 0 \end{array}\right] \qquad \begin{array}{l} \hat{C} = CT \\ \phantom{\hat{C}} = [\, 1 \quad 0|\quad 0\,] \end{array}$$

which is in the standard form for an unobservable state equation because the two–dimensional homogeneous state equation specified by

$$A_{11} = \left[\begin{array}{cc} 0 & 1 \\ -2 & -3 \end{array}\right] \qquad C_1 = [\, 1 \quad 0\,]$$

is seen easily to be observable. $\qquad\qquad\qquad\qquad\qquad\qquad \square$

4.5 POPOV-BELEVITCH-HAUTUS TESTS FOR OBSERVABILITY

Checking the rank of the observability matrix provides an *algebraic* test for observability. The Popov-Belevitch-Hautus eigenvector and rank tests for observability provide useful alternatives.

Theorem 4.7   *(Popov-Belevitch-Hautus Eigenvector Test for Observability). The state equation specified by the pair* (A, C) *is observable if and only if there exists no right eigenvector of* A *orthogonal to the rows of* C.

Proof.   By duality, the pair (A, C) is observable if and only if the pair (A^T, C^T) is controllable. By the Popov-Belevitch-Hautus eigenvector test for controllability, the pair (A^T, C^T) is controllable if and only if there exists no left eigenvector of A^T orthogonal to the columns of C^T, and equivalently, there exists no right eigenvector of A orthogonal to the rows of C. $\qquad\qquad \square$

Example 4.8   We again revisit the unobservable homogeneous state equation that was introduced in Example 4.2, that is,

$$\left[\begin{array}{c} \dot{x}_1(t) \\ \dot{x}_2(t) \\ \dot{x}_3(t) \end{array}\right] = \left[\begin{array}{ccc} 0 & 0 & -6 \\ 1 & 0 & -11 \\ 0 & 1 & -6 \end{array}\right] \left[\begin{array}{c} x_1(t) \\ x_2(t) \\ x_3(t) \end{array}\right]$$

$$y(t) = [\, 0 \quad 1 \quad -3\,] \left[\begin{array}{c} x_1(t) \\ x_2(t) \\ x_3(t) \end{array}\right]$$

The eigenvalues of A are $\lambda_1 = -1$, $\lambda_2 = -2$, and $\lambda_3 = -3$, with associated right eigenvectors

$$v_1 = \begin{bmatrix} 6 \\ 5 \\ 1 \end{bmatrix} \qquad v_2 = \begin{bmatrix} 3 \\ 4 \\ 1 \end{bmatrix} \qquad v_3 = \begin{bmatrix} 2 \\ 3 \\ 1 \end{bmatrix}$$

Of these, $Cv_3 = 0$, which confirms that the state equation is not observable. $\qquad\qquad\square$

Theorem 4.8 *(Popov-Belevitch-Hautus Rank Test for Observability). The state equation specified by the pair (A, C) is observable if and only if*

$$\text{rank}\begin{bmatrix} C \\ \lambda I - A \end{bmatrix} = n \quad \text{for all} \quad \lambda \in \mathbb{C}$$

Proof. By duality, the pair (A, C) is observable if and only if the pair (A^T, C^T) is controllable. By the Popov-Belevitch-Hautus rank test for controllability, the pair (A^T, C^T) is controllable if and only if

$$\text{rank}[\,\lambda I - A^T \quad C^T\,] = n \quad \text{for all} \quad \lambda \in \mathbb{C}$$

The desired conclusion follows from the fact that

$$\begin{aligned}
\text{rank}\begin{bmatrix} C \\ \lambda I - A \end{bmatrix} &= \text{rank}\begin{bmatrix} C \\ \lambda I - A \end{bmatrix}^T \\
&= \text{rank}\,[\, C^T \quad \lambda I - A^T \,] \\
&= \text{rank}\,[\, \lambda I - A^T \quad C^T \,] \qquad\qquad \square
\end{aligned}$$

4.6 MATLAB FOR OBSERVABILITY AND OBSERVER CANONICAL FORM

MATLAB for Observability

The MATLAB function that is useful for observability analysis is

Q = obsv(JbkR)   Calculate the observability matrix associated with the state equation data structure **JbkR**; only matrices **A** and **C** are used in the calculation.

MATLAB for Observer Canonical Form

A MATLAB function that is potentially useful for coordinate transformations and the observer canonical form realization is

canon   MATLAB function for canonical forms (use the
         companion switch for observer canonical form)
MATLAB **help** tells us that the **companion** observer canonical form computation is mathematically ill-conditioned and should be avoided if possible. We can manually compute the observer canonical form realization by using the coordinate transformation formula method derived in
Section 4.3. We also can obtain the observer canonical form by applying the duality properties (see Section 4.3), if we start with the controller
canonical form realization.

We also urge caution in the use of this function. We have found that
A_{OCF} has the correct structure but that B_{OCF} and C_{OCF} do not always
result in the required form for the observer canonical form defined in this
chapter.

Continuing MATLAB Example

Observability   We now assess the observability for the Continuing MAT-
LAB Example (rotational mechanical system). The following MATLAB code,
along with code from previous chapters, performs this assessment for the
Continuing MATLAB Example.

```
%- - - - - - - - - - - - - - - - - - - - - - - - - - - - - - - - - - - - -
%    Chapter 4.  Observability
%- - - - - - - - - - - - - - - - - - - - - - - - - - - - - - - - - - - - -

Q = obsv(JbkR);              % Calculate observability
                             % matrix Q
if (rank(Q) == size(A,1))    % Logic to assess
                             % observability
    disp('System is observable.');
else
    disp('System is NOT observable.');
end

Q1 = [C; C*A];               % Check Q via the formula
```

The result is

```
Q =
     1     0
     0     1
System is observable.
```

Coordinate Transformations and Observer Canonical Form For
the Continuing MATLAB Example, we compute the observer canonical form
state-space realization for the given open-loop system. The following MAT-
LAB code segment, along with code from previous chapters, performs this
computation:

```
%------------------------------------------------------
%  Chapter 4.  Coordinate Transformations and Observer
   % Canonical Form
%------------------------------------------------------

Qocf = inv(Pccfi);
Tocf = inv(Q)*Qocf;                  % Calculate OCF
                                     % transformation
   matrix
Aocf = inv(Tocf)*A*Tocf;             % Transform to OCF
                                     % via formula
Bocf = inv(Tocf)*B;
Cocf = C*Tocf;
Docf = D;

[JbkROCF,TOCF] = canon(JbkR,'companion'); % Compute OCF
                                     % using canon
AOCF = JbkROCF.a
BOCF = JbkROCF.b
COCF = JbkROCF.c
DOCF = JbkROCF.d
```

The following results are produced:

```
Tocf =
          0    1.0000
     1.0000   -4.0000

Aocf =
    -0.0000  -40.0000
     1.0000   -4.0000

Bocf =
     1
     0
```

```
Cocf =
    0      1

Docf = 0AOCF =

    0    -40
    1    -4

BOCF =
    1
    0

COCF =
    0      1

DOCF =
    0
```

In this case, the **canon** function yields the correct observer canonical form. Also, the observer canonical form obeys the correct duality properties with the controller canonical form as given in Chapter 3.

4.7 CONTINUING EXAMPLES FOR OBSERVABILITY AND OBSERVER CANONICAL FORM

Continuing Example 1: Two-Mass Translational Mechanical System

Observability Here we assess the observability for Continuing Example 1 (two-mass translational mechanical system), for both case a (multiple-input, multiple-output) and case b [input $u_2(t)$ and output $y_1(t)$].

Case a. The 8×4 observability matrix Q is

$$Q = \begin{bmatrix} C \\ CA \\ CA^2 \\ CA^3 \end{bmatrix}$$

$$
= \begin{bmatrix}
1 & 0 & 0 & 0 \\
0 & 0 & 1 & 0 \\
0 & 1 & 0 & 0 \\
0 & 0 & 0 & 1 \\
-15 & -0.75 & 5 & 0.25 \\
10 & 0.5 & -10 & -0.5 \\
13.75 & -14.31 & -6.25 & 4.69 \\
-12.5 & 9.38 & 7.50 & -9.63
\end{bmatrix}
$$

This observability matrix is of full rank, that is, rank$(Q) = 4$, which matches the system order $n = 4$. Therefore, the state equation is observable.

Case b. In case b, the system dynamics matrix A is identical to case a; however, since C is different due to the single output $y_1(t)$, we must again check for observability. The 4×4 observability matrix Q is

$$
Q = \begin{bmatrix}
C \\
CA \\
CA^2 \\
CA^3
\end{bmatrix}
$$

$$
= \begin{bmatrix}
1 & 0 & 0 & 0 \\
0 & 1 & 0 & 0 \\
-15 & -0.75 & 5 & 0.25 \\
13.75 & -14.31 & -6.25 & 4.69
\end{bmatrix}
$$

This observability matrix is of full rank, that is, rank$(Q) = 4$, which matches the system order $n = 4$. Also, the determinant of this square matrix Q is 25, confirming that Q is nonsingular. Therefore, the state equation is observable.

Now, let us return to the diagonal canonical form of Section 2.7. Since the system eigenvalues are distinct and C_{DCF} is fully populated (no zero elements), this means that the state equation is observable, which agrees with our conclusion above.

Coordinate Transformations and Observer Canonical Form Now we calculate the observer canonical form for case b. The observer canonical form coordinate transformation matrix is

$$
T_{\mathrm{OCF}} = Q^{-1} Q_{\mathrm{OCF}}
$$

$$= \begin{bmatrix} 0 & 0 & 0 & 1 \\ 0 & 0 & 1 & -1.25 \\ -0.01 & 0.2 & 0 & -2 \\ 0.2 & 0 & -2 & 1.5 \end{bmatrix}$$

in which Q was given above, and $Q_{OCF} = P_{CCF}^T$, where P_{CCF}^{-1} was given in Section 3.6. Using the coordinate transformations, we find the observer canonical form:

$$A_{OCF} = T_{OCF}^{-1} A T_{OCF} \qquad\qquad B_{OCF} = T_{OCF}^{-1} B$$

$$= \begin{bmatrix} 0 & 0 & 0 & -100 \\ 1 & 0 & 0 & -10 \\ 0 & 1 & 0 & -25.25 \\ 0 & 0 & 1 & -1.25 \end{bmatrix} \qquad = \begin{bmatrix} 0.25 \\ 0.0125 \\ 0 \\ 0 \end{bmatrix}$$

$$C_{OCF} = C T_{OCF} \qquad\qquad D_{OCF} = D$$

$$= [0 \quad 0 \quad 0 \quad 1] \qquad\qquad = 0$$

Again, this observer canonical form is the dual of the controller canonical form computed in Chapter 3, that is, $A_{OCF} = A_{CCF}^T$, $B_{OCF} = C_{CCF}^T$, $C_{OCF} = B_{CCF}^T$, and $D_{OCF} = D_{CCF}$. The reader is asked to identify in this observer canonical form the numerator and denominator polynomial coefficients from the single-input, single-output transfer function presented for case b in Chapter 2.

Continuing Example 2: Rotational Electromechanical System

Observability   This example assesses the observability for Continuing Example 2 (rotational electromechanical system). The 3×3 observability matrix Q

$$Q = \begin{bmatrix} C \\ CA \\ CA^2 \end{bmatrix} = \begin{bmatrix} 1 & 0 & 0 \\ 0 & 1 & 0 \\ 0 & 0 & 1 \end{bmatrix}$$

Clearly rank$(Q) = 3 = n$; also $|Q| = 1 \neq 0$. Therefore, this state equation is observable.

Again, since the system eigenvalues are distinct and C_{DCF} is fully populated (no zero elements) in the diagonal canonical form of Section 2.7, this confirms that the state equation is observable.

Coordinate Transformations and Observer Canonical Form

Here we calculate the observer canonical form for Continuing Example 2. The observer canonical form transformation matrix is

$$T_{OCF} = Q^{-1} Q_{OCF}$$

$$= \begin{bmatrix} 0 & 0 & 1 \\ 0 & 1 & -3 \\ 1 & -3 & 7 \end{bmatrix}$$

in which Q was given above, and $Q_{OCF} = P_{CCF}^T$, where P_{CCF}^{-1} was given in Section 3.6. Using this coordinate transformation, we find the observer canonical form:

$$A_{OCF} = T_{OCF}^{-1} A T_{OCF} \qquad B_{OCF} = T_{OCF}^{-1} B$$

$$= \begin{bmatrix} 0 & 0 & 0 \\ 1 & 0 & -2 \\ 0 & 1 & -3 \end{bmatrix} \qquad = \begin{bmatrix} 2 \\ 0 \\ 0 \end{bmatrix}$$

$$C_{OCF} = C T_{OCF} \qquad D_{OCF} = D$$

$$= [0 \quad 0 \quad 1] \qquad = 0$$

Again, this observer canonical form is the dual of the controller canonical form computed in Chapter 3 for Continuing Example 2. The reader is again asked to identify in this observer canonical form the numerator and denominator polynomial coefficients from the single-input, single-output transfer function presented for this example in Chapter 2.

4.8 HOMEWORK EXERCISES

Numerical Exercises

NE4.1 Assess the observability of the following systems, represented by the matrix pair (A, C).

a. $A = \begin{bmatrix} -4 & 0 \\ 0 & -5 \end{bmatrix} \qquad C = [1 \quad 1]$

b. $A = \begin{bmatrix} -4 & 0 \\ 0 & -5 \end{bmatrix} \qquad C = [1 \quad 0]$

c. $A = \begin{bmatrix} 0 & -10 \\ 1 & -2 \end{bmatrix}$ $\qquad C = [0 \quad 1]$

d. $A = \begin{bmatrix} 0 & 1 \\ -10 & -2 \end{bmatrix}$ $\qquad C = [1 \quad 2]$

e. $A = \begin{bmatrix} 2 & 0 \\ -1 & 1 \end{bmatrix}$ $\qquad C = [1 \quad 1]$

NE4.2 Compute the observer canonical form of the following systems ($D = 0$ for all cases).

a. $A = \begin{bmatrix} -4 & 0 \\ 0 & -5 \end{bmatrix}$ $\quad B = \begin{bmatrix} 1 \\ 1 \end{bmatrix}$ $\quad C = [1 \quad 1]$

b. $A = \begin{bmatrix} -4 & 0 \\ 0 & -5 \end{bmatrix}$ $\quad B = \begin{bmatrix} 1 \\ 0 \end{bmatrix}$ $\quad C = [1 \quad 0]$

c. $A = \begin{bmatrix} 0 & -10 \\ 1 & -2 \end{bmatrix}$ $\quad B = \begin{bmatrix} 1 \\ 2 \end{bmatrix}$ $\quad C = [0 \quad 1]$

d. $A = \begin{bmatrix} 0 & 1 \\ -10 & -2 \end{bmatrix}$ $\quad B = \begin{bmatrix} 0 \\ 1 \end{bmatrix}$ $\quad C = [1 \quad 2]$

e. $A = \begin{bmatrix} 2 & 0 \\ -1 & 1 \end{bmatrix}$ $\quad B = \begin{bmatrix} 1 \\ -1 \end{bmatrix}$ $\quad C = [1 \quad 1]$

NE4.3 Repeat **NE 4.1** using the Popov-Belevitch-Hautus tests for observability.

ANALYTICAL EXERCISES

AE4.1 Show that the n-dimensional controller canonical form realization of the transfer function

$$H(s) = \frac{b_{n-1}s^{n-1} + \cdots + b_1 s + b_0}{s^n + a_{n-1}s^{n-1} + \cdots + a_1 s + a_0}$$

is observable if and only if $H(s)$ has no pole-zero cancellations.

AE4.2 Show that the observability Gramian satisfies the matrix differential equation

$$\frac{d}{dt}M(t, t_f) + A^T M(t, t_f) + M(t, t_f)A + C^T C = 0$$

$$M(t_f, t_f) = 0$$

AE4.3 Show that the matrix exponential for

$$\begin{bmatrix} A & 0 \\ -C^T C & -A^T \end{bmatrix}$$

is given by

$$\begin{bmatrix} e^{At} & 0 \\ -e^{-A^T t} M(0, t) & e^{-A^T t} \end{bmatrix}$$

AE4.4 Show that the homogeneous single-output state equation characterized by

$$A = \begin{bmatrix} \mu & 1 & 0 & \cdots & 0 \\ 0 & \mu & 1 & \cdots & 0 \\ \vdots & \vdots & \vdots & \ddots & \vdots \\ 0 & 0 & 0 & \cdots & 1 \\ 0 & 0 & 0 & \cdots & \mu \end{bmatrix} \qquad C = [\, c_1 \quad c_2 \quad \cdots \quad c_{n-1} \quad c_n \,]$$

is observable if and only if $c_1 \neq 0$.

AE4.5 Show that the pair (A, C) is observable if and only if the only square matrix X that satisfies

$$AX = XA \qquad CX = 0$$

is $X = 0$.

AE4.6 Show that the pair (A, C) is observable if and only if the pair $(A - LC, HC)$ is observable for any $n \times p$ matrix L and any nonsingular $p \times p$ matrix H.

AE4.7 Show that the homogeneous multiple-output system

$$\begin{bmatrix} \dot{x}_1(t) \\ \dot{x}_2(t) \end{bmatrix} = \begin{bmatrix} A_{11} & A_{12} \\ A_{21} & A_{22} \end{bmatrix} \begin{bmatrix} x_1(t) \\ x_2(t) \end{bmatrix}$$

$$y(t) = [\, C_1 \quad 0 \,] \begin{bmatrix} x_1(t) \\ x_2(t) \end{bmatrix}$$

with C_1 nonsingular is observable if and only if the homogeneous system

$$\dot{x}_2(t) = A_{22} x_2(t)$$

$$y_2(t) = A_{12} x_2(t)$$

is observable.

Continuing MATLAB Exercises

CME4.1 For the CME1.1 system:

 a. Assess the system observability.

 b. Compute the observer canonical form. Compare with the controller canonical form from CME3.1b.

CME4.2 For the CME1.2 system:

 a. Assess the system observability.

 b. Compute the observer canonical form. Compare with the controller canonical form from CME3.2b.

CME4.3 For the CME1.3 system:

 a. Assess the system observability.

 b. Compute the observer canonical form. Compare with the controller canonical form from CME3.3b.

CME4.4 For the CME1.4 system:

 a. Assess the system observability.

 b. Compute the observer canonical form. Compare with the controller canonical form from CME3.4b.

Continuing Exercises

CE4.1a Determine if the CE1 system is observable for all three cases (cases from CE1.1b and numeric parameters from CE2.1a).

CE4.1b Calculate the observer canonical form realization numerically for the CE1 system, case iii only. Demonstrate the duality relationship between controller canonical form (CE3.1b) and observer canonical form. Assess the system observability by looking at the diagonal canonical form realization of CE2.1b; compare with your results from CE4.1a.

CE4.2a Determine if the CE2 system is observable for all three cases (cases from CE1.2b and numeric parameters from CE2.2a).

CE4.2b Calculate the observer canonical form realization numerically for the CE2 system, case i only. Demonstrate the duality relationship between controller canonical form (CE3.2b) and observer canonical form.

CE4.3a Determine if the CE3 system is observable for both cases (cases from CE1.3b and numeric parameters from CE2.3a).

CE4.3b Calculate the observer canonical form realization numerically for the CE3 system for both cases. Demonstrate the duality relationship between controller canonical form (CE3.3b) and observer canonical form. Assess the system observability by looking at the diagonal canonical form realization of CE2.3b; compare with your results from CE4.3a.

CE4.4a Determine if the CE4 system is observable (for the CE1.4b single-input, single-output case and with the numeric parameters from CE2.4a).

CE4.4b Calculate the observer canonical form realization numerically for the CE4 system. Assess the system observability by looking at the diagonal canonical form realization from CE 2.4b; compare with your results from CE4.4a.

CE4.5a Determine if the CE5 system is observable (for the CE1.5b single-input, single-output case and with the numeric parameters from CE2.5a).

CE4.5b Calculate the observer canonical form realization numerically for the CE5 system. Assess the system observability by looking at the diagonal canonical form realization from CE2.5b; compare with your results from CE4.5a.

5

MINIMAL REALIZATIONS

We have observed that a specified transfer function has more than one state-space realization because, starting with a particular realization, *any* valid coordinate transformation yields another state-space realization with the same dimension. Moreover, we have seen in a previous example that a given transfer function may have realizations of different dimension. This motivates the notion of a *minimal* realization of a particular transfer function, namely, a realization of least dimension out of all possible realizations.

It is clearly impractical to assess the minimality of a particular realization by comparing its dimension against that of all others. We therefore are interested in verifiable conditions for minimality. In this chapter we develop such conditions, first for single-input, single-output state equations having scalar transfer functions. In this case, there is an immediate link between minimality of a realization and possible pole-zero cancellations in the associated transfer function. Aided by this fact, along with the ability to write down state-space realizations by inspection of the transfer function, we strike an interesting connection between minimality of a state-space realization and its controllability and observability properties.

This analysis cannot be extended readily to multiple-input, multiple-output state equations because of several complications arising from the fact that the associated transfer functions are matrices of rational functions.

First, it is not immediately clear how to extend the notions of poles and zeros to the multiple-input, multiple-output case. Another reason is that the process of constructing a state-space realization from a transfer-function matrix is not nearly as straightforward as in the single-input, single-output case. Nevertheless, we derive the same result obtained in the single-input, single-output case for multiple-input, multiple-output state equations, albeit by different means.

Often in real-world problems, systems are modeled in the least possible size, and there are no pole/zero cancellations. Therefore, most of our continuing examples and exercises are already in minimal form, and hence there is nothing to present for these in this chapter. We present a MATLAB section to demonstrate the computation of minimal realizations.

5.1   MINIMALITY OF SINGLE-INPUT, SINGLE OUTPUT REALIZATIONS

Recall that the linear time-invariant state equation

$$\dot{x}(t) = Ax(t) + Bu(t)$$

$$y(t) = Cx(t) + Du(t) \tag{5.1}$$

is a *state-space realization* of a linear time-invariant system's input-output behavior if, given the associated transfer function $H(s)$, the coefficient matrices satisfy

$$C(sI - A)^{-1}B + D = H(s)$$

We consider a single-input, single-output system with scalar transfer function

$$H(s) = \frac{b(s)}{a(s)} = \frac{b_{n-1}s^{n-1} + \cdots + b_1 s + b_0}{s^n + a_{n-1}s^{n-1} + \cdots + a_1 s + a_0} \tag{5.2}$$

in which n defines the degree of the denominator polynomial $a(s)$ and hence the order of the transfer function. Also note that we have specified a *strictly proper* transfer function. This is essentially for convenience and does not affect any of the conclusions drawn in this section (as the reader may check). We call $H(s)$ *irreducible* if it has no pole-zero cancellations, and equivalently, its numerator and denominator polynomials have no common roots/factors.

Suppose that we have an n–dimensional realization of $H(s)$. Now it is clear that if the nth-order transfer function has one or more pole-zero cancellations, then, on canceling the corresponding factors in $a(s)$ and $b(s)$, we can write

$$H(s) = \frac{\tilde{b}(s)}{\tilde{a}(s)}$$

with $\tilde{n} = $ degree $\tilde{a}(s) < n$. It is clear that there is a state-space realization of $H(s)$ with dimension $\tilde{n}$ [say either controller canonical form or observer canonical form written down by inspection from the coefficients of $\tilde{a}(s)$ and $\tilde{b}(s)$], so the original n–dimensional realization is not minimal.

Conversely, suppose that we have an n–dimensional realization of the nth–order transfer function $H(s)$ that is not minimal, so there exists another realization of strictly lower dimension. Let $(\tilde{A}, \tilde{B}, \tilde{C})$ denote the coefficient matrices for this realization and $\tilde{n}$ denote its dimension. Since, by definition of the matrix adjoint, $\text{adj}(sI - \tilde{A})$ is an $\tilde{n} \times \tilde{n}$ matrix of polynomials of degree strictly less than $\tilde{n}$, it follows that

$$\tilde{b}(s) = \tilde{C} \, \text{adj}(sI - \tilde{A})\tilde{B}$$

is a scalar polynomial of degree strictly less than $\tilde{n}$. This along with

$$\tilde{a}(s) = |sI - \tilde{A}|$$

allows us to write

$$\begin{aligned} H(s) &= \tilde{C}(sI - \tilde{A})^{-1}\tilde{B} \\ &= \frac{\tilde{C} \, \text{adj}(sI - \tilde{A})\tilde{B}}{|sI - \tilde{A}|} \\ &= \frac{\tilde{b}(s)}{\tilde{a}(s)} \end{aligned}$$

We then conclude from the fact that degree $\tilde{a}(s) = \tilde{n} < n$ that $a(s)$ and $b(s)$ must have at least one common factor and that $H(s)$ is therefore not irreducible. The preceding argument can be summarized as follows.

Theorem 5.1 *Given the transfer function $H(s)$ in Equation (5.2) with $n = $ degree $a(s)$, an n–dimensional realization of $H(s)$ is minimal if and only if $H(s)$ is irreducible.*

Toward relating minimality of a realization to its controllability and observability properties, we consider the following example.

Example 5.1 For the third-order transfer function

$$H(s) = \frac{b(s)}{a(s)} = \frac{s^2 + 2s + 1}{s^3 + 6s^2 + 11s + 6}$$

the three–dimensional controller canonical form realization is specified by the coefficient matrices

$$A_{CCF} = \begin{bmatrix} 0 & 1 & 0 \\ 0 & 0 & 1 \\ -6 & -11 & -6 \end{bmatrix} \qquad B_{CCF} = \begin{bmatrix} 0 \\ 0 \\ 1 \end{bmatrix}$$

$$C_{CCF} = [1 \quad 2 \quad 1]$$

We know that this realization is automatically controllable. To check observability, we construct

$$\begin{bmatrix} C_{CCF} \\ C_{CCF}A_{CCF} \\ C_{CCF}A_{CCF}^2 \end{bmatrix} = \begin{bmatrix} 1 & 2 & 1 \\ -6 & -10 & -4 \\ 24 & 38 & 14 \end{bmatrix}$$

A direct calculation shows that the determinant of this observability matrix is zero, so the controller canonical form realization of this transfer function is not observable.

Alternatively, the three–dimensional observer canonical form realization is specified by the coefficient matrices

$$A_{OCF} = \begin{bmatrix} 0 & 0 & -6 \\ 1 & 0 & -11 \\ 0 & 1 & -6 \end{bmatrix} \qquad B_{OCF} = \begin{bmatrix} 1 \\ 2 \\ 1 \end{bmatrix}$$

$$C_{OCF} = [0 \quad 0 \quad 1]$$

and we know that this realization is automatically observable. Also, since the controller canonical form is not observable, we have by duality that the observer canonical form is not controllable.

Using results from previous chapters, we can conclude that

- All controllable three–dimensional realizations of $H(s)$ are not observable because every such realization can be related to the controller canonical form via a uniquely defined coordinate transformation and observability is unaffected by coordinate transformations.

- All observable three–dimensional realizations of $H(s)$ are not controllable because every such realization can be related to the observer canonical form via a uniquely defined coordinate transformation and controllability is unaffected by coordinate transformations.

Thus we conclude that there are no three–dimensional realizations of $H(s)$ that are both controllable and observable. Next, we observe that

$$H(s) = \frac{(s+1)^2}{(s+1)(s+2)(s+3)} = \frac{s+1}{(s+2)(s+3)}$$

which, on invoking Theorem 5.1, indicates that any three–dimensional realization of $H(s)$ is not minimal because $H(s)$ is not irreducible. In addition, since

$$H(s) = \frac{s+1}{(s+2)(s+3)} = \frac{s+1}{s^2+5s+6}$$

cannot be reduced further, we conclude that all two–dimensional realizations of $H(s)$ *are* minimal. The two–dimensional controller canonical form realization corresponding to the reduced transfer function is given by

$$A_{CCF} = \begin{bmatrix} 0 & 1 \\ -6 & -5 \end{bmatrix} \qquad B_{CCF} = \begin{bmatrix} 0 \\ 1 \end{bmatrix} \qquad C_{CCF} = \begin{bmatrix} 1 & 1 \end{bmatrix}$$

which is automatically controllable. From

$$\begin{bmatrix} C_{CCF} \\ C_{CCF}A_{CCF} \end{bmatrix} = \begin{bmatrix} 1 & 1 \\ -6 & -4 \end{bmatrix}$$

which is clearly nonsingular, we conclude that the two–dimensional controller canonical form realization is also observable. It follows by duality that the two–dimensional observer canonical form realization is controllable as well as observable.

We conclude from this analysis that

- All controllable two–dimensional realizations of the reduced $H(s)$ are also observable.
- All observable two–dimensional realizations of the reduced $H(s)$ are also controllable. □

This example suggests a link between minimality of a realization and joint controllability and observability. Toward establishing a concrete connection, we will make use of the following intermediate mathematical results:

Lemma 5.2 *Given $H(s)$ in Equation (5.2) with $n =$ degree $a(s)$, the n–dimensional controller canonical form realization is observable if and only if $H(s)$ is irreducible.*

Proof. The alert reader may recognize the connection between Lemma 5.2 and AE4.1. To prove this result, it suffices to verify that if $H(s)$ has a pole-zero cancellation, then the controller canonical form realization is not observable, and conversely, if the controller canonical form realization is not observable, then $H(s)$ has a pole-zero cancellation. For either implication, the explicit characterization of the eigenstructure of the companion matrix A_{CCF} suggests the use of the Popov-Belevich-Hautus eigenvector test for observability. This, together with the explicit connection between the controller canonical form and the numerator and denominator polynomials of $H(s)$, makes for quick work. $\square$

With Lemma 5.2 in hand, we have by duality:

Lemma 5.3 *Given $H(s)$ in Equation (5.2) with $n =$ degree $a(s)$, the n–dimensional observer canonical form realization is controllable if and only if $H(s)$ is irreducible.*

We require one last mathematical result:

Lemma 5.4 *If there is one n–dimensional realization of $H(s)$ that is both controllable and observable, then all n–dimensional realizations of $H(s)$ are both controllable and observable.*

Proof. Let (A, B, C) represent an n–dimensional realization of $H(s)$ that is both controllable and observable, and let $(\hat{A}, \hat{B}, \hat{C})$ represent another n–dimensional realization. In the time domain, we therefore have

$$Ce^{At}B = h(t) = \hat{C}e^{\hat{A}t}\hat{B} \quad \text{for all} \quad t \geq 0$$

Repeated differentiation and subsequent evaluation at $t = 0$ yields

$$CA^kB = \hat{C}\hat{A}^k\hat{B} \quad \text{for all} \quad k \geq 0$$

which can be packaged into the following $n \times n$ matrix identity:

$$\begin{bmatrix} CB & CAB & \cdots & CA^{n-1}B \\ CAB & CA^2B & \cdots & CA^{n-2}B \\ \vdots & \vdots & \ddots & \vdots \\ CA^{n-1}B & CA^nB & \cdots & CA^{2n-2}B \end{bmatrix}$$

$$
= \begin{bmatrix} \hat{C}\hat{B} & \hat{C}\hat{A}\hat{B} & \cdots & \hat{C}\hat{A}^{n-1}\hat{B} \\ \hat{C}\hat{A}\hat{B} & \hat{C}\hat{A}^2\hat{B} & \cdots & \hat{C}\hat{A}^n\hat{B} \\ \vdots & \vdots & \ddots & \vdots \\ \hat{C}\hat{A}^{n-1}\hat{B} & \hat{C}\hat{A}^n\hat{B} & \cdots & \hat{C}\hat{A}^{2n-2}\hat{B} \end{bmatrix}
$$

Each side can be factored as

$$
\begin{bmatrix} C \\ CA \\ \vdots \\ CA^{n-1} \end{bmatrix} \begin{bmatrix} B & AB & \cdots & A^{n-1}B \end{bmatrix} = \begin{bmatrix} \hat{C} \\ \hat{C}\hat{A} \\ \vdots \\ \hat{C}\hat{A}^{n-1} \end{bmatrix} \begin{bmatrix} \hat{B} & \hat{A}\hat{B} & \cdots & \hat{A}^{n-1}\hat{B} \end{bmatrix}
$$

in which each factor on both left and right sides is an $n \times n$ matrix. By assumption, the realization (A, B, C) is controllable and observable, so each factor on the left hand side is nonsingular. Consequently, the product on the left-hand side is nonsingular and so the same is true for the product on the right-hand side. This, in turn, implies that each factor on the right-hand side is nonsingular, so $(\hat{A}, \hat{B}, \hat{C})$ also must specify a controllable and observable n-dimensional realization. Since the second realization $(\hat{A}, \hat{B}, \hat{C})$ was selected arbitrarily, we conclude that all n-dimensional realizations of $H(s)$ must be both controllable and observable. □

We now can formally establish a link between minimality of a realization and joint controllability and observability.

Theorem 5.5 *An n-dimensional realization of the transfer function $H(s)$ in Equation (5.2) with $n =$ degree $a(s)$ is minimal if and only if it is both controllable and observable.*

Proof. If the n-dimensional realization is minimal, then $H(s)$ is irreducible by Theorem 5.1. Following Lemma 5.2, this implies that the n-dimensional controller canonical form realization is both controllable and observable. Consequently, by Lemma 5.4, our minimal n-dimensional realization also must be both controllable and observable. Conversely, if we start with an n-dimensional realization of $H(s)$ that is both controllable and observable, then the same must be true for the n-dimensional controller canonical form realization. This implies that $H(s)$ is irreducible, which, in turn, implies that the n-dimensional realization is minimal. □

5.2   MINIMALITY OF MULTIPLE-INPUT, MULTIPLE OUTPUT REALIZATIONS

Our goal in this section is to show that Theorem 5.5 can be extended to the multiple-input, multiple-output case.

Theorem 5.6   *Suppose that the linear state equation (5.1) is a realization of the $p \times m$ transfer function $H(s)$. Then it is a minimal realization if and only if it is both controllable and observable.*

Proof.   We first show that joint controllability/observability implies minimality. Suppose that the linear state equation (5.1) is an n–dimensional realization of $H(s)$ that is *not* minimal. Then there is a realization

$$\dot{\tilde{x}}(t) = \tilde{A}\tilde{x}(t) + \tilde{B}u(t)$$

$$y(t) = \tilde{C}\tilde{x}(t) + \tilde{D}u(t)$$

of dimension $\tilde{n} < n$. Since both are realizations of $H(s)$, we equate the impulse responses in the time domain, that is,

$$Ce^{At}B + D\delta(t) = \tilde{C}e^{\tilde{A}t}\tilde{B} + \tilde{D}\delta(t) \quad \text{for all } t \geq 0$$

which implies that $D = \tilde{D}$ and $\tilde{C}e^{\tilde{A}t}\tilde{B} = Ce^{At}B$, for all $t \geq 0$. Repeated differentiation of the latter identity and evaluation at $t = 0$ gives

$$CA^k B = \tilde{C}\tilde{A}^k \tilde{B} \text{ for all } k \geq 0$$

Arranging these data for $k = 0, 1, \ldots, 2n - 2$ into matrix form yields

$$\begin{bmatrix} CB & CAB & \cdots & CA^{n-1}B \\ CAB & CA^2B & \cdots & CA^nB \\ \vdots & \vdots & \ddots & \vdots \\ CA^{n-1}B & CA^nB & \cdots & CA^{2n-2}B \end{bmatrix}$$

$$= \begin{bmatrix} \tilde{C}\tilde{B} & \tilde{C}\tilde{A}\tilde{B} & \cdots & \tilde{C}\tilde{A}^{n-1}\tilde{B} \\ \tilde{C}\tilde{A}\tilde{B} & \tilde{C}\tilde{A}^2\tilde{B} & \cdots & \tilde{C}\tilde{A}^n\tilde{B} \\ \vdots & \vdots & \ddots & \vdots \\ \tilde{C}\tilde{A}^{n-1}\tilde{B} & \tilde{C}\tilde{A}^n\tilde{B} & \cdots & \tilde{C}\tilde{A}^{2n-2}\tilde{B} \end{bmatrix}$$

Each side can be factored to yield

$$
\begin{bmatrix} C \\ CA \\ \vdots \\ CA^{n-1} \end{bmatrix}
\begin{bmatrix} B & AB & \cdots & A^{n-1}B \end{bmatrix}
=
\begin{bmatrix} \tilde{C} \\ \tilde{C}\tilde{A} \\ \vdots \\ \tilde{C}\tilde{A}^{n-1} \end{bmatrix}
\begin{bmatrix} \tilde{B} & \tilde{A}\tilde{B} & \cdots & \tilde{A}^{n-1}\tilde{B} \end{bmatrix}
$$

The right-hand side is the product of a $pn \times \tilde{n}$ matrix and an $\tilde{n} \times mn$ matrix, each of which has rank no greater than $\tilde{n}$ (because $pn > \tilde{n}$ and $mn > \tilde{n}$). Consequently, the rank of each product is upper bounded by $\tilde{n} < n$. Applying a general matrix rank relationship to the left-hand side product, written compactly as QP, we obtain

$$
\text{rank}(QP) = \text{rank}\, P - \dim(\text{Ker } Q \cap \text{Im} P)
$$

$$
\leq \tilde{n}
$$

$$
< n
$$

We claim that this implies that the linear state equation (5.1) cannot be both controllable and observable because

- If the pair (A, B) is controllable (rank $P = n$), then necessarily $\dim(\text{Ker } Q \cap \text{Im} P) \geq 1$, implying that $\dim(\text{Ker } Q) \geq 1$, so the pair (A, C) is not observable.
- If the pair (A, C) is observable (rank $Q = n$, equivalently nullity $Q = 0$), then necessarily, $\dim(\text{Ker } Q \cap \text{Im} P) = 0$, implying that rank $P < n$ so that the pair (A, B) is not controllable.

We therefore have shown that joint controllability and observability implies minimality.

We next show that minimality implies joint controllability and observability. Suppose that the linear state equation (5.1) is an n–dimensional realization of $H(s)$ that is *not* controllable, so

$$
\text{rank}[\, B \quad AB \quad \cdots \quad A^{n-1}B \,] = q < n
$$

From Section 3.4, we know that there exists a nonsingular $n \times n$ matrix T such that

$$
\hat{A} = T^{-1}AT \qquad\qquad \hat{B} = T^{-1}B
$$

$$
= \begin{bmatrix} A_{11} & A_{12} \\ 0 & A_{22} \end{bmatrix} \qquad\qquad = \begin{bmatrix} B_1 \\ 0 \end{bmatrix}
$$

in which the pair (A_{11}, B_1) specifies a controllable q–dimensional state equation. In addition, we partition $\hat{C} = CT$ conformably as $\hat{C} = [\,C_1 \quad C_2\,]$. This, along with $\hat{D} = D$, yields

$$
\begin{aligned}
H(s) &= C(sI - A)^{-1}B + D \\
&= \hat{C}(sI - \hat{A})^{-1}\hat{B} + \hat{D} \\
&= [\,C_1 \quad C_2\,]
\begin{bmatrix} sI - A_{11} & -A_{12} \\ 0 & sI - A_{22} \end{bmatrix}^{-1}
\begin{bmatrix} B_1 \\ 0 \end{bmatrix} + D \\
&= [\,C_1 \quad C_2\,]
\begin{bmatrix} (sI - A_{11})^{-1} & (sI - A_{11})^{-1}A_{12}(sI - A_{22})^{-1} \\ 0 & (sI - A_{22})^{-1} \end{bmatrix} \\
&\quad \times \begin{bmatrix} B_1 \\ 0 \end{bmatrix} + D \\
&= C_1(sI - A_{11})^{-1}B_1 + D
\end{aligned}
$$

It follows that

$$
\dot{z}_1(t) = A_{11}z_1(t) + B_1u(t)
$$
$$
y(t) = C_1z_1(t) + Du(t)
$$

is a q–dimensional state-space realization of $H(s)$. Since $q < n$, we conclude that the original n–dimensional realization (5.1) is not minimal.

A similar argument can be made assuming initially that the linear state equation (5.1) is not observable using the discussion in Section 4.4 regarding the standard form for unobservable state equations. We therefore conclude that a realization that is either not controllable or not observable (or both) is not minimal.   □

5.3 MATLAB **FOR MINIMAL REALIZATIONS**

The following command is useful for computing minimal realizations via pole/zero cancellations using MATLAB.

```
MinSysName = minreal(SysName)
```

Given the linear time-invariant system data structure **SysName**, the **minreal** function produces a linear time-invariant system data structure **MinSysName** in which all possible pole/zero cancellations are made. For state-space models, this function produces a minimal realization **MinSys-Name** of **SysName** by removing all uncontrollable or unobservable modes.

The Continuing MATLAB Example (rotational mechanical system) is already a minimal realization as developed in Chapters 1 and 2, so there is nothing to do for that example in this chapter. Therefore, we instead employ MATLAB to compute a minimal realization of the system given in Example 5.1. The following MATLAB code performs this computation. Again, it is not related to the Continuing MATLAB Example, and hence this code stands alone and does not appear in the complete Continuing MATLAB Example m-file in Appendix C.

```
%- - - - - - - - - - - - - - - - - - - - - - - - - - - - - - - - - - - - - - -
%   Chapter 5.  Minimal Realizations
%- - - - - - - - - - - - - - - - - - - - - - - - - - - - - - - - - - - - - - -

Accf = [0 1 0; 0 0 1; -6 -11 -6];   % Define CCF system
                                     % for Example 5.1
Bccf = [0; 0; 1];
Cccf = [1 2 1];
Dccf = 0;

CCF = ss(Accf,Bccf,Cccf,Dccf);

minCCF = minreal(CCF)                % Compute the
                                     % minimal
                                     % realization
```

This m-file yields the following output, a two–dimensional minimal realization for the given three–dimensional controller canonical form system from Example 5.1:

```
1 state removed.

a =
              x1        x2
    x1     -2.789    -10.98
    x2    -0.01517   -2.211

b =
              u1
    x1    0.7887
    x2    0.2113
```

```
c =
        x1      x2
  y1  0.634   2.366
```

```
d =
       u1
  y1    0
```

Continuous-time model.

MATLAB tells us that one state was removed corresponding to the cancellation of the common factor $(s + 1)$ in both the numerator and denominator of the transfer function $H(s)$. The resulting two–dimensional state-space realization has no discernible structure, but it is minimal. That is, it is both controllable and observable, and there are no further possible pole/zero cancellations in the reduced second-order transfer function.

5.4 HOMEWORK EXERCISES

Numerical Exercises

NE5.1 Compute a minimal realization for each of the following state equations.

a. $A = \begin{bmatrix} 0 & 1 \\ -2 & -3 \end{bmatrix}$ $B = \begin{bmatrix} 0 \\ 1 \end{bmatrix}$ $C = [1 \quad 1]$

b. $A = \begin{bmatrix} 0 & 1 & 0 \\ 0 & 0 & 1 \\ -15 & -17 & -7 \end{bmatrix}$ $B = \begin{bmatrix} 0 \\ 0 \\ 1 \end{bmatrix}$ $C = [5 \quad 4 \quad 1]$

c. $A = \begin{bmatrix} 0 & 1 & 0 \\ 0 & 0 & 1 \\ -15 & -17 & -7 \end{bmatrix}$ $B = \begin{bmatrix} 0 \\ 0 \\ 1 \end{bmatrix}$ $C = [3 \quad 1 \quad 0]$

d. $A = \begin{bmatrix} 0 & 1 & 0 & 0 \\ 0 & 0 & 1 & 0 \\ 0 & 0 & 0 & 1 \\ -250 & -255 & -91 & -15 \end{bmatrix}$ $B = \begin{bmatrix} 0 \\ 0 \\ 0 \\ 1 \end{bmatrix}$

$C = [25 \quad 8 \quad 1 \quad 0]$

e. $A = \begin{bmatrix} 0 & 0 & 0 & -250 \\ 1 & 0 & 0 & -255 \\ 0 & 1 & 0 & -91 \\ 0 & 0 & 1 & -15 \end{bmatrix}$ $B = \begin{bmatrix} 24 \\ 9 \\ 2 \\ 0 \end{bmatrix}$

$C = \begin{bmatrix} 0 & 0 & 0 & 1 \end{bmatrix}$

Analytical Exercises

AE 5.1 Show that the state equation represented by the triple (A, B, C) is a minimal realization if and only if the state equation represented by the triple $(A - BKC, B, C)$ is a minimal realization for any $m \times p$ gain matrix K. In other words, minimality is invariant with respect to *output feedback* $u(t) = -Ky(t)$.

AE 5.2 Show that the single-input, single-output state equation

$$\dot{x}(t) = Ax(t) + Bu(t)$$
$$y(t) = Cx(t) + Du(t)$$

is minimal if and only if A and

$$\begin{bmatrix} A & B \\ C & D \end{bmatrix}$$

have no eigenvalue in common.

AE 5.3 Consider a single-input, single-output n—dimensional state-space realization represented by (A, B, C), and assume that $n \geq 2$. Show that if (A, B, C) is a minimal realization, then A and the product BC do not commute with respect to matrix multiplication.

Continuing MATLAB Exercises

CME5.3 For the system given in CME1.3, compute a minimal state-space realization.

CME5.4 For the system given in CME1.4, compute a minimal state-space realization.

Continuing Exercises

CE5.5 You should have found in CE 3.5 and CE 4.5 that the system given in CE1.5 is controllable but not observable. Therefore, the original state-space realization is not minimal. Compute a minimal state-space realization for this system.

6

STABILITY

For the linear time-invariant state equation

$$\begin{aligned} \dot{x}(t) &= Ax(t) + Bu(t) \\ y(t) &= Cx(t) + Du(t) \end{aligned} \qquad x(0) = x_0 \qquad (6.1)$$

we consider two types of stability in this chapter. The first is an internal notion of stability and involves the qualitative behavior of the zero-input state response, i.e., the response of a related homogeneous state equation that depends solely on the initial state. We introduce the fundamental concepts in a more general setting involving homogeneous nonlinear state equations but specialize to the linear case for which explicit stability criteria are available that involve the eigenvalues of the system dynamics matrix A. We also present an energy-based stability analysis that has its origins in the work of the Russian mathematician A. M. Lyapunov more than a century ago. The power of Lyapunov stability theory lies in its applicability to nonlinear systems; however, we focus on its specialization to linear systems.

The second type of stability we study in this chapter focuses on external, or input-output, behavior. In particular, we characterize state equations for which the zero-state output response is a bounded signal for every bounded input signal. This is referred to, not surprisingly, as *bounded-input, bounded-output stability*. Although these internal and external stability properties are fundamentally different in character, as they pertain to different response components, they are related, as we will show. This chapter concludes with an overview of stability analysis using MATLAB featuring our Continuing MATLAB Example and Continuing Examples 1 and 2.

6.1 INTERNAL STABILITY

In this section we focus on the stability of equilibrium states for homogeneous state equations. We begin our analysis with the nonlinear state equation

$$\dot{x}(t) = f[x(t)] \qquad x(0) = x_0 \tag{6.2}$$

for which equilibrium states are constant $n \times 1$ vectors $\tilde{x}$ that satisfy $f(\tilde{x}) = 0$. Stability of an equilibrium state refers to the qualitative behavior of trajectories that start in the vicinity of the equilibrium state. A nonlinear state equation can have multiple isolated equilibrium states each with different stability properties. This is why we refer to stability of a particular equilibrium state rather than the state equation itself. The stability-related notions we consider are illustrated in Figure 6.1. Point *a* represents an *unstable* equilibrium. A ball perfectly balanced atop the curved surface on the left will remain at rest if undisturbed. However, the slightest perturbation will cause the ball to roll away from that rest position. Point *b* illustrates a *stable* equilibrium in the sense that the ball will move only a small distance from a rest position on the flat surface when experiencing a perturbation that is suitably small. Finally, point *c* depicts an asymptotically stable equilibrium. Suppose that the ball is initially at rest nestled in the curved surface on the right. For a reasonably small perturbation, the ball will not stray too far from the rest position and, in addition, eventually will return to the rest position.

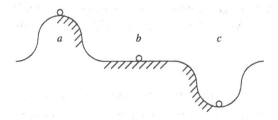

FIGURE 6.1 Equilibrium states.

Armed with this intuition, we now give precise stability definitions of an equilibrium state. Because a nonzero equilibrium state can be translated to the origin by change of variables with an accompanying modification to the state equation, we can assume without loss of generality that the equilibrium state under scrutiny is at the origin, that is, $\tilde{x} = 0 \in \mathbb{R}^n$.

Definition 6.1   *The equilibrium state $\tilde{x} = 0$ of Equation (6.2) is*

- *Stable if, given any $\varepsilon > 0$ there corresponds a $\delta > 0$ such that $\|x_0\| < \delta$ implies that $\|x(t)\| < \varepsilon$ for all $t \geq 0$.*
- *Unstable if it is not stable.*
- *Asymptotically stable if it is stable and it is possible to choose $\delta > 0$ such that $\|x_0\| < \delta$ implies that $\lim_{t \to \infty} \|x(t)\| = 0$. Specifically, given any $\varepsilon > 0$, there exists $T > 0$ for which the corresponding trajectory satisfies $\|x(t)\| \leq \varepsilon$ for all $t \geq T$.*
- *Globally asymptotically stable if it is stable and $\lim_{t \to \infty} \|x(t)\| = 0$ for any initial state. Specifically, given any $M > 0$ and $\varepsilon > 0$, there exists $T > 0$ such that $\|x_0\| < M$ implies that the corresponding trajectory satisfies $\|x(t)\| \leq \varepsilon$ for all $t \geq T$.*
- *Exponentially stable if there exist positive constants $\delta, k,$ and λ such that $\|x_0\| < \delta$ implies that $\|x(t)\| < k e^{-\lambda t} \|x_0\|$ for all $t \geq 0$.*
- *Globally exponentially stable if there exist positive constants k and λ such that $\|x(t)\| \leq k e^{-\lambda t} \|x_0\|$ for all $t \geq 0$ for all initial states.*

The wording in these definitions is a bit subtle, but the basic ideas are conveyed in Figure 6.2. An equilibrium state is stable if the state trajectory can be made to remain as close as desired to the equilibrium state for all time by restricting the initial state to be sufficiently close to the equilibrium state. An unstable equilibrium state does not necessarily involve trajectories that diverge arbitrarily far from the equilibrium; rather only that there is some bound on $\|x(t)\|$ that cannot be achieved for all $t \geq 0$ by at least one trajectory no matter how small the initial deviation $\|x_0\|$. Asymptotic stability requires, in addition to stability, that trajectories converge to the equilibrium state over time with no further constraint on the rate of convergence. By comparison, exponential stability is a stronger stability property. As an illustration, consider the one-dimensional state equation

$$\dot{x}(t) = -x^3(t)$$

which, for any initial state, has the solution $x(t) = x_0 / \sqrt{1 + 2x_0^2 t}$ that asymptotically converges to the equilibrium $\tilde{x} = 0$ over time. However, the rate of convergence is slower than any decaying exponential bound.

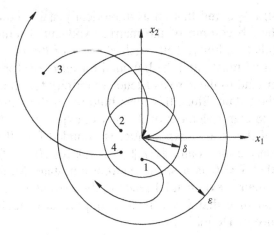

FIGURE 6.2 Stability of an equilibrium state.

Our ultimate focus is on the homogeneous linear time-invariant state equation

$$\dot{x}(t) = Ax(t) \qquad x(0) = x_0 \qquad (6.3)$$

for which $\tilde{x} = 0 \in \mathbb{R}^n$ is seen easily to be an equilibrium state. It is possible to show by exploiting the linearity of the solution to (6.3) in the initial state that the preceding stability definitions can be reformulated as follows:

Definition 6.2 *The equilibrium state* $\tilde{x} = 0$ *of Equation* (6.3) *is*

- **Stable** *if there exists a finite positive constant* γ *such that for any initial state* x_0 *the corresponding trajectory satisfies* $\|x(t)\| \leq \gamma \|x_0\|$ *for all* $t \geq 0$.
- **Unstable** *if it is not stable.*
- **(Globally) asymptotically stable** *if given any* $\mu > 0$ *there exists* $T > 0$ *such that for any initial state* x_0 *the corresponding trajectory satisfies* $\|x(t)\| \leq \mu \|x_0\|$ *for all* $t \geq T$.
- **(Globally) exponentially stable** *if there exist positive constants* k *and* λ *such that that for any initial state* x_0 *the corresponding trajectory satisfies* $\|x(t)\| \leq ke^{-\lambda t}\|x_0\|$ *for all* $t \geq 0$.

Since the trajectory of Equation (6.3) is given by $x(t) = e^{At}x_0$, we see from the choice $x_0 = e_i$, the ith standard basis vector, that a stable equilibrium state implies that the ith column of the matrix exponential is bounded for all $t \geq 0$ and that an asymptotically stable equilibrium state implies that the ith column of the matrix exponential tends to the zero vector as t tends to infinity. Thus each element of the ith column of the matrix exponential must either be bounded for all $t \geq 0$ for a stable equilibrium state or tend to the zero as t tends to infinity for an asymptotically stable equilibrium state. Since this must hold for each column, these conclusions apply to every element of the matrix exponential. Conversely, if each element of the matrix exponential is bounded for all $t \geq 0$, then it is possible to derive the bound $\|x(t)\| \leq \mu \|x_0\|$ for all $t \geq 0$, from which we conclude that $\tilde{x} = 0$ is a stable equilibrium state. Similarly, if each element of the matrix exponential tends to the zero as t tends to infinity, then we can conclude that $\tilde{x} = 0$ is an asymptotically stable equilibrium state in the sense of Definition 6.2.

We further investigate the behavior of elements of the matrix exponential using the Jordan canonical form of A. As summarized in Appendix B, for any square matrix A there exists a nonsingular matrix T yielding

$$J = T^{-1}AT$$

for which J is block diagonal, and each block has the form

$$J_k(\lambda) = \begin{bmatrix} \lambda & 1 & 0 & \cdots & 0 \\ 0 & \lambda & 1 & \cdots & 0 \\ 0 & 0 & \lambda & \ddots & 0 \\ \vdots & \vdots & \vdots & \ddots & 1 \\ 0 & 0 & 0 & \cdots & \lambda \end{bmatrix} \quad (k \times k)$$

representing one of the Jordan blocks associated with the eigenvalue λ. Since we can write $A = T J T^{-1}$, we then have $e^{At} = T e^{Jt} T^{-1}$. Consequently, boundedness or asymptotic properties of elements of e^{At} can be inferred from corresponding properties of the elements of e^{Jt}. Since J is block diagonal, so is e^{Jt}. Specifically, for each submatrix of the form $J_k(\lambda)$ on the block diagonal of J, e^{Jt} will contain the diagonal block

$$e^{J_k(\lambda)t} = e^{\lambda t} \begin{bmatrix} 1 & t & \frac{1}{2}t^2 & \cdots & \frac{1}{(k-1)!}t^{k-1} \\ 0 & 1 & t & \cdots & \frac{1}{(k-2)!}t^{k-2} \\ 0 & 0 & 1 & \ddots & \vdots \\ \vdots & \vdots & \vdots & \ddots & t \\ 0 & 0 & 0 & \cdots & 1 \end{bmatrix} \quad (k \times k)$$

We also point out that all the Jordan blocks associated with a particular eigenvalue are scalar if and only if the associated geometric and algebraic multiplicities of that eigenvalue are equal. Furthermore, when $J_1(\lambda) = \lambda$, we have $e^{J_1(\lambda)t} = e^{\lambda t}$. With this preliminary analysis in place, we are prepared to establish the following result:

Theorem 6.3 *The equilibrium state $\tilde{x} = 0$ of Equation (6.3) is:*

- *Stable if and only if all eigenvalues of A have a nonpositive real part and the geometric multiplicity of any eigenvalue with zero real part equals the associated algebraic multiplicity.*
- *(Globally) asymptotically stable if and only if every eigenvalue of A has strictly negative real part.*

Proof (stability.) We see that each Jordan block has a matrix exponential containing bounded terms provided that $\text{Re}(\lambda) < 0$, in which case terms of the form $t^j e^{\lambda t}$ are bounded for all $t \geq 0$ for any power of t (and in fact decay to zero as t tends to infinity), or that whenever $\text{Re}(\lambda) = 0$, the size of any corresponding Jordan block is 1×1 so that $|e^{J_1(\lambda)t}| = |e^{\lambda t}| = e^{\text{Re}(\lambda)t} \equiv 1$. As noted earlier, each Jordan block associated with an eigenvalue λ has size 1 when and only when the geometric and algebraic multiplicities are equal. Conversely, if there exists an eigenvalue with $\text{Re}(\lambda) > 0$, the matrix exponential of each associated Jordan block contains unbounded terms, or if there exists an eigenvalue with $\text{Re}(\lambda) = 0$ and a Jordan block of size 2 or greater, the associated matrix exponential will contain terms of the form $t^j e^{\lambda t}$ with $j \geq 1$ having magnitude $|t^j e^{\lambda t}| = t^j e^{\text{Re}(\lambda)t} = t^j$ that also grow without bound despite the fact that $|e^{\lambda t}| = e^{\text{Re}(\lambda)t} \equiv 1$.

(Asymptotic stability.) Each Jordan block has a matrix exponential containing elements that are either zero or asymptotically tend to zero as t tends to infinity provided that $\text{Re}(\lambda) < 0$. Conversely, in addition to the preceding discussion, even if there exists an eigenvalue with $\text{Re}(\lambda) = 0$ having scalar Jordan blocks, $|e^{J_1(\lambda)}| = |e^{\lambda t}| = e^{\text{Re}(\lambda)t} \equiv 1$, which does not tend asymptotically to zero as t tends to infinity. $\qquad\square$

We note that if A has strictly negative real-part eigenvalues, then it must be nonsingular. Consequently, $\tilde{x} = 0$ is the only equilibrium state for the homogeneous linear state equation (6.3) because it is the only solution to the homogeneous linear equation $Ax = 0$. It is therefore customary to refer to Equation (6.3) as an asymptotically stable system in this case. The eigenvalue criteria provided by Theorem 6.3 are illustrated in Figure 6.3. Case 1 depicts strictly negative real-part eigenvalues corresponding to an

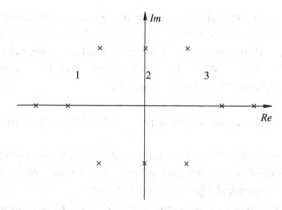

FIGURE 6.3 Eigenvalue locations in the complex plane.

asymptotically stable system. Case 2 indicates nonrepeated eigenvalues on the imaginary axis that, since the geometric and algebraic multiplicities are each 1 in this case, indicates a stable system. Finally, case 3 shows one or more eigenvalues with positive real-part that corresponds to an unstable system.

We remark that stability in the sense of Definition 6.2 and Theorem 6.3 is commonly referred to as *marginal stability*. We will adopt this practice in situations where we wish to emphasize the distinction between stability and asymptotic stability.

Energy-Based Analysis

Here we establish intuitive connections between the types of stability defined earlier and energy-related considerations. Consider a physical system for which total energy can be defined as a function of the system state. If an equilibrium state corresponds to a (local) minimum of this energy function, and if the energy does not increase along any trajectory that starts in the vicinity of the equilibrium state, it is reasonable to conclude that the trajectory remains close to the equilibrium state, thus indicating a stable equilibrium state. If the system dissipates energy along any trajectory that starts near the equilibrium state so that the system energy converges to the local minimum, we expect this to correspond to asymptotic convergence of the trajectory to the equilibrium, thereby indicating an asymptotically stable equilibrium. We observe that conclusions regarding the stability of an equilibrium state have been inferred from the time rate of change of the system energy along trajectories. The following example presents a quantitative illustration of these ideas.

Example 6.1   We consider the second-order linear translational mechanical system that was introduced in Example 1.1, which for zero external applied force is governed by

$$m\,\ddot{y}(t) + c\dot{y}(t) + ky(t) = 0$$

in which $y(t)$ represents the displacement of both the mass and spring from rest. The state variables were chosen previously as the mass/spring displacement $x_1(t) = y(t)$ and the mass velocity $x_2(t) = \dot{y}(t)$, yielding the homogeneous state equation

$$\begin{bmatrix} \dot{x}_1(t) \\ \dot{x}_2(t) \end{bmatrix} = \begin{bmatrix} 0 & 1 \\ -\dfrac{k}{m} & -\dfrac{c}{m} \end{bmatrix} \begin{bmatrix} x_1(t) \\ x_2(t) \end{bmatrix}$$

We recall that these state variables are related to energy stored in this system. The spring displacement characterizes the potential energy stored in the spring, and the mass velocity characterizes the kinetic energy stored in the mass. We therefore can express the total energy stored in the system by the function

$$E(x_1, x_2) = \frac{1}{2}kx_1^2 + \frac{1}{2}mx_2^2$$

We observe that the system energy is positive whenever $[x_1, x_2]^T \neq [0, 0]^T$ and attains the minimum value of zero at the equilibrium state $[\tilde{x}_1, \tilde{x}_2]^T = [0, 0]^T$.

On evaluating the energy function along a system trajectory, we can compute the time derivative

$$\frac{d}{dt}E[x_1(t), x_2(t)] = \frac{d}{dt}\left[\frac{1}{2}kx_1^2(t) + \frac{1}{2}mx_2^2(t)\right]$$

$$= kx_1(t)\dot{x}_1(t) + mx_2(t)\dot{x}_2(t)$$

$$= kx_1(t)\left[x_2(t)\right] + mx_2(t)\left[-\frac{k}{m}x_1(t) - \frac{c}{m}x_2(t)\right]$$

$$= -cx_2^2(t)$$

where we have invoked the chain rule and have used the state equation to substitute for the state-variable derivatives.

For zero damping ($c = 0$) we have $dE/dt \equiv 0$, so the total system energy is constant along any trajectory. This corresponds to a perpetual exchange between the potential energy stored in the spring and the kinetic energy stored in the mass. This also indicates that $[\tilde{x}_1, \tilde{x}_2]^T = [0, 0]^T$ is a stable equilibrium in the sense of Definition 6.2. Specifically, since

$$\frac{1}{2} \min\{k, m\}[x_1^2(t) + x_2^2(t)]$$

$$\leq \frac{1}{2} kx_1^2(t) + \frac{1}{2} mx_2^2(t)$$

$$= \frac{1}{2} kx_1^2(0) + \frac{1}{2} mx_2^2(0) \leq \frac{1}{2} \max\{k, m\}[x_1^2(0) + x_2^2(0)]$$

we have the norm bound on the state trajectory $x(t) = [x_1(t), x_2(t)]^T$, that is,

$$\|x(t)\| \leq \sqrt{\frac{\max\{k, m\}}{\min\{k, m\}}} \|x(0)\| \quad \text{for all } t \geq 0$$

which suggests an obvious choice for the positive constant γ in Definition 6.2.

For positive damping ($c > 0$), we have $dE/dt < 0$ along any trajectory for which the mass velocity is not identically zero. A trajectory for which $x_2(t) = \dot{y}(t)$ is identically zero corresponds to identically zero acceleration and a constant displacement. Since such a trajectory also must satisfy the equations of motion we see, on substituting $\dot{x}_2(t) = \ddot{y}(t) \equiv 0$, $\dot{x}_1(t) = x_2(t) = \dot{y}(t) \equiv 0$, and $x_1(t) = y(t) \equiv y_0$ (constant), that $ky_0 = 0$. Consequently, the only trajectory for which the mass velocity is identically zero corresponds to the equilibrium state $x(t) \equiv \tilde{x} = [0, 0]^T$. We conclude that the total energy in the system is strictly decreasing along all other trajectories and converges to zero as time tends to infinity. We expect that this should correspond to asymptotic convergence of the trajectory to the equilibrium state $[\tilde{x}_1, \tilde{x}_2]^T = [0, 0]^T$ as time tends to infinity. To see this, convergence of the total energy to zero implies that given any $\mu > 0$, there is a $T > 0$ for which

$$E[x_1(t), x_2(t)] \leq \mu^2 \frac{\min\{k, m\}}{\max\{k, m\}} E[x_1(0), x_2(0)] \quad \text{for all } t \geq T$$

Using this and a previous bound, we have

$$\|x(t)\| \leq \sqrt{\frac{2E[x_1(t), x_2(t)]}{\min\{k, m\}}}$$

$$\leq \mu \sqrt{\frac{2E[x_1(0), x_2(0)]}{\max\{k, m\}}} \leq \mu\|x(0)\| \quad \text{for all } t \geq T$$

thereby verifying that $x(t) \equiv \tilde{x} = [0, 0]^T$ is an asymptotically stable equilibrium state.

For negative damping ($c < 0$), we have $dE/dt > 0$ along any trajectory for which the mass velocity is not identically zero. The same reasoning applied earlier indicates that the total energy in the system is strictly increasing along any trajectory other than $x(t) \equiv \tilde{x} = [0, 0]^T$. It can be argued that any initial state other than the equilibrium state yields a trajectory that diverges infinitely far away from the origin as time tends to infinity.

Simulation results and eigenvalue computations bear out these conclusions. For $m = 1$ kg, $k = 10$ N/m, and $c = 0$ N-s/m along with the initial state $x(0) = x_0 = [1, 2]^T$, the state-variable time responses are shown in Figure 6.4a, the phase portrait [$x_2(t) = \dot{x}_1(t)$ plotted versus

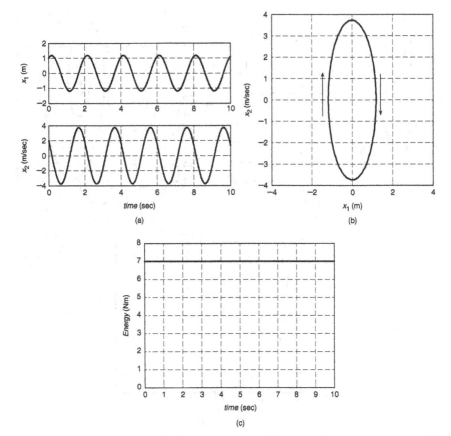

FIGURE 6.4   (a) State-variable responses; (b) phase portrait; (c) energy response for a marginally-stable equilibrium.

$x_1(t)$ parameterized by time t] is shown in Figure 6.4b, and the time response of the total system energy is shown in Figure 6.4c. In this case, we see oscillatory state-variable time responses, an elliptical phase portrait, and constant total energy. The system eigenvalues are $\lambda_{1,2} = \pm j3.16$, purely imaginary (zero real part). Since they are distinct, the geometric multiplicity equals the algebraic multiplicity for each.

With $c = 1$ N-s/m and all other parameters unchanged, the state-variable time responses are shown in Figure 6.5a, the phase portrait is shown in Figure 6.5b, and the time response of the total system energy is shown in Figure 6.5c. In this case, each state-variable time response decays to zero as time tends to infinity, as does the total energy response. The phase portrait depicts a state trajectory that spirals in toward the equilibrium state at

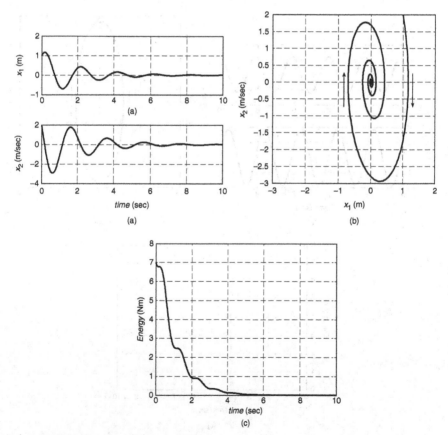

FIGURE 6.5 (a) State-variable responses; (b) phase portrait; (c) energy response for an asymptotically-stable equilibrium.

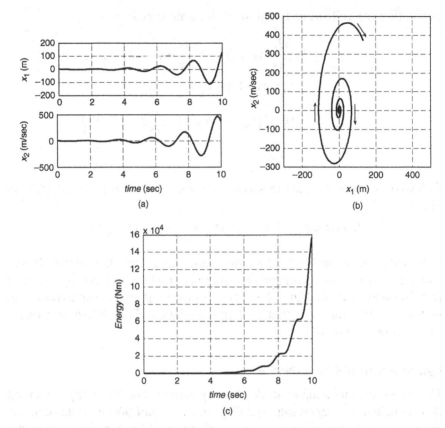

FIGURE 6.6 (a) State-variable responses; (b) phase portrait; (c) energy response for an unstable equilibrium.

the origin. The system eigenvalues are $\lambda_{1,2} = -0.50 \pm j3.12$, each with negative real part.

Finally, with the damping coefficient changed to $c = -1$ N-s/m, the state-variable time responses are shown in Figure 6.6a, the phase portrait is shown in Figure 6.6b, and the time response of the total system energy is shown in Figure 6.6c. Here each state variable time response grows in amplitude and the total energy increases with time. The phase portrait depicts a state trajectory that spirals away from the equilibrium state at the origin. The system eigenvalues are $\lambda_{1,2} = +0.50 \pm j3.12$, each with positive real part.

An extremely appealing feature of the preceding energy-based analysis is that stability of the equilibrium state can be determined directly from the time derivative of the total energy function along trajectories of the system. Computation of this time derivative can be interpreted as first computing the following function of the state variables

$$\dot{E}(x_1, x_2) \triangleq \frac{\partial E}{\partial x_1}(x_1, x_2)\dot{x}_1 + \frac{\partial E}{\partial x_2}(x_1, x_2)\dot{x}_2$$

$$= (kx_1)\dot{x}_1 + (mx_2)\dot{x}_2$$

$$= (kx_1)(x_2) + (mx_2)\left(-\frac{k}{m}x_1 - \frac{c}{m}x_2\right)$$

$$= -cx_2^2$$

followed by evaluating along a system trajectory $x(t) = [x_1(t), x_2(t)]^T$ to obtain

$$\dot{E}[x_1(t), x_2(t)] = -cx_2^2(t) = \frac{d}{dt}E[x_1(t), x_2(t)]$$

In addition, properties of the total energy time derivative along all system trajectories and accompanying stability implications can be deduced from properties of the function $\dot{E}(x_1, x_2)$ over the entire state space. An important consequence is that explicit knowledge of the trajectories themselves is not required.

Lyapunov Stability Analysis

The Russian mathematician A. M. Lyapunov (1857–1918) observed that conclusions regarding stability of an equilibrium state can be drawn from a more general class of energy-like functions. For the nonlinear state equation (6.2), we consider real-valued functions $V(x) = V(x_1, x_2, \ldots, x_n)$ with continuous partial derivatives in each state variable that are *positive definite*, meaning that $V(0) = 0$ and $V(x) > 0$ for all $x \neq 0$ at least in a neighborhood of the origin. This generalizes the property that the total energy function has a local minimum at the equilibrium. To analyze the time derivative of the function $V(x)$ along trajectories of Equation (6.2), we define

$$\dot{V}(x) = \frac{\partial V}{\partial x_1}(x)\dot{x}_1 + \frac{\partial V}{\partial x_2}(x)\dot{x}_2 + \cdots + \frac{\partial V}{\partial x_n}(x)\dot{x}_n$$

$$= \left[\frac{\partial V}{\partial x_1}(x) \frac{\partial V}{\partial x_2}(x) \cdots \frac{\partial V}{\partial x_2}(x)\right]\begin{bmatrix} \dot{x}_1 \\ \dot{x}_2 \\ \vdots \\ \dot{x}_n \end{bmatrix}$$

$$= \frac{\partial V}{\partial x}(x)f(x)$$

Thus $\dot{V}(x)$ is formed from the inner product of the gradient of $V(x)$ and the nonlinear map $f(x)$ that defines the system dynamics. The fundamental discovery of Lyapunov is that the equilibrium $\tilde{x} = 0$ is

- Stable if $\dot{V}(x)$ is *negative semidefinite*; that is, $\dot{V}(x) \leq 0$ for all x in a neighborhood of the origin.
- Asymptotically stable if $\dot{V}(x)$ is *negative definite*; that is, $\dot{V}(x) < 0$ for all $x \neq 0$ in a neighborhood of the origin.

A positive-definite function $V(x)$ for which $\dot{V}(x)$ is at least negative semidefinite is called a *Lyapunov function*. The preceding result is extremely powerful because stability of an equilibrium can be determined directly from the system dynamics without explicit knowledge of system trajectories. Consequently, this approach is referred to as *Lyapunov's direct method*. This is extremely important in the context of nonlinear systems because system trajectories, i.e., solutions to the nonlinear state equation (6.2), in general are not available in closed form.

We observe that Lyapunov's direct method only provides a sufficient condition for (asymptotic) stability of an equilibrium in terms of an unspecified Lyapunov function for which, in general, there are no systematic ways to construct. As a consequence, if a particular positive-definite function $V(x)$ fails to have $\dot{V}(x)$ negative semidefinite, we cannot conclude immediately that the origin is unstable. Similarly, if, for a Lyapunov function $V(x)$, $\dot{V}(x)$ fails to be negative definite, we cannot rule out asymptotic stability of the origin. On the other hand, a so-called converse theorem exists for asymptotic stability that, under additional hypotheses, guarantees the existence of a Lyapunov function with a negative-definite $\dot{V}(x)$. For a thorough treatment of Lyapunov stability analysis, we refer the interested reader to Khalil (2002).

For the linear state equation (6.3), Lyapunov stability analysis can be made much more explicit. First, we can focus on energy-like functions that are quadratic forms given by

$$V(x) = x^T P x = \sum_{i,j=1}^{n} p_{ij} x_i x_j \tag{6.4}$$

in which the associated matrix $P = [p_{ij}]$ is symmetric without loss of generality so that its elements satisfy $p_{ij} = p_{ji}$. We note here that P does not refer to the controllability matrix introduced in Chapter 3. A quadratic form is a positive-definite function over all of $\mathbb{R}^n$ if and only if P is a positive-definite symmetric matrix. A symmetric $n \times n$ matrix P is positive definite if and only if every eigenvalue of P is real and positive.

Consequently, the eigenvalues of a symmetric positive-definite matrix can be ordered via

$$0 < \lambda_{\min}(P) = \lambda_1 \leq \lambda_2 \leq \cdots \leq \lambda_n = \lambda_{\max}(P)$$

and the associated quadratic form satisfies the so-called Rayleigh-Ritz inequality

$$\lambda_{\min}(P)x^T x \leq x^T P x \leq \lambda_{\max}(P)x^T x \text{ for all } x \in \mathbb{R}^n$$

Another useful characterization of positive definiteness of a symmetric $n \times n$ matrix $P = [p_{ij}]$ is that its n leading principal minors defined as the submatrix determinants

$$p_{11} \quad \begin{vmatrix} p_{11} & p_{12} \\ p_{12} & p_{22} \end{vmatrix} \quad \begin{vmatrix} p_{11} & p_{12} & p_{13} \\ p_{12} & p_{22} & p_{23} \\ p_{13} & p_{23} & p_{33} \end{vmatrix} \quad \cdots \quad \begin{vmatrix} p_{11} & p_{12} & \cdots & p_{1n} \\ p_{12} & p_{22} & \cdots & p_{2n} \\ \vdots & \vdots & \ddots & \vdots \\ p_{1n} & p_{2n} & \cdots & p_{nn} \end{vmatrix}$$

are all positive. This is referred to as *Sylvester's criterion*. It follows directly that a quadratic form $x^T P x$ and associated symmetric matrix P are negative definite if and only if $-P$ is a positive-definite matrix.

The gradient of the quadratic form $V(x) = x^T P x$ is $(\partial V/\partial x)(x) = 2x^T P$, as can be verified from the summation in Equation (6.4). Using this and the linear dynamics in Equation (6.3), we can compute $\dot{V}(x)$ according to

$$\dot{V}(x) = \frac{\partial V}{\partial x}(x) f(x)$$

$$= (2x^T P)(Ax)$$

$$= x^T A^T P x + x^T P A x$$

$$= x^T [A^T P + PA] x$$

in which we also have used the fact that $x^T A^T P x = x^T P A x$ because these are scalar quantities related by the transpose operation. We observe that $\dot{V}(x)$ is also a quadratic form expressed in terms of the symmetric matrix $A^T P + PA$. Therefore, a sufficient condition for asymptotic stability of the equilibrium state $\tilde{x} = 0$ is the existence of a symmetric positive-definite matrix P for which $A^T P + PA$ is negative definite. The following result links the existence of such a matrix to the eigenvalue condition for asymptotic stability established in Theorem 6.3.

Theorem 6.4 *For any symmetric positive definite matrix Q, the Lyapunov matrix equation*

$$A^T P + P A = -Q \tag{6.5}$$

has a unique symmetric positive definite solution P if and only if every eigenvalue of A has strictly negative real part.

Proof. For necessity, suppose that, given a symmetric positive-definite matrix Q, there exists a unique symmetric positive-definite solution P to the Lyapunov matrix equation (6.5). Let λ be any eigenvalue of A and $v \in \mathbb{R}^n$ be a corresponding (right) eigenvector. Premultiplying Equation (6.5) by v^* and postmultiplying by v yields

$$
\begin{aligned}
-v^* Q v &= v^*[A^T P + P A]v \\
&= v^* A^T P v + v^* P A v \\
&= \bar{\lambda} v^* P v + \lambda v^* P v \\
&= (\bar{\lambda} + \lambda) v^* P v \\
&= 2 \operatorname{Re}(\lambda) v^* P v
\end{aligned}
$$

Since $v \neq 0$ (because it is an eigenvector) and P and Q are positive definite matrices, these quadratic forms satisfy $v^* P v > 0$ and $v^* Q v > 0$. This gives

$$\operatorname{Re}(\lambda) = -\frac{1}{2}\frac{v^* Q v}{v^* P v} < 0$$

from which we conclude that every eigenvalue of A necessarily has a strictly negative real part.

For sufficiency, suppose that every eigenvalue of A has a strictly negative real part and consider, for any symmetric positive definite matrix Q,

$$P = \int_0^\infty e^{A^T t} Q e^{At} dt$$

Since elements of e^{At} are sums of terms of the form $M t^{k-1} e^{\operatorname{Re}(\lambda)t} \cos(\operatorname{Im}(\lambda)t + \theta)$ that decay to zero exponentially fast, this improper integral converges to a finite limit, so P is well defined. Also, P is symmetric by inspection. Next, for any $x \in \mathbb{R}^n$,

$$x^T P x = \int_0^\infty (e^{At}x)^T Q(e^{At}x) dt \geq 0$$

because the integrand is nonnegative for all $t \geq 0$. Moreover, equality holds when and only when the integrand is identically zero which, by positive definiteness of Q implies that $e^{At}x \equiv 0$. Since e^{At} is nonsingular for all t, it follows that equality holds if and only if $x = 0$, which implies that P is positive definite.

Next, P satisfies the Lyapunov matrix equation because

$$A^T P + PA = \int_0^\infty [A^T e^{A^T t} Q e^{At} + e^{A^T t} Q e^{At} A] dt$$

$$= \int_0^\infty \frac{d}{dt} [e^{A^T t} Q e^{At}] dt$$

$$= e^{A^T t} Q e^{At} \big|_0^\infty$$

$$= 0 - Q$$

$$= -Q$$

Finally, to show that P is the only solution to the Lyapunov matrix equation, suppose that $\overline{P}$ is another solution. Subtracting the Lyapunov matrix equations satisfied by each solution gives

$$A^T(P - \overline{P}) + (P - \overline{P})A = 0$$

Premultiplying by $e^{A^T t}$ and postmultiplying by e^{At} yields

$$0 = e^{A^T t}[A^T(P - \overline{P}) + (P - \overline{P})A]e^{At}$$

$$= \frac{d}{dt}[e^{A^T t}(P - \overline{P})e^{At}]$$

for all $t \geq 0$. Integrating this result from 0 to ∞ produces

$$0 = \int_0^\infty \frac{d}{dt}[e^{A^T t}(P - \overline{P})e^{At}]dt$$

$$= e^{A^T t}(P - \overline{P})e^{At} \big|_0^\infty$$

$$= -(P - \overline{P})$$

Thus $\overline{P} = P$, so P is the only, hence unique, solution to the Lyapunov matrix equation (6.5). $\square$

Example 6.2   We investigate the solution to the Lyapunov matrix equation for the system dynamics matrix

$$A = \begin{bmatrix} 0 & 1 \\ -2 & -3 \end{bmatrix}$$

In general, the Lyapunov matrix equation effectively represents $(n^2 + n)/2$ equations in as many unknowns because P, Q, and $A^T P + PA$ are each symmetric. We take $Q = I$ for simplicity (obviously a symmetric positive-definite matrix) and proceed to solve the Lyapunov matrix equation for

$$P = \begin{bmatrix} p_{11} & p_{12} \\ p_{12} & p_{22} \end{bmatrix} = P^T$$

We then assess whether or not it is positive definite. Direct substitutions give

$$\begin{bmatrix} 0 & -2 \\ 1 & -3 \end{bmatrix} \begin{bmatrix} p_{11} & p_{12} \\ p_{12} & p_{22} \end{bmatrix} + \begin{bmatrix} p_{11} & p_{12} \\ p_{12} & p_{22} \end{bmatrix} \begin{bmatrix} 0 & 1 \\ -2 & -3 \end{bmatrix}$$

$$= \begin{bmatrix} -2p_{12} & -2p_{22} \\ p_{11} - 3p_{12} & p_{12} - 3p_{22} \end{bmatrix} + \begin{bmatrix} -2p_{12} & p_{11} - 3p_{12} \\ -2p_{22} & p_{12} - 3p_{22} \end{bmatrix}$$

$$= \begin{bmatrix} -4p_{12} & p_{11} - 3p_{12} - 2p_{22} \\ p_{11} - 3p_{12} - 2p_{22} & 2p_{12} - 6p_{22} \end{bmatrix}$$

$$= \begin{bmatrix} -1 & 0 \\ 0 & -1 \end{bmatrix}$$

The (1,2) and (2,1) elements of either side yield the same equation which is why for $n = 2$ we extract $(n^2 + n)/2 = 3$ equations in the three unknowns p_{11}, p_{12}, and p_{22} and repackage as

$$\begin{bmatrix} 0 & -4 & 0 \\ 1 & -3 & -2 \\ 0 & 2 & -6 \end{bmatrix} \begin{bmatrix} p_{11} \\ p_{12} \\ p_{22} \end{bmatrix} = \begin{bmatrix} -1 \\ 0 \\ -1 \end{bmatrix}$$

The 3×3 coefficient on the left is nonsingular, so

$$\begin{bmatrix} p_{11} \\ p_{12} \\ p_{22} \end{bmatrix} = \begin{bmatrix} 0 & -4 & 0 \\ 1 & -3 & -2 \\ 0 & 2 & -6 \end{bmatrix}^{-1} \begin{bmatrix} -1 \\ 0 \\ -1 \end{bmatrix} = \begin{bmatrix} 1.25 \\ 0.25 \\ 0.25 \end{bmatrix} \quad \text{and}$$

$$P = \begin{bmatrix} p_{11} & p_{12} \\ p_{12} & p_{22} \end{bmatrix}$$

$$= \begin{bmatrix} 1.25 & 0.25 \\ 0.25 & 0.25 \end{bmatrix}$$

Now we employ Sylvester's criterion to test for the positive definiteness of P: We must check the leading two principal minors

$$|p_{11}| = +1.25 \text{ and } \begin{vmatrix} p_{11} & p_{12} \\ p_{12} & p_{22} \end{vmatrix} = +0.25$$

each of which is positive, so P is positive definite. Hence A must have negative real-part eigenvalues and, by Theorem 6.4, define an asymptotically stable homogeneous linear state equation.

The characteristic polynomial A is $\lambda^2 + 3\lambda + 2 = (\lambda + 1)(\lambda + 2)$ from which the eigenvalues are $\lambda_{1,2} = -1, -2$ each of which is real and negative. $\qquad\square$

Exponential Stability

We have demonstrated previously that a nonlinear state equation can have an asymptotically stable equilibrium state that is not exponentially stable. In contrast, asymptotic stability and exponential stability are equivalent for the linear time-invariant case. This fact is not terribly surprising given that when the matrix A has negative real-part eigenvalues, elements of the matrix exponential tend to zero with an exponential rate of decay. Here we apply the preceding Lyapunov analysis to derive an explicit exponential bound on the norm of the state trajectory in Definition 6.2.

Suppose that A has negative real-part eigenvalues so that with $Q = I$, the Lyapunov matrix equation

$$A^T P + PA = -Q$$

has a unique symmetric positive-definite solution P. With $V(x) = x^T P x$ we have

$$\dot{V}(x) = x^T(A^T P + PA)x = -x^T x$$

For any initial state $x(0) = x_0$, the time derivative of $V[x(t)] = x^T(t) Px(t)$ along the resulting trajectory is

$$\frac{d}{dt} V[x(t)] = \dot{V}[x(t)]$$

$$= -x(t)^T x(t)$$

$$\leq -\frac{1}{\lambda_{\max}(P)} x(t)^T P x(t)$$

$$= -\frac{1}{\lambda_{\max}(P)} V[x(t)]$$

where we have used the Rayliegh-Ritz inequality. This differential inequality can be used to bound $V[x(t)]$ as follows: Define

$$w(t) = \frac{d}{dt} V[x(t)] + \frac{1}{\lambda_{max}(P)} V[x(t)]$$

from which it follows that $w(t) \le 0$ for all $t \ge 0$. The scalar linear ordinary differential equation

$$\frac{d}{dt} V[x(t)] = -\frac{1}{\lambda_{max}(P)} V[x(t)] + w(t)$$

has the unique solution

$$V[x(t)] = e^{-\frac{1}{\lambda_{max}(P)}t} V(x_0) + \int_0^t e^{-\frac{1}{\lambda_{max}(P)}(t-\tau)} w(\tau)d\tau$$

For any $t \ge 0$, the integral term is nonpositive because the exponential term in the integrand is strictly positive and $w(\tau) \le 0$. Thus

$$V[x(t)] \le e^{-\frac{1}{\lambda_{max}(P)}t} V(x_0) \qquad \text{for all} \qquad t \ge 0$$

This, together with another application of the Rayliegh-Ritz inequality, yields

$$\lambda_{min}(P)x^T(t)x(t) \le V[x(t)] \le e^{-\frac{1}{\lambda_{max}(P)}t} V(x_0)$$

$$\le \lambda_{max}(P)e^{-\frac{1}{\lambda_{max}(P)}t} x_0^T x_0$$

Dividing through by $\lambda_{min}(P) > 0$ and taking square roots gives, using $\|x\| = \sqrt{x^T x}$,

$$\|x(t)\| \le \sqrt{\frac{\lambda_{max}(P)}{\lambda_{min}(P)}} e^{-\frac{1}{2\lambda_{max}(P)}t} \|x_0\| \qquad \text{for all} \qquad t \ge 0$$

This exponentially decaying bound is of the form given in Definition 6.2 with

$$k = \sqrt{\frac{\lambda_{max}(P)}{\lambda_{min}(P)}} \quad \text{and} \quad \lambda = \frac{1}{2\lambda_{max}(P)}$$

6.2 BOUNDED-INPUT, BOUNDED-OUTPUT STABILITY

Thus far in this chapter, we have been concerned with internal stability. This section discusses a type of external stability called *bounded-input, bounded-output stability*. As mentioned at the outset, bounded-input, bounded-output stability pertains to the zero-state output response. In this section we study bounded-input, bounded-output stability, and in the next section we relate asymptotic and bounded-input, bounded-output stability.

A vector-valued signal $u(t)$ is bounded if there is a finite, positive constant ν for which

$$\|u(t)\| \leq \nu \qquad \text{for all} \qquad t \geq 0$$

If such an upper bound exits, we denote the least upper bound, or *supremum*, by

$$\sup_{t \geq 0} \|u(t)\|$$

When $\|u(t)\|$ cannot be bounded above in this way, we write

$$\sup_{t \geq 0} \|u(t)\| = \infty$$

The supremum of $\|u(t)\|$ over the infinite interval $[0, \infty)$ is in general different from a maximum value because the latter must be achieved at some $t \in [0, \infty)$ and therefore may not exist. For example, consider the bounded scalar signal $u(t) = 1 - e^{-t}$, for which
$$\sup_{t \geq 0} \|u(t)\| = \sup_{t \geq 0} (1 - e^{-t}) = 1$$

but a maximum value for $\|u(t)\|$ is never attained on $t \in [0, \infty)$.

Definition 6.5 *The linear state equation* (6.1) *is called* **bounded-input, bounded-output stable** *if there exists a finite constant η such that for any input $u(t)$ the zero-state output response satisfies*

$$\sup_{t \geq 0} \|y(t)\| \leq \eta \sup_{t \geq 0} \|u(t)\|$$

This definition is not terribly useful as a test for bounded-input, bounded-output stability because it requires an exhaustive search over all bounded input signals, which is, of course, impossible. The next result is of interest because it establishes a necessary and sufficient test for bounded-input, bounded-output stability involving the system's impulse response that in principle requires a single computation. The theorem is cast in the multiple-input, multiple-output context and, as such, involves a choice of vector norm for both the input space $\mathbb{R}^m$ and the output space $\mathbb{R}^p$ and a corresponding induced matrix norm on the set of $p \times m$ matrices (see Appendix B, Section 9).

Theorem 6.6   *The linear state equation* (6.1) *is bounded-input, bounded-output stable if and only if the impulse response matrix* $H(t) = Ce^{At}B + D\delta(t)$ *satisfies*

$$\int_0^\infty \|H(\tau)\| d\tau < \infty$$

Proof.   To show sufficiency, suppose that this integral is finite, and set

$$\eta = \int_0^\infty \|H(\tau)\| d\tau$$

Then, for all $t \geq 0$, the zero-state output response satisfies

$$\|y(t)\| = \|\int_0^t H(\tau)u(t-\tau)d\tau\|$$

$$\leq \int_0^t \|H(\tau)u(t-\tau)\| d\tau$$

$$\leq \int_0^t \|H(\tau)\| \|u(t-\tau)\| d\tau$$

$$\leq \int_0^t \|H(\tau)\| d\tau \sup_{0 \leq \sigma \leq t} \|u(\sigma)\|$$

$$\leq \int_0^\infty \|H(\tau)\| d\tau \sup_{t \geq 0} \|u(t)\|$$

$$= \eta \sup_{t \geq 0} \|u(t)\|$$

from which the bound in Definition 6.5 follows from the definition of supremum. Since the input signal was arbitrary, we conclude that the system (6.1) is bounded-input, bounded-output stable.

To show that bounded-input, bounded-output stability implies

$$\int_0^\infty \|H(\tau)\| d\tau < \infty$$

we prove the contrapositive. Assume that for any finite $\eta > 0$ there exists a $T > 0$ such that

$$\int_0^T \|H(\tau)\| d\tau > \eta$$

It follows that there exists an element $H_{ij}(\tau)$ of $H(\tau)$ and a $T_{ij} > 0$ such that

$$\int_0^{T_{ij}} |H_{ij}(\tau)| \, d\tau > \eta$$

Consider the bounded input defined by $u(t) \equiv 0 \in \mathbb{R}^>$ for all $t > T_{ij}$, and on the interval $[0, T_{ij}]$ every component of $u(t)$ is zero except for $u_j(t)$, which is set to

$$u_j(t) = \begin{cases} -1 & H_{ij}(T_{ij} - t) > 0 \\ 0 & H_{ij}(T_{ij} - t) = 0 \\ 1 & H_{ij}(T_{ij} - t) < 0 \end{cases}$$

Then $\|u(t)\| \le 1$ for all $t \ge 0$, but the ith component of the zero-state output response satisfies

$$\begin{aligned} y_i(T_{ij}) &= \int_0^{T_{ij}} H_{ij}(T_{ij} - \sigma) u_j(\sigma) d\sigma \\ &= \int_0^{T_{ij}} |H_{ij}(T_{ij} - \sigma)| d\sigma \\ &= \int_0^{T_{ij}} |H_{ij}(\tau)| d\tau \\ &> \eta \\ &\ge \eta \sup_{t \ge 0} \|u(t)\| \end{aligned}$$

Since $\|y(T_{ij})\| \ge |y_i(T_{ij})|$ and η was arbitrary, we conclude that the system (6.1) is not bounded-input, bounded-output stable. $\qquad\square$

6.3 BOUNDED-INPUT, BOUNDED-OUTPUT STABILITY VERSUS ASYMPTOTIC STABILITY

It is reasonable to expect, since elements of the impulse-response matrix involve linear combinations of elements of the matrix exponential via premultiplication by C and postmultiplication by B, that asymptotic stability implies bounded-input, bounded-output stability. This indeed is true. However, the converse in general is not true as illustrated by the following example.

Example 6.3 Consider the following two-dimensional state equation:

$$\begin{bmatrix} \dot{x}_1(t) \\ \dot{x}_2(t) \end{bmatrix} = \begin{bmatrix} 0 & 1 \\ 1 & 0 \end{bmatrix} \begin{bmatrix} x_1(t) \\ x_2(t) \end{bmatrix} + \begin{bmatrix} -1 \\ 1 \end{bmatrix} u(t)$$

$$y(t) = \begin{bmatrix} 0 & 1 \end{bmatrix} \begin{bmatrix} x_1(t) \\ x_2(t) \end{bmatrix}$$

The characteristic polynomial is

$$|sI - A| = \begin{vmatrix} s & -1 \\ -1 & s \end{vmatrix}$$

$$= s^2 - 1$$

$$= (s + 1)(s - 1)$$

indicating that the eigenvalues of A are $\lambda_{1,2} = -1, +1$ and according to Theorem 6.3, the state equation is not asymptotically stable as a result. This can be seen by inspection of the matrix exponential

$$e^{At} = \begin{bmatrix} \dfrac{1}{2}(e^t + e^{-t}) & \dfrac{1}{2}(e^t - e^{-t}) \\ \dfrac{1}{2}(e^t - e^{-t}) & \dfrac{1}{2}(e^t + e^{-t}) \end{bmatrix}$$

The growing exponential term e^t associated with the positive eigenvalue causes every element of e^{At} to diverge as t increases. The transfer function is

$$H(s) = C(sI - A)^{-1} B$$

$$= \begin{bmatrix} 0 & 1 \end{bmatrix} \begin{bmatrix} s & -1 \\ -1 & s \end{bmatrix}^{-1} \begin{bmatrix} -1 \\ 1 \end{bmatrix}$$

$$= \begin{bmatrix} 0 & 1 \end{bmatrix} \frac{\begin{bmatrix} s & 1 \\ 1 & s \end{bmatrix}}{s^2 - 1} \begin{bmatrix} -1 \\ 1 \end{bmatrix}$$

$$= \frac{s - 1}{(s + 1)(s - 1)}$$

$$= \frac{1}{(s + 1)}$$

from which the impulse response $h(t) = e^{-t}$, $t \geq 0$, satisfies

$$\int_0^\infty |h(\tau)| d\tau = 1$$

so, by Theorem 6.6, the system is bounded-input, bounded-output stable.

In this example, we observe that the state equation is not minimal (it is observable but not controllable). Thus the transfer function $H(s)$ necessarily must have a pole-zero cancellation. In this case, the unstable pole at $s = 1$ is canceled by a zero at the same location, yielding a first-order transfer function with a stable pole at $s = -1$. In the time domain, the unstable exponential term e^t that appears in the matrix exponential is missing from the impulse response. This unstable pole-zero cancellation therefore leads to a bounded-input, bounded-output-stable state equation that is not asymptotically stable. If we are concerned only with input-output behavior, we might be tempted to think that bounded-input, bounded-output stability is good enough. The problem lies in the fact that bounded-input, bounded-output stability characterizes only the zero-state response, and it may happen that an initial state yields a zero-input response component resulting in a complete output response that does not remain bounded. For this example, the initial state $x(0) = [1, 1]^T$ yields the zero-input response component $y_{zi}(t) = e^t$, so for, say, a unit step input, the complete response is

$$
\begin{aligned}
y(t) &= y_{zi}(t) + y_{zs}(t) \\
&= e^t + (1 - e^{-t})
\end{aligned}
\qquad t \geq 0
$$

that, despite the bounded zero-state response component, diverges with increasing t. □

In the single-input, single-output case, we see that the situation encountered in the preceding example can be avoided by disallowing pole-zero cancellations of any type, i.e., by requiring an irreducible transfer function associated with the given state equation. This, as we have seen, is equivalent to minimality of the realization. It turns out that while the transfer-function interpretation is difficult to extend to the multiple-input, multiple-output case, the minimality condition remains sufficient for bounded-input, bounded-output stability to imply asymptotic stability.

Theorem 6.7   *For the linear state equation* (6.1):

1. Asymptotic stability always implies bounded-input, bounded-output stability.

2. If Equation (6.1) is minimal, then bounded-input, bounded-output stability implies asymptotic stability.

Proof.   As argued previously, asymptotic stability is equivalent to exponential stability. Thus there exist finite positive constants k and λ such that

$$\|e^{At}x(0)\| \le ke^{-\lambda t}\|x(0)\| \qquad \text{for all} \qquad x(0) \in \mathbb{R}^n$$

Thus, for any $x(0) \ne 0$,

$$\frac{\|e^{At}x(0)\|}{\|x(0)\|} \le ke^{-\lambda t}$$

thereby establishing an upper bound on the left-hand-side ratio for each $t \ge 0$. By definition of supremum, the induced matrix norm therefore has the bound

$$\|e^{At}\| = \sup_{x(0)\ne 0}\frac{\|e^{At}x(0)\|}{\|x(0)\|} \le ke^{-\lambda t} \qquad \text{for all} \qquad t \ge 0$$

Using this, familiar bounding arguments give

$$\int_0^\infty \|H(\tau)\|d\tau \le \int_0^\infty \left[\|Ce^{A\tau}B\| + \|D\delta(t)\|\right]d\tau$$

$$\le \|C\|\|B\|\int_0^\infty ke^{-\lambda \tau}d\tau + \|D\|$$

$$= \frac{k\|C\|\|B\|}{\lambda} + \|D\|$$

$$\le \infty$$

and so state equation (6.1) is bounded-input, bounded-output stable by Theorem 6.6.

Next, we show via a contradiction argument that under the minimality hypothesis, bounded-input, bounded-output stability implies asymptotic stability. Suppose that Equation (6.1) is bounded-input, bounded-output stable and yet there is an eigenvalue λ_i of A with $\text{Re}(\lambda_i) \ge 0$. Let m_i denote the associated algebraic multiplicity so that, based on the partial fraction expansion of $(sI - A)^{-1}$, we see that e^{At} will contain terms of the form

$$R_{i1}e^{\lambda_i t} + R_{i2}te^{\lambda_i t} + \cdots + R_{im_i}\frac{t^{m_i-1}}{(m_i - 1)!}e^{\lambda_i t}$$

in which $R_{im_i} \ne 0$ because of the assumed algebraic multiplicity, along with similar terms associated with the other distinct eigenvalues of A. Bounded-input, bounded-output stability implies, via Theorem 6.6, that

$$\lim_{t\to\infty} Ce^{At}B = 0 \in \mathbb{R}^{p\times m}$$

and since exponential time functions associated with distinct eigenvalues are linearly independent, the only way the nondecaying terms in e^{At} will be absent in $Ce^{At}B$ is if

$$CR_{ij}B = 0 \qquad j = 1, 2, \ldots, m_i$$

In particular,

$$
\begin{aligned}
0 &= CR_{im_i}B \\
&= C[(s - \lambda_i)^{m_i}(sI - A)^{-1}]|_{s=\lambda_i}B \\
&= [(s - \lambda_i)^{m_i}C(sI - A)^{-1}B]|_{s=\lambda_i}
\end{aligned}
$$

This allows us to conclude that

$$
\begin{aligned}
&\begin{bmatrix} C \\ \lambda_i I - A \end{bmatrix} R_{im_i}[\, B \quad \lambda_i I - A \,] \\
&= \begin{bmatrix} C \\ sI - A \end{bmatrix} [(s - \lambda_i)^{m_i}(sI - A)^{-1}][\, B \quad sI - A \,]\Bigg|_{s=\lambda_i} \\
&= \begin{bmatrix} (s - \lambda_i)_i^m C(sI - A)^{-1}B & (s - \lambda_i)^{m_i}C \\ (s - \lambda_i)^{m_i}B & (s - \lambda_i)^{m_i}(sI - A) \end{bmatrix}\Bigg|_{s=\lambda_i} \\
&= \begin{bmatrix} 0 & 0 \\ 0 & 0 \end{bmatrix}
\end{aligned}
$$

Now, minimality of the linear state equation (6.1) implies that the leftmost factor has full-column rank n by virtue of the Popov-Belevich-Hautus rank rest for observability. Consequently, it is possible to select n linearly independent rows from the leftmost factor to give a nonsingular matrix M_O. Similarly, that the rightmost factor has full-row rank n by the Popov-Belevich-Hautus rank rest for controllability, and it is possible to select n linearly independent columns from the rightmost factor to yield a nonsingular matrix M_C. This corresponds to selecting n rows and n columns from the left-hand-side product. An identical selection from the right hand side yields the identity

$$M_O R_{im_i} M_C = 0$$

which, because of the nonsingularity of M_O and M_C, implies that $R_{im_i} = 0$. This contradicts $R_{im_i} \neq 0$, which enables us to conclude that under the minimality hypothesis, bounded-input, bounded-output stability implies asymptotic stability. $\qquad\square$

6.4 MATLAB **FOR STABILITY ANALYSIS**

The following MATLAB function is useful for Lyapunov stability analysis:
> **lyap(A',Q)** Solve **A'P+PA=-Q** for matrix **P**, given a positive-definite matrix **Q**.

Note that the MATLAB function **lyap** directly solves the problem **AP+PA'=-Q**, so we must give the transpose of matrix **A** (that is, **A'**) as an input to this function.

We now assess the stability of the Continuing MATLAB Example (rotational mechanical system) via Lyapunov stability analysis. The following MATLAB code segment performs this computation and logic:

```
%---------------------------------------------------------
% Chapter 6. Lyapunov Stability Analysis
%---------------------------------------------------------

if (real(Poles(1))==0 | real(Poles(2))==0) % lyap
                                            % will fail
  if (real(Poles(1)) < =0 | real(Poles(2)) < =0)
    disp('System is marginally stable.');
  else
    disp('System is unstable.');
  end
else                        % lyap will succeed
    Q = eye(2);             % Given positive definite
                            % matrix
    P = lyap(A',Q);         % Solve for P
    pm1 = det(P(1,1));      % Sylvester's method to see
                            % if P is positive definite
    pm2 = det(P(1:2,1:2));
    if (pm1>0 & pm2>0)      % Logic to assess stability
                            % condition
    disp('System is asymptotically stable.');
  else
    disp('System is unstable.');
  end
end

figure;                     % Plot phase portraits to
                            % enforce stability analysis
plot(Xo(:,1),Xo(:,2),'k'); grid; axis('square');
  axis([-1.5 1.5 -2 1]);
```

```
set(gca,'FontSize',18);
xlabel('\itx_1 (rad)'); ylabel('\itx_2 (rad/s)');
```

This code segment yields the following output plus the phase-portrait plot of Figure 6.7:

```
R =
    5.1750 0.0125
    0.0125 0.1281

pm1 =   5.1750
pm2 =   0.6628

System is asymptotically stable.
```

Figure 6.7 plots velocity $x_2(t)$ versus displacement $x_1(t)$. Since this is an asymptotically stable system, the phase portrait spirals in from the given initial state $x(0) = [0.4, 0.2]^{\mathrm{T}}$ to the system equilibrium state $\tilde{x} = [0, 0]^{\mathrm{T}}$. This is another view of the state variable responses shown in Figure 2.3.

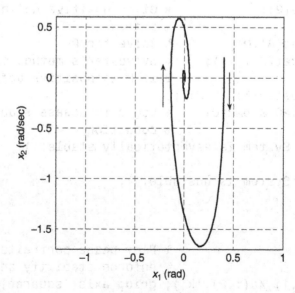

FIGURE 6.7   Phase-portrait for the Continuing MATLAB Example.

6.5   CONTINUING EXAMPLES: STABILITY ANALYSIS

Continuing Example 1: Two-Mass Translational Mechanical System

Here we assess stability properties of the system of Continuing Example 1 (two-mass translational mechanical system). Asymptotic stability is a fundamental property of a system; it is only dependent on the system dynamics matrix A and not on matrices B, C, or D. Asymptotic is therefore independent of the various possible combinations of input and output choices. Therefore, in this section there is only one stability analysis; it is the same for case a (multiple-input, multiple-output), case b [single input $u_2(t)$ and single output $y_1(t)$], and all other possible input-output combinations. However, we will employ two methods to get the same result, eigenvalue analysis and Lyapunov stability analysis.

Eigenvalue Analysis   The four open-loop system eigenvalues for Continuing Example 1, found from the eigenvalues of A, are $s_{1,2} = -0.5 \pm 4.44i$ and $s_{3,4} = -0.125 \pm 2.23i$. Thus this open-loop system is asymptotically stable because the real parts of the four eigenvalues are strictly negative.

Lyapunov Analysis   The Lyapunov matrix equation is given in Equation (6.5). The stability analysis procedure is as follows: For a given-positive definite matrix Q (I_4 is a good choice), solve for P in this equation. If P turns out to be positive definite, then the system is asymptotically stable. The solution for P is

$$P = \begin{bmatrix} 15.76 & 0.29 & 1.93 & 0.38 \\ 0.29 & 1.78 & -0.19 & 1.09 \\ 1.93 & -0.19 & 9.16 & -0.04 \\ 0.38 & 1.09 & -0.04 & 1.46 \end{bmatrix}$$

Now we must check the positive definiteness of P using Sylvester's criterion. The four principal minors of P are the following four submatrix determinants:

$$|15.76| \qquad \begin{vmatrix} 15.76 & 0.29 \\ 0.29 & 1.78 \end{vmatrix}$$

$$\begin{vmatrix} 15.76 & 0.29 & 1.93 \\ 0.29 & 1.78 & -0.19 \\ 1.93 & -0.19 & 9.16 \end{vmatrix} \qquad \begin{vmatrix} 15.76 & 0.29 & 1.93 & 0.38 \\ 0.29 & 1.78 & -0.19 & 1.09 \\ 1.93 & -0.19 & 9.16 & -0.04 \\ 0.38 & 1.09 & -0.04 & 1.46 \end{vmatrix}$$

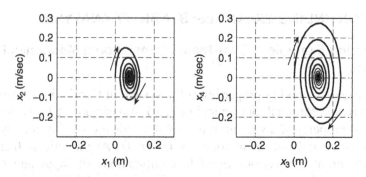

FIGURE 6.8 Phase portraits for Continuing Example 1, case *a*.

These determinants evaluate to 15.76, 27.96, 248.58, and 194.88, respectively. All four principal minors are positive, and therefore, P is positive definite. Therefore, this system is asymptotically stable, and consequently, bounded-input, bounded-output stable.

To reinforce these stability results, Figure 6.8 presents the phase portraits for Continuing Example 1, case a, with two step inputs of 20 and 10 N, respectively, and zero initial conditions.

Figure 6.8 plots velocity $x_2(t)$ versus displacement $x_1(t)$ on the left and velocity $x_4(t)$ versus displacement $x_3(t)$ on the right. Since this is an asymptotically stable system, the phase portraits both spiral in from zero initial conditions on all state variables to the steady-state state vector $x_{ss} = [0.075, 0, 0.125, 0]^T$. Since each plot in Figure 6.8 is both plotted on the same scale, we see that mass 2 undergoes higher amplitude displacement and velocity motions than mass 1.

To further reinforce the stability results, Figure 6.9 presents the phase portraits for Continuing Example 1, case b, with initial conditions $x(0) = [0.1, 0, 0.2, 0]^T$ and zero input.

Figure 6.9 again plots velocity $x_2(t)$ versus displacement $x_1(t)$ on the left and velocity $x_4(t)$ versus displacement $x_3(t)$ on the right. Since this is an asymptotically stable system, the phase portraits both spiral in from the given nonzero initial displacements (initial velocities are zero) to the zero equilibrium values for all state variables. Since its initial displacement is double that of mass 1, we see that mass 2 undergoes higher amplitude displacement and velocity motions than mass 1.

Continuing Example 2: Rotational Electromechanical System

Now we assess stability properties of the system of Continuing Example 2 (rotational electromechanical system). We will attempt to employ two methods, eigenvalue analysis and Lyapunov stability analysis.

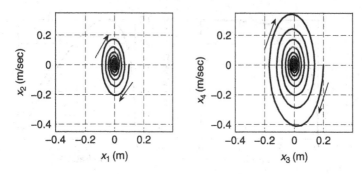

FIGURE 6.9 Phase Portraits for Continuing Example 1, case *b*.

Eigenvalue Analysis If all real parts of all system eigenvalues are strictly negative, the system is stable. If just one real part of an eigenvalue is zero (and the real parts of the remaining system eigenvalues are zero or strictly negative), the system is marginally stable. If just one real part of an eigenvalue is positive (regardless of the real parts of the remaining system eigenvalues), the system is unstable.

From Chapter 2, the three open-loop system eigenvalues for Continuing Example 2, found from the eigenvalues of A, are $s_{1,2,3} = 0, -1, -2$. Thus this open-loop system is marginally stable, i.e., stable but not asymptotically stable, because there is a nonrepeated zero eigenvalue, and the rest are real and negative. This system is not bounded-input, bounded-output stable because when a constant voltage is applied, the output shaft angle will increase linearly without bound in steady-state motion. This does not pose a problem as this is how a DC servomotor is supposed to behave.

Lyapunov Analysis MATLAB cannot solve the Lyapunov matrix equation (the MATLAB error is: *Solution does not exist or is not unique*) because of the zero eigenvalue in the system dynamics matrix A. This indicates that the system is not asymptotically stable.

These marginal-stability results can be further demonstrated by the phase portrait plots of Figure 6.10. Figures 6.10 plot motor shaft angular velocity $x_2(t)$ versus angular displacement $x_1(t)$ on the left and angular acceleration $x_3(t)$ versus angular velocity $x_2(t)$ on the right. The angular displacement $x_1(t)$ grows linearly without bound as the angular velocity $x_2(t)$ approaches a constant steady-state value of 1 rad/s, so the phase portrait on the left diverges from the origin parallel to the horizontal axis. In addition, the angular acceleration $x_3(t)$ approaches a constant steady-state value of 0 rad/s^2, so the phase portrait on the right converges to the point $[1, 0]^T$.

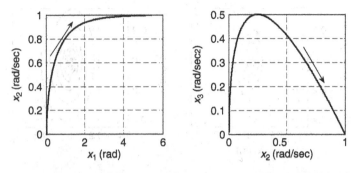

FIGURE 6.10 Phase portraits for Continuing Example 2.

6.6 HOMEWORK EXERCISES

Numerical Exercises

NE6.1 Assess the stability properties of the following systems, represented by the given system dynamics matrices A. Use the eigenvalue test for stability.

a. $A = \begin{bmatrix} 0 & 1 \\ -14 & -4 \end{bmatrix}$

b. $A = \begin{bmatrix} 0 & 1 \\ -14 & 4 \end{bmatrix}$

c. $A = \begin{bmatrix} 0 & 1 \\ 0 & -4 \end{bmatrix}$

d. $A = \begin{bmatrix} 0 & 1 \\ -14 & 0 \end{bmatrix}$

NE6.2 Repeat NE6.1 using an energy-based approach with phase portraits.

NE6.3 Repeat NE6.1 using Lyapunov stability analysis. Will the calculations work in each case? Why or why not?

NE6.4 For the single-input, single output system with transfer function

$$H(s) = \frac{s^2 - s - 2}{s^3 + 2s^2 - 4s - 8}$$

a. Is the system *bounded-input, bounded-output* stable?

b. Obtain a realization in controller canonical form. Is this realization observable? Is this realization asymptotically stable?

c. Find a minimal realization of $H(s)$.

NE6.5 For the single-input, single output system with transfer function

$$H(s) = \frac{s^2 + s - 2}{s^3 + 2s^2 - 4s - 8}$$

a. Is the system *bounded-input, bounded-output* stable?

b. Obtain a realization in observer canonical form. Is this realization controllable? Is this realization asymptotically stable?

c. Find a minimal realization of $H(s)$.

Analytical Exercises

AE6.1 If $A = -A^T$, show that the homogeneous linear state equation

$$\dot{x}(t) = Ax(t)$$

is stable but not asymptotically stable.

AE6.2 Given matrices A and Q, let P satisfy the Lyapunov matrix equation

$$A^T P + P A = -Q$$

Show that for all $t \geq 0$,

$$P = e^{A^T t} Q e^{At} + \int_0^t e^{A^T \tau} Q e^{A \tau} d\tau$$

AE6.3 Show that the eigenvalues of A have real part less than $-\mu$ if and only if for every symmetric positive-definite matrix Q there exists a unique symmetric positive-definite solution to

$$A^T P + P A + 2\mu P = -Q$$

AE6.4 Suppose that A has negative real-part eigenvalues, and let P denote the unique symmetric-positive definite solution to the Lyapunov matrix equation

$$A^T P + P A = -I$$

Show that the perturbed homogeneous linear state equation

$$\dot{x}(t) = (A + \Delta A)x(t)$$

is asymptotically stable if the perturbation matrix satisfies the spectral norm bound

$$\|\Delta A\| < \frac{1}{2\lambda_{\max}(P)}$$

AE6.5 Suppose that the pair (A, B) is controllable and that A has negative real-part eigenvalues. Show that the Lyapunov matrix equation

$$AW + WA = -BB^T$$

has a symmetric positive definite solution W.

AE6.6 Suppose that the pair (A, B) is controllable and that the Lyapunov matrix equation

$$AW + WA = -BB^T$$

has a symmetric positive-definite solution W. Show that A has negative real-part eigenvalues.

AE6.7 Consider a bounded-input, bounded-output stable single-input, single-output state equation with transfer function $H(s)$. For positive constants λ and μ, show that the zero-state response $y(t)$ to the input $u(t) = e^{-\lambda t}, t \geq 0$ satisfies

$$\int_0^\infty y(t)e^{-\mu t}dt = \frac{1}{\lambda + \mu}H(\mu)$$

Under what conditions can this relationship hold if the state equation is *not* bounded-input, bounded-output stable?

Continuing MATLAB Exercises

CME6.1 For the system given in CME1.1:
   a. Assess system stability using Lyapunov analysis. Compare this result with eigenvalue analysis.
   b. Plot phase portraits to reinforce your stability results.

CME6.2 For the system given in CME1.2:
   a. Assess system stability using Lyapunov analysis. Compare this result with eigenvalue analysis.
   b. Plot phase portraits to reinforce your stability results.

CME6.3 For the system given in CME1.3:
   a. Assess system stability condition using Lyapunov analysis. Compare this result with eigenvalue analysis.
   b. Plot phase portraits to reinforce your stability results.

CME6.4 For the system given in CME1.4:

    a. Assess system stability condition using Lyapunov analysis. Compare this result with eigenvalue analysis.

    b. Plot phase portraits to reinforce your stability results.

Continuing Exercises

CE6.1 Using Lyapunov analysis, assess the stability properties of the CE1 system; any case will do—since the A matrix is identical for all input/output cases, the stability condition does not change. Check your results via eigenvalue analysis. Plot phase portraits to reinforce your results.

CE6.2 Using Lyapunov analysis, assess the stability properties of the CE2 system; any case will do—because the A matrix is identical for all input-output cases, stability does not change. Lyapunov stability analysis will not succeed (why?); therefore, assess system stability via eigenvalue analysis. Plot phase portraits to reinforce your results.

CE6.3 Using Lyapunov analysis, assess the stability properties of the CE3 system; either case will do—since the A matrix is identical for all input-output cases, stability does not change. Lyapunov stability analysis will not succeed (why?); therefore, assess system stability via eigenvalue analysis. Plot phase portraits to reinforce your results.

CE6.4 Using Lyapunov analysis, assess the stability properties of the CE4 system. Lyapunov stability analysis will not succeed (why?); therefore, assess system stability via eigenvalue analysis. Plot phase portraits to reinforce your results.

CE6.5 Using Lyapunov analysis, assess the stability properties of the CE5 system. Lyapunov stability analysis will not succeed (why?); therefore, assess system stability via eigenvalue analysis. Plot phase portraits to reinforce your results.

7

DESIGN OF LINEAR STATE FEEDBACK CONTROL LAWS

Previous chapters, by introducing fundamental state-space concepts and analysis tools, have now set the stage for our initial foray into state-space methods for control system design. In this chapter, our focus is on the design of state feedback control laws that yield desirable closed-loop performance in terms of both transient and steady-state response characteristics. The fundamental result that underpins much of this chapter is that controllability of the open-loop state equation is both necessary and sufficient to achieve arbitrary closed-loop eigenvalue placement via state feedback. Furthermore, explicit feedback gain formulas for eigenvalue placement are available in the single-input case. To support the design of state feedback control laws, we discuss important relationship between the eigenvalue locations of a linear state equation and its dynamic response characteristics.

For state equations that are not controllable and so arbitrary eigenvalue placement via state feedback is not possible, we investigate whether state feedback can be used to at least stabilize the closed-loop state equation. This leads to the concept of stabilizability. Following that, we discuss techniques for improving steady-state performance first by introducing an additional gain parameter in the state feedback law, followed by the incorporation of integral-error compensation into our state feedback structure.

Linear State-Space Control Systems, by Robert L. Williams II and Douglas A. Lawrence
Copyright © 2007 John Wiley & Sons, Inc.

The chapter concludes by illustrating the use of MATLAB for shaping the dynamic response and state feedback control law design in the context of our Continuing MATLAB Example and Continuing Examples 1 and 2.

7.1   STATE FEEDBACK CONTROL LAW

We begin this section with the linear time-invariant state equation

$$\dot{x}(t) = Ax(t) + Bu(t)$$
$$y(t) = Cx(t) \tag{7.1}$$

which represents the open-loop system or *plant* to be controlled. Our focus is on the application of state feedback control laws of the form

$$u(t) = -Kx(t) + r(t) \tag{7.2}$$

with the goal of achieving desired performance characteristics for the closed-loop state equation

$$\dot{x}(t) = (A - BK)x(t) + Br(t)$$
$$y(t) = Cx(t) \tag{7.3}$$

The effect of state feedback on the open-loop block diagram of Figure 1.1 is shown in Figure 7.1.

The state feedback control law (7.2) features a constant state feedback gain matrix K of dimension $m \times n$ and a new external reference input $r(t)$ necessarily having the same dimension $m \times 1$ as the open-loop input $u(t)$, as well as the same physical units. Later in this chapter we will modify

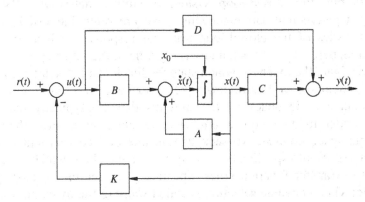

FIGURE 7.1   Closed-loop system block diagram.

the state feedback control law to include a gain matrix multiplying the reference input. The state feedback control law can be written in terms of scalar components as

$$
\begin{bmatrix} u_1(t) \\ u_2(t) \\ \vdots \\ u_m(t) \end{bmatrix} = - \begin{bmatrix} k_{11} & k_{12} & \cdots & k_{1n} \\ k_{21} & k_{22} & \cdots & k_{2n} \\ \vdots & \vdots & \ddots & \vdots \\ k_{m1} & k_{m2} & \cdots & k_{mn} \end{bmatrix} \begin{bmatrix} x_1(t) \\ x_2(t) \\ \vdots \\ x_n(t) \end{bmatrix} + \begin{bmatrix} r_1(t) \\ r_2(t) \\ \vdots \\ r_m(t) \end{bmatrix}
$$

For the single-input, single-output case, the feedback gain K is a $1 \times n$ row vector, the reference input $r(t)$ is a scalar signal, and the state feedback control law has the form

$$
u(t) = - \begin{bmatrix} k_1 & k_2 & \cdots & k_n \end{bmatrix} \begin{bmatrix} x_1(t) \\ x_2(t) \\ \vdots \\ x_2(t) \end{bmatrix} + r(t)
$$

$$
= -k_1 x_1(t) - k_2 x_2(t) - \cdots - k_n x_n(t) + r(t)
$$

If the external reference input is absent, the state feedback control law is called a *regulator* that is designed to deliver desirable transient response for nonzero initial conditions and/or attenuate disturbances to maintain the equilibrium state $\tilde{x} = 0$.

7.2 SHAPING THE DYNAMIC RESPONSE

In addition to closed-loop asymptotic stability (see Chapter 6), which requires that the closed-loop system dynamics matrix $A - BK$ have strictly negative real-part eigenvalues, we are often interested in other characteristics of the closed-loop transient response, such as rise time, peak time, percent overshoot, and settling time of the step response. Before we investigate the extent to which state feedback can influence the closed-loop eigenvalues, we first review topics associated with transient response performance of feedback control systems that are typically introduced in an undergraduate course emphasizing classical control theory. In our state-space context, we seek to translate desired transient response characteristics into specifications on system eigenvalues, which are closely related to transfer function poles. Specifying desired closed-loop system behavior via eigenvalue selection is called *shaping* the dynamic response. Control system engineers often use dominant first- and second-order subsystems as approximations in the design process, along with criteria that justify such approximations for higher-order systems. Our discussion follows this approach.

Eigenvalue Selection for First-Order Systems

Figure 7.2 shows unit step responses of typical of first- through fourth-order systems.

For a first-order system, we can achieve desired transient behavior via specifying a single eigenvalue. Figure 7.2 (*top left*) shows a standard first-order system step response. All stable first-order systems driven by unit step inputs behave this way, with transient response governed by a single decaying exponential involving the time constant τ. After three time constants, the first-order unit step response is within 95 percent of its steady-state value. A smaller time constant responds more quickly, whereas a larger time constant responds more slowly. On specifying a desired time constant, the associated characteristic polynomial and eigenvalue are

$$\lambda + \frac{1}{\tau} \quad \text{and} \quad \lambda_1 = -\frac{1}{\tau}$$

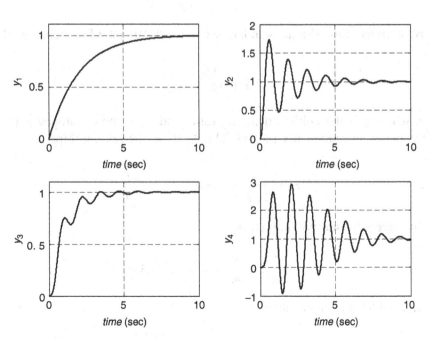

FIGURE 7.2 First- through fourth-order system unit step responses.

Eigenvalue Selection for Second-Order Systems

For a second-order system, we can achieve desired transient behavior via specifying a pair of eigenvalues. To illustrate, we consider the linear translational mechanical system of Example 1.1 (see Figure 1.2) with applied force $f(t)$ as the input and mass displacement $y(t)$ as the output. We identify this with a standard second-order system by redefining the input via $u(t) = f(t)/k$. The new input $u(t)$ can be interpreted as a commanded displacement. This will normalize steady-state value of the unit step response to 1.0. With this change, the state equation becomes

$$\begin{bmatrix} \dot{x}_1(t) \\ x_2(t) \end{bmatrix} = \begin{bmatrix} 0 & 1 \\ -\dfrac{k}{m} & -\dfrac{c}{m} \end{bmatrix} \begin{bmatrix} x_1(t) \\ x_2(t) \end{bmatrix} + \begin{bmatrix} 0 \\ \dfrac{k}{m} \end{bmatrix} u(t)$$

$$y(t) = \begin{bmatrix} 1 & 0 \end{bmatrix} \begin{bmatrix} x_1(t) \\ x_2(t) \end{bmatrix}$$

with associated transfer function

$$H(s) = \frac{\dfrac{k}{m}}{s^2 + \dfrac{c}{m}s + \dfrac{k}{m}}$$

We compare this with the standard second-order transfer function, namely,

$$\frac{\omega_n^2}{s^2 + 2\xi\omega_n + \omega_n^2}$$

in which ξ is the unitless damping ratio, and ω_n is the undamped natural frequency in radians per second. This leads to the relationships

$$\xi = \frac{c}{2\sqrt{km}} \quad \text{and} \quad \omega_n = \sqrt{\frac{k}{m}}$$

The characteristic polynomial is

$$\lambda^2 + \frac{c}{m}\lambda + \frac{k}{m} = \lambda^2 + 2\xi\omega_n\lambda + \omega_n^2$$

from which the eigenvalues are

$$\lambda_{1,2} = -\xi\omega_n \pm \omega_n\sqrt{\xi^2 - 1}$$

TABLE 7.1   Damping Ratio versus Step Response Characteristics

Case	Damping Ratio	Eigenvalues	Unit Step Response
Overdamped	$\xi > 1$	Real and distinct	Slowest transient response
Critically damped	$\xi = 1$	Real and equal	Fastest transient response without overshoot
Underdamped	$0 < \xi < 1$	Complex conjugate pair	Faster transient response but with overshoot and oscillation
Undamped	$\xi = 0$	Imaginary pair	Undamped oscillation
Unstable	$\xi < 0$	At least one with positive real part	Unbounded response

To study the relationship between these eigenvalues and system transient response, we identify five distinct cases in Table 7.1, determined by the dimensionless damping ratio ξ for a fixed undamped natural frequency ω_n.

We next relate step response characteristics to the system eigenvalues for the most interesting of these cases: the underdamped case characterized by $0 < \xi < 1$. In this case, the complex conjugate eigenvalues are given by

$$\lambda_{1,2} = -\xi\omega_n \pm j\omega_d$$

in which $\omega_d = \omega_n\sqrt{1 - \xi^2}$ is the damped natural frequency in radians per second. The unit step response for a standard second-order system in the underdamped case is

$$y(t) = 1 - \frac{e^{-\xi\omega_n t}}{\sqrt{1 - \xi^2}} \sin(\omega_d t + \theta)$$

in which the phase angle is given by $\theta = \cos^{-1}(\xi)$ and therefore is referred to as the *damping angle*. This response features a sinusoidal component governed by the damped natural frequency and damping angle that is damped by a decaying exponential envelope related to the negative real part of the eigenvalues. A response of this type is plotted in Figure 7.3.

For the underdamped case, there are four primary performance characteristics (see Figure 7.3) associated with the unit step response that either directly or approximately can be related to the damping ratio and undamped natural frequency. The performance characteristics definitions and formulas are found in Dorf and Bishop (2005). Rise time t_R is defined as the elapsed time between when the response first reaches 10 percent of

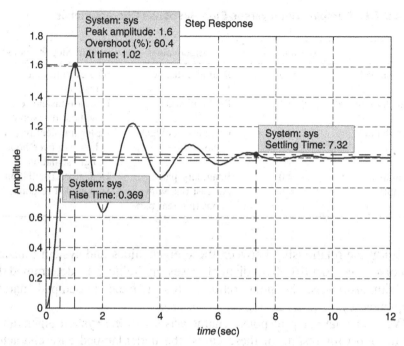

FIGURE 7.3   Unit step response for Example 7.1.

the steady-state value to when the response first reaches 90 percent of the steady-state value. For a damping ratio in the range $0.3 < \xi < 0.8$, rise time can be approximated by

$$t_R \cong \frac{2.16\xi + 0.60}{\omega_n}$$

Peak time t_P is the time at which the peak response value is reached and is given exactly by

$$t_P = \frac{\pi}{\omega_n\sqrt{1 - \xi^2}} = \frac{\pi}{\omega_d}$$

Percent overshoot PO characterizes the relationship between the peak value and steady-state value according to

$$PO = \frac{\text{peak value} - \text{steady-state value}}{\text{steady-state value}} \times 100\%$$

and can be computed exactly using

$$PO = 100e^{-\xi\pi/\sqrt{1-\xi^2}}$$

Settling time t_S is typically defined to be the time at which the response enters and remains within a ± 2 percent band about the steady-state value and can be approximated by

$$t_S \cong \frac{4}{\xi \omega_n}$$

The swiftness of the response is related to the rise time and peak time; the deviation between the response and its steady-state value is related to the percent overshoot and settling time.

Example 7.1   We return to the linear translational mechanical system for the parameter values $m = 1$ kg, $c = 1$ N-s/m, and $k = 10$ N/m. The undamped natural frequency and dimensionless damping ratio are

$$\omega_n = \sqrt{\frac{k}{m}} = \sqrt{\frac{10}{1}} = 3.16 \text{ rad/s}$$

$$\xi = \frac{c}{2\sqrt{km}} = \frac{1}{2\sqrt{10(1)}} = 0.158$$

and the damped natural frequency is

$$\omega_d = \omega_n \sqrt{1 - \xi^2} = 3.12 \text{ rad/s}$$

The characteristic polynomial is

$$\lambda^2 + 2\xi \omega_n \lambda + \omega_n^2 = \lambda^2 + \lambda + 10$$

yielding the complex conjugate eigenvalues

$$\lambda_{1,2} = -0.5 \pm 3.12i$$

The unit step response is given by

$$y(t) = 1 - 1.01e^{-0.5t} \sin(3.12t + 80.9°)$$

We calculate directly from ξ and ω_n the step response characteristics: $t_R = 0.30$ s (because $\xi < 0.3$ this estimate may be somewhat inaccurate), $t_P = 1.01$ s, $PO = 60.5$ percent, and $t_S = 8$ s.

A plot of the unit step response is given is Figure 7.3, with rise time, peak time, percent overshoot, and settling time displayed. This plot was obtained from MATLAB using the **step** command and right-clicking in the resulting figure window to add the performance specifications (Characteristics → choose Peak Response, Settling Time, Rise Time, and Steady State in turn. MATLAB will mark the response with a dot, which the user may click to display the numerical values). We see that with the exception of rise time (0.30 s estimated versus 0.37 s from the plot), the formula values agree well with the MATLAB results values labeled on Figure 7.3 (settling time is also inaccurate, 8 s from the equation versus 7.32 s as measured from the MATLAB generated response). □

Example 7.2 The unit step response in Figure 7.3 is typical of lightly damped systems such as large-scale space structures. We consider the application of a proportional-derivative (PD) control law given by

$$u(t) = -k_P y(t) - k_D \dot{y}(t) + Gr(t)$$

$$= -\begin{bmatrix} k_P & k_D \end{bmatrix} \begin{bmatrix} x_1(t) \\ x_2(t) \end{bmatrix} + Gr(t)$$

which, as shown, also has the form of a state feedback control law. The closed-loop system that results is given by

$$\begin{bmatrix} \dot{x}_1(t) \\ x_2(t) \end{bmatrix} = \begin{bmatrix} 0 & 1 \\ -\left(\dfrac{k}{m} + k_P\right) & -\left(\dfrac{c}{m} + k_D\right) \end{bmatrix} \begin{bmatrix} x_1(t) \\ x_2(t) \end{bmatrix} + \begin{bmatrix} 0 \\ \dfrac{k}{m} \end{bmatrix} Gr(t)$$

$$y(t) = \begin{bmatrix} 1 & 0 \end{bmatrix} \begin{bmatrix} x_1(t) \\ x_2(t) \end{bmatrix}$$

Our objective is to improve the step response characteristics by reducing the percent overshoot and settling time via adjusting the proportional and derivative gains k_P and k_D. In particular, we specify a desired percent overshoot of 4 percent and a settling time of 2 s. The gain G will be chosen so that the closed-loop unit step response will have the same steady-state value as the open-loop unit step response.

In terms of these performance specifications, we calculate the desired closed-loop damping ratio ξ' from the percent-overshoot formula according to

$$\xi' = \frac{\left| \ln\left(\dfrac{PO}{100}\right) \right|}{\sqrt{\pi^2 + \left[\ln\left(\dfrac{PO}{100}\right)\right]^2}} = \frac{\left| \ln\left(\dfrac{4}{100}\right) \right|}{\sqrt{\pi^2 + \left[\ln\left(\dfrac{4}{100}\right)\right]^2}} = 0.716$$

Using this, we calculate the desired closed-loop undamped natural frequency from the settling-time formula via

$$\omega'_n = \frac{4}{\xi' t_S} = 2.79 \text{ rad/s}$$

The desired closed-loop damped natural frequency is

$$\omega'_d = \omega'_n \sqrt{1 - (\xi')^2} = 1.95 \text{ rad/s}$$

The proportional and derivative gains then can be determined by comparing the closed-loop characteristic polynomial of the closed-loop state equation with the desired closed-loop characteristic polynomial:

$$\lambda^2 + (2\xi\omega_n + k_D)\lambda + (\omega_n^2 + k_P) = \lambda^2 + 2\xi'\omega'_n\lambda + {\omega'_n}^2 = \lambda^2 + 4\lambda + 7.81$$

which leads to

$$k_P = {\omega'_n}^2 - \omega_n^2 = -2.20 \quad k_D = 2\xi'\omega'_n - 2\xi\omega_n = 3.00$$

The input gain G is determined from the relationship

$$G = \frac{{\omega'_n}^2}{\omega_n^2} = 0.781$$

which yields the closed-loop transfer function

$$\frac{\frac{k}{m}G}{s^2 + (\frac{c}{m} + k_D)s + (\frac{k}{m} + k_P)} = \frac{\omega_n^2 G}{s^2 + (2\xi\omega_n + k_D)s + (\omega_n^2 + k_P)}$$

$$= \frac{{\omega'_n}^2}{s^2 + 2\xi'\omega'_n s + {\omega'_n}^2} = \frac{7.81}{s^2 + 4s + 7.81}$$

The closed-loop eigenvalues then are:

$$\lambda_{1,2} = -2 \pm j1.95$$

and the closed-loop unit step response is

$$y(t) = 1 - 1.43e^{-2t} \sin(1.95t + 44.3°)$$

The MATLAB **step** command gives us the four performance specifications: $t_R = 0.78$ s, $t_P = 1.60$ s, $PO = 4$ percent, and $t_S = 2.12$ s. The actual settling time differs slightly from the desired settling time because of the approximation in the settling-time formula. Figure 7.4 shows a

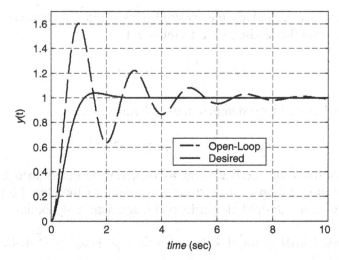

FIGURE 7.4 Closed-loop versus open-loop system unit step responses for Example 7.2.

comparison of the unit step response for the improved closed-loop system with the unit step response of the open-loop system from Figure 7.3.

In Figure 7.4 and Table 7.2 we see that the step response of the closed-loop system has a slower rise and peak time compared with the open-loop system, but the percent overshoot and settling time are much improved. Values given below for rise time and settling time are from MATLAB-generated step responses rather than the approximate formulas.   □

Inequality Constraints   When specifying desired transient response behavior, the control system engineer typically specifies bounds on the step response characteristics rather than exact values. For first-order systems, this usually involves an upper bound on the time required for the step response to reach steady state, which yields an upper bound for the desired time constant. For second-order systems, this typically involves

TABLE 7.2   Comparison of Open-Loop and Closed-Loop Step Response Characteristics

Specification	Open-Loop Response	Closed-Loop Response
t_R (sec)	0.37	0.78
t_P (sec)	1.02	1.60
PO (%)	60.4	4
t_S (sec)	7.32	2.12

any combination of upper bounds the on rise time, peak time, percent overshoot, and settling time of the step response. This yields bounds on the desired damping ratio and undamped natural frequency that, in turn, can be translated into regions in the complex plane that specify desired eigenvalue locations. This more general viewpoint will be demonstrated in the following example for the second-order case.

Example 7.3 In this example we characterize acceptable eigenvalue locations in the complex plane so that the following second-order performance specifications are satisfied:

$$ PO \leq 4\% \qquad t_S \leq 2 \text{ s} \qquad t_P \leq 0.5 \text{ s} $$

The formula for percent overshoot indicates that percent overshoot decreases as damping ratio increases. Therefore, an upper bound on percent overshoot $PO \leq PO_{max}$ corresponds to a lower bound on damping ratio $\xi \geq \xi_{min}$. From Example 7.2, we know that a damping ratio of $\xi_{min} = 0.716$ will yield a percent overshoot equal to $PO_{max} = 4$ percent. Therefore, in order to achieve a percent overshoot less than 4 percent, we require a damping ratio greater than 0.716. We previously defined the damping angle $\theta = \cos^{-1}(\xi)$, which appeared in the expression for the unit step response of the standard underdamped second-order system. From this relationship, we see that a lower bound on the damping ratio corresponds to an upper bound on the damping angle $\theta \leq \theta_{max} = \cos^{-1}(\xi_{min})$. To relate this to allowable eigenvalue locations, we see from basic trigonometry that

$$ \tan(\theta) = \frac{\sqrt{1 - \xi^2}}{\xi} = \frac{\omega_n\sqrt{1 - \xi^2}}{\xi\omega_n} = \frac{\omega_d}{\xi\omega_n} $$

Recalling that eigenvalues of an underdamped second-order system are $\lambda_{1,2} = -\xi\omega_n \pm j\omega_d$, in which $\omega_d = \omega_n\sqrt{1 - \xi^2}$ is the damped natural frequency, we therefore interpret the damping angle θ as the angle made by the radial line from the origin in the complex plane passing through the eigenvalue $\lambda_1 = -\xi\omega_n + j\omega_d$ measured with respect to the negative real axis. By symmetry, $-\theta$ is the angle made by the radial line from the origin in the complex plane passing through the conjugate eigenvalue $\lambda_2 = -\xi\omega_n - j\omega_d$, also measured with respect to the negative real axis. An upper bound on the damping angle characterizes a sector in the left half of the complex plane bounded by the pair of radial lines corresponding to the maximum allowable damping angle. In this example, a minimum allowable damping ratio of 0.716 corresponds to a maximum allowable damping angle of $\cos^{-1}(0.716) = 44.3°$. The associated sector specifying acceptable eigenvalue locations is bounded by the radial lines making angles of $\pm 44.3°$ with respect to the negative real axis.

The formula that approximates settling time shows that settling time is inversely proportional to the product $\xi \omega_n$ that is directly related to the real part of the complex-conjugate eigenvalues. Therefore, an upper bound on settling time $t_S \leq t_{S, \max}$ corresponds to eigenvalues that lie to the left of a vertical line passing through the point $-4/t_{S, \max}$ on the negative real axis. With a specified upper bound on settling time of 2 s, eigenvalues must lie to the left of the vertical line passing through the point $-4/2 = -2$ on the negative real axis.

The formula for peak time shows that peak time is inversely proportional to the damped natural frequency ω_d that characterizes the imaginary part of the complex-conjugate eigenvalues. Thus an upper bound on peak time $t_P \leq t_{P, \max}$ yields a lower bound on ω_d. This corresponds to an eigenvalue that lies above the horizontal line passing through $j(\pi/t_{P, \max})$ on the positive imaginary axis and a conjugate eigenvalue that lies below the horizontal line passing through $-j(\pi/t_{P, \max})$ on the negative imaginary axis. With a specified upper bound on peak time of 0.5 s, these horizontal lines pass through $\pm j(\pi/0.5) = \pm j2\pi$.

The region in the complex plane characterizing allowable eigenvalue locations with respect to all three constraints is given by the intersection of the individual regions. The shaded region in Figure 7.5 shows allowable eigenvalue locations for this example along with the bounding lines for the individual regions.

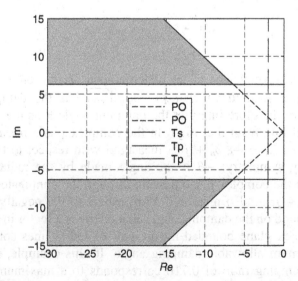

FIGURE 7.5 Allowable eigenvalue locations for Example 7.3.

In this example we see that only the percent-overshoot and peak-time constraints are active. That is, the settling time will be less than 2 s for all eigenvalues satisfying the other two constraints. $\qquad\square$

Higher-Order Systems

We typically encounter systems of order or dimension greater than two. When we approximate higher-order systems by a dominant first- or second-order model and specify the associated dominant eigenvalue locations that meet various transient response specifications, we eventually need to specify locations for the remaining eigenvalues. The general rule of thumb we adopt is to augment the dominant eigenvalues with the requisite number of additional eigenvalues that are 10 times further to the left than the dominant eigenvalues in the complex plane. In so doing, we expect that the higher-order system response will be dominated by that of the first- or second-order approximate model.

Figure 7.6 shows the effect of augmenting a dominant first-order system eigenvalue ($\lambda_1 = -0.5$) with additional real negative eigenvalues at least 10 times higher, for second- through fourth-order systems. The eigenvalues associated with each case are given in Table 7.3. The second- through fourth-order step responses are similar to the desired dominant first-order step response. The settling time increases slightly as the system order increases.

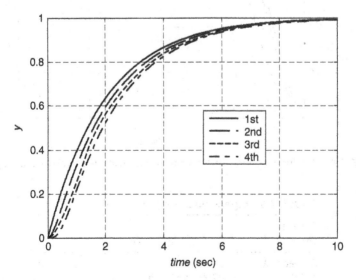

FIGURE 7.6 Dominant first-order system versus second- through fourth-order systems.

TABLE 7.3   Eigenvalues for Figure 7.6

System Order	Eigenvalues
First	$\lambda_1 = -0.5$
Second	$\lambda_{1,2} = -0.5, -5$
Third	$\lambda_{1,2,3} = -0.5, -5, -6$
Fourth	$\lambda_{1,2,3,4} = -0.5, -5, -6, -7$

Figure 7.7 shows the effect of augmenting dominant second-order system eigenvalues ($s_{1,2} = -2 \pm 1.95i$ from Example 7.2) with additional real negative eigenvalues at least 10 times further to the left in the complex plane for third- and sixth-order systems. The eigenvalues associated with each case are given in Table 7.4. The third- and sixth-order step responses are similar to the desired dominant second-order step response. The rise time increases slightly as the system order increases.

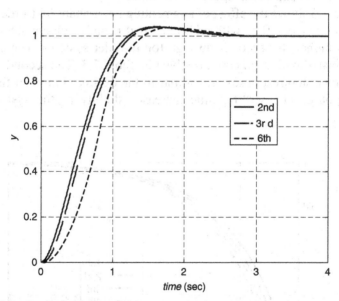

FIGURE 7.7   Dominant second-order system versus third- and sixth-order systems.

TABLE 7.4   Eigenvalues for Figure 7.7

System Order	Eigenvalues
Second	$\lambda_{1,2} = -2 \pm 1.95i$
Third	$\lambda_{1,2,3} = -2 \pm 1.95i, -20$
Sixth	$\lambda_{1,2,3,4,5,6} = -2 \pm 1.95i, -20, -21, -22, -23$

ITAE Method for Shaping the Dynamic Response

The ITAE (integral of time multiplying the absolute value of error), method attempts to accomplish dynamic shaping by penalizing the error, in our case the deviation between the unit step response and the steady-state value that yields the transient response component, more heavily later in the response as time increases (Graham and Lathrop, 1953). The ITAE objective function is

$$\text{ITAE} = \int_0^\infty t\,|e(t)|\,dt$$

Minimizing the ITAE objective function yields a step response with relatively small overshoot and relatively little oscillation. The ITAE is just one possible objective function; others have been proposed and used in practice. For first- through sixth-order systems, the characteristic polynomials given in Table 7.5 correspond to systems that minimize the ITAE criterion (Dorf and Bishop, 2005). In each case, one must specify the desired natural frequency ω_n (higher values correspond to faster response). Then the desired eigenvalues are obtained by finding the roots of the appropriate characteristic polynomial for the particular system order.

Figure 7.8 shows the unit step responses for the ITAE first- through sixth-order systems described by transfer functions of the form

$$H_k(s) = \frac{\omega_n^k}{d_k(s)}$$

in which the index k denotes the system order, and the denominator polynomial is taken from Table 7.5. The independent axis in Figure 7.8 is the normalized time $\omega_n t$, which is unitless. Note that beyond the first-order case, some overshoot is involved in optimizing the ITAE objective function.

This section has presented several approaches to translate transient response specifications into desired eigenvalue locations. These methods

TABLE 7.5 ITAE Characteristic Polynomials

System Order	Characteristic Polynomial
First	$s + \omega_n$
Second	$s^2 + 1.4\,\omega_n s + \omega_n^2$
Third	$s^3 + 1.75\,\omega_n s^2 + 2.15\,\omega_n^2 s + \omega_n^3$
Fourth	$s^4 + 2.1\,\omega_n s^3 + 3.4\,\omega_n^2 s^2 + 2.7\,\omega_n^3 s + \omega_n^4$
Fifth	$s^5 + 2.8\,\omega_n s^4 + 5.0\,\omega_n^2 s^3 + 5.5\,\omega_n^3 s^2 + 3.4\,\omega_n^4 s + \omega_n^5$
Sixth	$s^6 + 3.25\,\omega_n s^5 + 6.6\,\omega_n^2 s^4 + 8.6\,\omega_n^3 s^3 + 7.45\,\omega_n^4 s^2 + 3.95\,\omega_n^5 s + \omega_n^6$

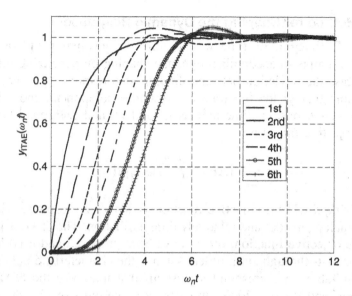

FIGURE 7.8   ITAE unit step responses.

have focused on so-called all-pole system models. It is well known from classical control that the presence of zeros (roots of the transfer function numerator polynomial) will change the system's transient behavior. One approach to remedy this is to use a prefilter to reduce the effect of these zeros.

7.3   CLOSED-LOOP EIGENVALUE PLACEMENT VIA STATE FEEDBACK

The following result establishes a connection between the ability to arbitrarily place the closed-loop eigenvalues by proper choice of the state feedback gain matrix K and controllability of the open-loop state equation, i.e., the pair (A, B).

Theorem 7.1   *For any symmetric set of n complex numbers $\{\mu_1, \mu_2, \ldots, \mu_n\}$, there exists a state feedback gain matrix K such that $\sigma(A - BK) = \{\mu_1, \mu_2, \ldots, \mu_n\}$ if and only if the pair (A, B) is controllable.*

Proof.   Here we prove that controllability of the pair (A, B) is necessary for arbitrary eigenvalue placement via state feedback. That controllability of the pair (A, B) is sufficient for arbitrary eigenvalue placement via state feedback will be established constructively in the single-input case by deriving formulas for the requisite state feedback gain vector. The multiple-input case is considerably more complicated, and we will be content to refer the interested reader to treatments available in the literature.

For necessity, we prove the contrapositive. Suppose the pair (A, B) is not controllable so that

$$\text{rank}\begin{bmatrix} B & AB & A^2B & \cdots & A^{n-1}B \end{bmatrix} = q < n$$

and there exists a nonsingular $n \times n$ matrix T for which

$$\hat{A} = T^{-1}AT \qquad\qquad \hat{B} = T^{-1}B$$

$$= \begin{bmatrix} A_{11} & A_{12} \\ 0 & A_{22} \end{bmatrix} \qquad = \begin{bmatrix} B_1 \\ 0 \end{bmatrix}$$

where A_{11} is of dimension $q \times q$ and B_1 is $q \times m$. For any state feedback gain matrix K, let

$$\hat{K} = KT$$

$$= \begin{bmatrix} K_1 & K_2 \end{bmatrix}$$

where K_1 is of dimension $m \times q$ and K_2 is $m \times (n - q)$. Then

$$\hat{A} - \hat{B}\hat{K} = \begin{bmatrix} A_{11} & A_{12} \\ 0 & A_{22} \end{bmatrix} - \begin{bmatrix} B_1 \\ 0 \end{bmatrix}\begin{bmatrix} K_1 & K_2 \end{bmatrix}$$

$$= \begin{bmatrix} A_{11} - B_1K_1 & A_{12} - B_1K_2 \\ 0 & A_{22} \end{bmatrix}$$

in which we note that the bottom-block row is unaffected by the state feedback. The characteristic polynomial is

$$\left| sI - (\hat{A} - \hat{B}\hat{K}) \right| = \begin{vmatrix} sI - (A_{11} - B_1K_1) & -(A_{12} - B_1K_2) \\ 0 & sI - A_{22} \end{vmatrix}$$

$$= |sI - (A_{11} - B_1K_1)| \cdot |sI - A_{22}|$$

Thus the n eigenvalues of $\hat{A} - \hat{B}\hat{K}$ are the q eigenvalues of $A_{11} - B_1K_1$ along with the $n - q$ eigenvalues of A_{22}, which represents the uncontrollable subsystem. Since

$$\hat{A} - \hat{B}\hat{K} = T^{-1}AT - T^{-1}BKT$$

$$= T^{-1}(A - BK)T$$

the same conclusion holds for the eigenvalues of $A - BK$. Since for any feedback gain matrix K the eigenvalues of A_{22} are always among the closed-loop eigenvalues, we conclude that arbitrary eigenvalue placement is *not* possible. Equivalently, if arbitrary eigenvalue placement is possible by state feedback, then the pair (A, B) necessarily must be controllable.

$\square$

Feedback Gain Formula for Controller Canonical Form

We know from Chapter 3 that the controller canonical form specifies a controllable single-input state equation. The construction of a feedback gain vector that achieves arbitrary eigenvalue placement proceeds as follows: The coefficient matrices for controller canonical form are given below:

$$A_{\text{CCF}} = \begin{bmatrix} 0 & 1 & 0 & \cdots & 0 \\ 0 & 0 & 1 & \cdots & 0 \\ \vdots & \vdots & \vdots & \ddots & \vdots \\ 0 & 0 & 0 & \cdots & 1 \\ -a_0 & -a_1 & -a_2 & \cdots & -a_{n-1} \end{bmatrix} \quad B_{\text{CCF}} = \begin{bmatrix} 0 \\ 0 \\ \vdots \\ 0 \\ 1 \end{bmatrix}$$

and we recall that since A_{CCF} is a companion matrix, its characteristic polynomial is written down by inspection:

$$|sI - A_{\text{CCF}}| = s^n + a_{n-1}s^{n-1} + \cdots + a_2 s^2 + a_1 s + a_0$$

We denote a feedback gain vector by

$$K_{\text{CCF}} = \begin{bmatrix} \delta_0 & \delta_1 & \delta_2 & \cdots & \delta_{n-1} \end{bmatrix}$$

which yields the closed-loop system dynamics matrix

$$A_{\text{CCF}} - B_{\text{CCF}} K_{\text{CCF}}$$

$$= \begin{bmatrix} 0 & 1 & 0 & \cdots & 0 \\ 0 & 0 & 1 & \cdots & 0 \\ \vdots & \vdots & \vdots & \ddots & \vdots \\ 0 & 0 & 0 & \cdots & 1 \\ -a_0 - \delta_0 & -a_1 - \delta_1 & -a_2 - \delta_2 & \cdots & -a_{n-1} - \delta_{n-1} \end{bmatrix}$$

We see that $A_{\text{CCF}} - B_{\text{CCF}} K_{\text{CCF}}$ is also in companion form, so its characteristic polynomial can be written down by inspection of the bottom row:

$$|sI - A_{\text{CCF}} + B_{\text{CCF}} K_{\text{CCF}}| = s^n + (a_{n-1} + \delta_{n-1})s^{n-1} + \cdots + (a_2 + \delta_2)s^2$$
$$+ (a_1 + \delta_1)s + (a_0 + \delta_0)$$

Beginning with an arbitrary symmetric set of complex numbers $\{\mu_1, \mu_2, \ldots, \mu_n\}$ that represents the desired closed-loop eigenvalues, we define the associated closed-loop characteristic polynomial

$$\alpha(s) = (s - \mu_1)(s - \mu_2) \cdots (s - \mu_n)$$
$$= s^n + \alpha_{n-1}s^{n-1} + \cdots + \alpha_2 s^2 + \alpha_1 s + \alpha_0$$

It is important to note that because the roots of a polynomial uniquely determine and are uniquely determined by the polynomial coefficients, specifying n desired closed-loop eigenvalues is equivalent to specifying the n coefficients $\alpha_0, \alpha_1, \alpha_2, \ldots, \alpha_{n-1}$. In terms of these parameters, the crux of the problem is to determine K_{CCF} so that the characteristic polynomial of $A_{CCF} - B_{CCF}K_{CCF}$ matches the desired closed-loop characteristic polynomial $\alpha(s)$. Equating

$$|sI - A_{CCF} + B_{CCF}K_{CCF}| = s^n + (a_{n-1} + \delta_{n-1})s^{n-2} + \cdots + (a_2 + \delta_2)s^2$$
$$+ (a_1 + \delta_1)s + (a_0 + \delta_0)$$
$$= s^n + \alpha_{n-1}s^{n-1} + \cdots + \alpha_2 s^2 + \alpha_1 s + \alpha_0$$

yields, on comparing coefficients of like powers of s, the relationships

$$a_0 + \delta_0 = \alpha_0 \quad a_1 + \delta_1 = \alpha_1 \quad a_2 + \delta_2 = \alpha_2 \quad \cdots \quad a_{n-1} + \delta_{n-1} = \alpha_{n-1}$$

so that

$$\delta_0 = \alpha_0 - a_0 \quad \delta_1 = \alpha_1 - a_1 \quad \delta_2 = \alpha_2 - a_2 \quad \cdots \quad \delta_{n-1} = \alpha_{n-1} - a_{n-1}$$

and the state feedback gain vector is then given by

$$K_{CCF} = \begin{bmatrix} (\alpha_0 - a_0) & (\alpha_1 - a_1) & (\alpha_2 - a_2) & \cdots & (\alpha_{n-1} - a_{n-1}) \end{bmatrix}$$

Example 7.4 We consider the following three-dimensional state equation given in controller canonical form specified by the coefficient matrices

$$A_{CCF} = \begin{bmatrix} 0 & 1 & 0 \\ 0 & 0 & 1 \\ -18 & -15 & -2 \end{bmatrix} \quad B_{CCF} = \begin{bmatrix} 0 \\ 0 \\ 1 \end{bmatrix} \quad C_{CCF} = \begin{bmatrix} 1 & 0 & 0 \end{bmatrix}$$

The open-loop characteristic polynomial, is by inspection,

$$a(s) = s^3 + a_2 s^2 + a_1 s + a_0 = s^3 + 2s^2 + 15s + 18$$

which yields the open-loop eigenvalues $\lambda_{1,2,3} = -1.28, -0.36 \pm j3.73$. This open-loop system exhibits a typical third-order lightly damped step response, as shown in Figure 7.9 below.

This open-loop system is already asymptotically stable, but we are interested in designing a state feedback control law to improve the transient response performance. We first specify a pair of dominant eigenvalues to yield a percent overshoot of 6 percent and a settling time of 3 s. The associated damping ratio and undamped natural frequency are $\xi = 0.67$ and $\omega_n = 2.00$ rad/s. The resulting dominant second-order eigenvalues are $\lambda_{1,2} = -1.33 \pm j1.49$. The open-loop system is third-order, so we need to specify a third desired eigenvalue, which we choose to be negative, real, and 10 further to the left of the dominant second-order eigenvalues in the complex plane: $\lambda_3 = -13.33$. Thus the desired characteristic polynomial is

$$\alpha(s) = s^3 + \alpha_2 s^2 + \alpha_1 s + \alpha_0 = s^3 + 16s^2 + 39.55s + 53.26$$

This leads immediately to the state feedback gain vector

$$
\begin{aligned}
K_{\text{CCF}} &= \left[(\alpha_0 - a_0) \quad (\alpha_1 - a_1) \quad (\alpha_2 - a_2) \right] \\
&= \left[(53.26 - 18) \quad (39.55 - 15) \quad (16 - 2) \right] \\
&= \left[35.26 \quad 24.55 \quad 14.00 \right]
\end{aligned}
$$

The state feedback control law

$$u(t) = -K_{\text{CCF}} \, x_{\text{CCF}}(t) + r(t)$$

yields the closed-loop state equation specified by the coefficient matrices

$$
A_{\text{CCF}} - B_{\text{CCF}} K_{\text{CCF}} = \begin{bmatrix} 0 & 1 & 0 \\ 0 & 0 & 1 \\ -53.26 & -39.55 & -16 \end{bmatrix} \quad B_{\text{CCF}} = \begin{bmatrix} 0 \\ 0 \\ 1 \end{bmatrix}
$$

$$C_{\text{CCF}} = \begin{bmatrix} 1 & 0 & 0 \end{bmatrix}$$

which is also in controller canonical form. Figure 7.9 shows a comparison of the open-loop and closed-loop output responses to a unit step input. The closed-loop transient response is greatly improved, and the achieved percent overshoot (5.9 percent) and settling time (3.09 s) are in close agreement with the design specifications. Note, however, that neither unit step response achieves a steady-state value of 1.0. For the closed-loop system, this means that the steady-state output does not match the reference

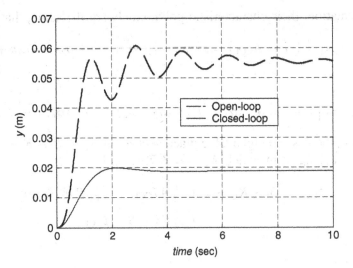

FIGURE 7.9 Open-loop versus closed-loop unit step responses for Example 7.4.

input, yielding, in this case, substantial steady-state error. Methods to correct this will be addressed later in this chapter.   □

Bass-Gura Formula

We generalize the preceding feedback gain construction for the controller canonical form to the case of an arbitrary controllable single-input state equation. Our plan of attack is to use the explicitly defined state coordinate transformation linking a controllable single-input state equation to its associated controller canonical form.

As before, we let $\{\mu_1, \mu_2, \ldots, \mu_n\}$ be a symmetric set of n complex numbers representing the desired closed-loop eigenvalues, which uniquely determines the desire closed-loop characteristic polynomial

$$\alpha(s) = (s - \mu_1)(s - \mu_2) \cdots (s - \mu_n)$$
$$= s^n + \alpha_{n-1} s^{n-1} + \cdots + \alpha_2 s^2 + \alpha_1 s + \alpha_0$$

For the controllable pair (A, B), Section 3.4 indicates that the state coordinate transformation $x(t) = T_{\text{CCF}} \, x_{\text{CCF}}(t)$ with

$$T_{\text{CCF}} = P P_{\text{CCF}}^{-1}$$

transforms the original state equation to its controller canonical form. Since the system dynamics matrices A and A_{CCF} are related by a similarity transformation, their characteristic polynomials are the same, that is,

$$|sI - A| = |sI - A_{CCF}| = s^n + a_{n-1}s^{n-2} + \cdots + a_2s^2 + a_1s + a_0$$

With

$$K_{CCF} = \begin{bmatrix} (\alpha_0 - a_0) & (\alpha_1 - a_1) & (\alpha_2 - a_2) & \cdots & (\alpha_{n-1} - a_{n-1}) \end{bmatrix}$$

achieving the desired eigenvalue placement for $A_{CCF} - B_{CCF}K_{CCF}$, we set $K = K_{CCF}T_{CCF}^{-1}$, from which $K_{CCF} = KT_{CCF}$ and

$$A_{CCF} - B_{CCF}K_{CCF} = T_{CCF}^{-1}AT_{CCF} - T_{CCF}^{-1}B(KT_{CCF})$$
$$= T_{CCF}^{-1}(A - BK)T_{CCF}$$

Thus

$$|sI - A + BK| = |sI - A_{CCF} + B_{CCF}K_{CCF}|$$
$$= s^n + \alpha_{n-1}s^{n-1} + \cdots + \alpha_2s^2 + \alpha_1s + \alpha_0$$

so that $K = K_{CCF}T_{CCF}^{-1}$ achieves the desired eigenvalue placement for the closed-loop state equation in the original coordinates.

Putting everything together yields the *Bass-Gura formula*, namely,

$$K = K_{CCF}T_{CCF}^{-1}$$
$$= \begin{bmatrix} (\alpha_0 - a_0) & (\alpha_1 - a_1) & (\alpha_2 - a_2) & \cdots & (\alpha_{n-1} - a_{n-1}) \end{bmatrix} \begin{bmatrix} PP_{CCF}^{-1} \end{bmatrix}^{-1}$$
$$= \begin{bmatrix} (\alpha_0 - a_0) & (\alpha_1 - a_1) & (\alpha_2 - a_2) & \cdots & (\alpha_{n-1} - a_{n-1}) \end{bmatrix}$$

$$\times \begin{bmatrix} P \begin{bmatrix} a_1 & a_2 & \cdots & a_{n-1} & 1 \\ a_2 & a_3 & \cdots & 1 & 0 \\ \vdots & \vdots & \ddots & \vdots & \vdots \\ a_{n-1} & 1 & \cdots & 0 & 0 \\ 1 & 0 & \cdots & 0 & 0 \end{bmatrix} \end{bmatrix}^{-1} \qquad (7.4)$$

which depends only on the controllable open-loop pair (A, B), the open-loop characteristic polynomial, and the desired closed-loop characteristic polynomial, as determined by the desired closed-loop eigenvalues. We observe that nonsingularity of the controllability matrix P is required to implement this formula. Finally, we note that by providing a formula for the state feedback gain vector that achieves arbitrary eigenvalue placement for a controllable single-input state equation, we have proved the sufficiency part of Theorem 7.1, at least in the single-input case.

Example 7.5   We demonstrate application of the Bass-Gura formula for the three-dimensional single-input state equation with coefficient matrices

$$A = \begin{bmatrix} 0 & 0 & 0 \\ 0 & 1 & 0 \\ 0 & 0 & 2 \end{bmatrix} \qquad B = \begin{bmatrix} 1 \\ 1 \\ 1 \end{bmatrix}$$

We take the desired closed-loop eigenvalues to be $\{\mu_1, \mu_2, \mu_3\} = \{-1, -1, -1\}$. We break the feedback gain vector computation into several steps.

We first check the controllability of the pair (A, B) by computing

$$P = \begin{bmatrix} B & AB & A^2 B \end{bmatrix}$$
$$= \begin{bmatrix} 1 & 0 & 0 \\ 1 & 1 & 1 \\ 1 & 2 & 4 \end{bmatrix}$$

which has $|P| = 2 \neq 0$, so this state equation is controllable. We note that the same conclusion follows by inspection because the state equation is in diagonal canonical form, in which the eigenvalues of A are distinct and the elements of B are all nonzero.

Next, we form the open-loop and desired closed-loop characteristic polynomials. Here, A is a diagonal matrix with its eigenvalues displayed on the main diagonal $\lambda_{1,2,3} = 0, 1, 2$. The open-loop characteristic polynomial and associated coefficients are

$$\begin{aligned} a(s) &= (s - 0)(s - 1)(s - 2) \\ &= s^3 - 3s^2 + 2s + 0 \end{aligned} \qquad a_2 = -3, \quad a_1 = 2, \quad a_0 = 0$$

The open-loop system is unstable because there are two positive real eigenvalues. The closed-loop desired characteristic polynomial and associated coefficients are

$$\begin{aligned} \alpha(s) &= (s + 1)^3 \\ &= s^3 + 3s^2 + 3s + 1 \end{aligned} \qquad \alpha_2 = 3, \quad \alpha_1 = 3, \quad \alpha_0 = 1$$

From this we can directly compute

$$\begin{aligned} K_{CCF} &= \begin{bmatrix} (\alpha_0 - a_0) & (\alpha_1 - a_1) & (\alpha_2 - a_2) \end{bmatrix} \\ &= \begin{bmatrix} (1 - 0) & (3 - 2) & (3 - (-3)) \end{bmatrix} \\ &= \begin{bmatrix} 1 & 1 & 6 \end{bmatrix} \end{aligned}$$

We next calculate T_{CCF} and T_{CCF}^{-1}:

$$T_{CCF} = PP_{CCF}^{-1}$$

$$= \begin{bmatrix} 1 & 0 & 0 \\ 1 & 1 & 1 \\ 1 & 2 & 4 \end{bmatrix} \begin{bmatrix} 2 & -3 & 1 \\ -3 & 1 & 0 \\ 1 & 0 & 0 \end{bmatrix} \qquad T_{CCF}^{-1} = \begin{bmatrix} \frac{1}{2} & -1 & \frac{1}{2} \\ 0 & -1 & 1 \\ 0 & -1 & 2 \end{bmatrix}$$

$$= \begin{bmatrix} 2 & -3 & 1 \\ 0 & -2 & 1 \\ 0 & -1 & 1 \end{bmatrix}$$

Finally, we calculate the state feedback gain vector

$$K = K_{CCF}T_{CCF}^{-1}$$

$$= \begin{bmatrix} 1 & 1 & 6 \end{bmatrix} \begin{bmatrix} \frac{1}{2} & -1 & \frac{1}{2} \\ 0 & -1 & 1 \\ 0 & -1 & 2 \end{bmatrix}$$

$$= \begin{bmatrix} \frac{1}{2} & -8 & \frac{27}{2} \end{bmatrix}$$

As a final check, we see that

$$A - BK = \begin{bmatrix} -\frac{1}{2} & 8 & -\frac{27}{2} \\ -\frac{1}{2} & 9 & -\frac{27}{2} \\ -\frac{1}{2} & 8 & -\frac{23}{2} \end{bmatrix}$$

has eigenvalues $\{-1, -1, -1\}$ as desired.

Figure 7.10 compares the MATLAB generated state variable step responses for open and closed-loop state equations each for zero initial state. Because the open-loop state equation is in diagonal canonical form, we can easily compute

$$x_1(t) = t \qquad x_2(t) = e^t - 1 \qquad x_3(t) = \frac{1}{2}(e^{2t} - 1)$$

which grow without bound as time increases. The stabilizing effect of the state feedback law is evident from the closed-loop responses.   □

Ackermann's Formula

Here we present *Ackermann's formula* for computing the state feedback gain vector. In terms of the desired closed-loop characteristic polynomial

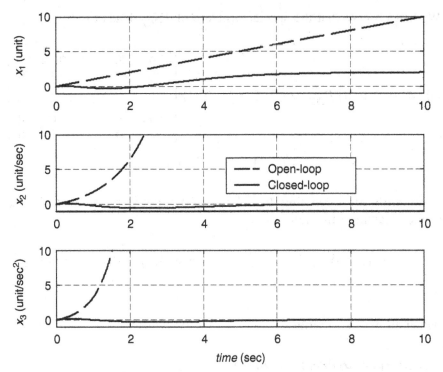

FIGURE 7.10   Open-loop versus closed-loop step response for Example 7.5.

$\alpha(s)$, the state feedback gain vector is given by

$$K = \begin{bmatrix} 0 & 0 & \cdots & 0 & 1 \end{bmatrix} P^{-1}\alpha(A)$$

Here $P = \begin{bmatrix} B & AB & A^2B & \cdots & A^{n-1}B \end{bmatrix}$ is the controllability matrix for the controllable pair (A, B) with inverse P^{-1} and $\alpha(A)$ represents

$$\alpha(A) = A^n + \alpha_{n-1}A^{n-1} + \cdots + \alpha_2 A^2 + \alpha_1 A + \alpha_0 I$$

which yields an $n \times n$ matrix. To verify Ackermann's formula, it is again convenient to begin with a controllable pair in controller canonical form, for which we must check that

$$K_{\text{CCF}} = \begin{bmatrix} (\alpha_0 - a_0) & (\alpha_1 - a_1) & (\alpha_2 - a_2) & \cdots & (\alpha_{n-1} - a_{n-1}) \end{bmatrix}$$

$$= \begin{bmatrix} 0 & 0 & \cdots & 0 & 1 \end{bmatrix} P_{\text{CCF}}^{-1}\alpha(A_{\text{CCF}})$$

We first observe that the special structure of P_{CCF}^{-1} leads to

$$
\begin{bmatrix} 0 & 0 & \cdots & 0 & 1 \end{bmatrix} P_{CCF}^{-1} = \begin{bmatrix} 0 & 0 & \cdots & 0 & 1 \end{bmatrix}
\begin{bmatrix}
a_1 & a_2 & \cdots & a_{n-1} & 1 \\
a_2 & a_3 & \cdots & 1 & 0 \\
\vdots & \vdots & \ddots & \vdots & \vdots \\
a_{n-1} & 1 & \cdots & 0 & 0 \\
1 & 0 & \cdots & 0 & 0
\end{bmatrix}
$$

$$
= \begin{bmatrix} 1 & 0 & \cdots & 0 & 0 \end{bmatrix}
$$

Now, with $a(s)$ denoting the characteristic polynomial of A_{CCF}, the Cayley-Hamilton theorem gives

$$
a(A_{CCF}) = A_{CCF}^n + a_{n-1} A_{CCF}^{n-1} + \cdots + a_2 A_{CCF}^2 + a_1 A_{CCF} + a_0 I
$$
$$
= 0 \quad (n \times n)
$$

which allows us to write

$$
\begin{aligned}
\alpha(A_{CCF}) &= \alpha(A_{CCF}) - a(A_{CCF}) \\
&= (A_{CCF}^n + \alpha_{n-1} A_{CCF}^{n-1} + \cdots + \alpha_2 A_{CCF}^2 + \alpha_1 A_{CCF} + \alpha_0 I) \\
&\quad - (A_{CCF}^n + a_{n-1} A_{CCF}^{n-1} + \cdots + a_2 A_{CCF}^2 + a_1 A_{CCF} + a_0 I) \\
&= (\alpha_{n-1} - a_{n-1}) A_{CCF}^{n-1} + \cdots + (\alpha_2 - a_2) A_{CCF}^2 \\
&\quad + (\alpha_1 - a_1) A_{CCF} + (\alpha_0 - a_0) I
\end{aligned}
$$

Next, the special structure of A_{CCF} yields

$$
\begin{bmatrix} 1 & 0 & 0 & \cdots & 0 & 0 \end{bmatrix} I = \begin{bmatrix} 1 & 0 & 0 & \cdots & 0 & 0 \end{bmatrix}
$$
$$
\begin{bmatrix} 1 & 0 & 0 & \cdots & 0 & 0 \end{bmatrix} A_{CCF} = \begin{bmatrix} 0 & 1 & 0 & \cdots & 0 & 0 \end{bmatrix}
$$
$$
\begin{bmatrix} 1 & 0 & 0 & \cdots & 0 & 0 \end{bmatrix} A_{CCF}^2 = \begin{bmatrix} 0 & 1 & 0 & \cdots & 0 & 0 \end{bmatrix} A_{CCF}
$$
$$
= \begin{bmatrix} 0 & 0 & 1 & \cdots & 0 & 0 \end{bmatrix}
$$

$$\vdots$$

$$
\begin{bmatrix} 1 & 0 & 0 & \cdots & 0 & 0 \end{bmatrix} A_{CCF}^{n-1} = \begin{bmatrix} 0 & 0 & 0 & \cdots & 1 & 0 \end{bmatrix} A_{CCF}^{n-2}
$$
$$
= \begin{bmatrix} 0 & 0 & 0 & \cdots & 0 & 1 \end{bmatrix}
$$

By combining these intermediate results, we see that

$$\begin{bmatrix} 0 & 0 & \cdots & 0 & 1 \end{bmatrix} P_{\text{CCF}}^{-1} \alpha(A_{\text{CCF}})$$

$$= \begin{bmatrix} 1 & 0 & \cdots & 0 & 0 \end{bmatrix} [(\alpha_{n-1} - a_{n-1}) A_{\text{CCF}}^{n-1} + \cdots + (\alpha_2 - a_2) A_{\text{CCF}}^2$$

$$+ (\alpha_1 - a_1) A_{\text{CCF}} + (\alpha_0 - a_0) I]$$

$$= (\alpha_0 - a_0) \begin{bmatrix} 1 & 0 & \cdots & 0 & 0 \end{bmatrix} I$$

$$+ (\alpha_1 - a_1) \begin{bmatrix} 1 & 0 & \cdots & 0 & 0 \end{bmatrix} A_{\text{CCF}}$$

$$+ (\alpha_2 - a_2) \begin{bmatrix} 1 & 0 & \cdots & 0 & 0 \end{bmatrix} A_{\text{CCF}}^2$$

$$+$$

$$\vdots$$

$$+ (\alpha_{n-1} - a_{n-1}) \begin{bmatrix} 1 & 0 & \cdots & 0 & 0 \end{bmatrix} A_{\text{CCF}}^{n-1}$$

$$= \begin{bmatrix} (\alpha_0 - a_0) & (\alpha_1 - a_1) & (\alpha_2 - a_2) & \cdots & (\alpha_{n-1} - a_{n-1}) \end{bmatrix}$$

$$= K_{\text{CCF}}$$

as required. Now, to address the general case, we let (A, B) denote a controllable pair for which $T_{\text{CCF}} = P P_{\text{CCF}}^{-1}$ is such that $A_{\text{CCF}} = T_{\text{CCF}}^{-1} A T_{\text{CCF}}$. It is straightforward to verify that for any integer $k \geq 0$, $A_{\text{CCF}}^k = (T_{\text{CCF}}^{-1} A T_{\text{CCF}})^k = T_{\text{CCF}}^{-1} A^k T_{\text{CCF}}$, which allows us to write

$$\alpha(A_{\text{CCF}}) = A_{\text{CCF}}^n + \alpha_{n-1} A_{\text{CCF}}^{n-1} + \cdots + \alpha_2 A_{\text{CCF}}^2 + \alpha_1 A_{\text{CCF}} + \alpha_0 I$$

$$= (T_{\text{CCF}}^{-1} A^n T_{\text{CCF}}) + \alpha_{n-1} (T_{\text{CCF}}^{-1} A^{n-1} T_{\text{CCF}})$$

$$+ \cdots + \alpha_2 (T_{\text{CCF}}^{-1} A^2 T_{\text{CCF}}) + \alpha_1 (T_{\text{CCF}}^{-1} A T_{\text{CCF}}) + \alpha_0 (T_{\text{CCF}}^{-1} T_{\text{CCF}})$$

$$= T_{\text{CCF}}^{-1} (A^n + \alpha_{n-1} A^{n-1} + \cdots + \alpha_2 A^2 + \alpha_1 A + \alpha_0 I) T_{\text{CCF}}$$

$$= T_{\text{CCF}}^{-1} \alpha(A) T_{\text{CCF}}$$

We then have

$$K = K_{\text{CCF}} T_{\text{CCF}}^{-1}$$

$$= \begin{bmatrix} 0 & 0 & \cdots & 0 & 1 \end{bmatrix} P_{\text{CCF}}^{-1} \alpha(A_{\text{CCF}}) T_{\text{CCF}}^{-1}$$

$$= \begin{bmatrix} 0 & 0 & \cdots & 0 & 1 \end{bmatrix} P_{\text{CCF}}^{-1} (T_{\text{CCF}}^{-1} \alpha(A) T_{\text{CCF}}) T_{\text{CCF}}^{-1}$$

$$= \begin{bmatrix} 0 & 0 & \cdots & 0 & 1 \end{bmatrix} P_{\text{CCF}}^{-1} T_{\text{CCF}}^{-1} \alpha(A)$$

$$= \begin{bmatrix} 0 & 0 & \cdots & 0 & 1 \end{bmatrix} P_{\text{CCF}}^{-1} (P P_{\text{CCF}}^{-1})^{-1} \alpha(A)$$

$$= \begin{bmatrix} 0 & 0 & \cdots & 0 & 1 \end{bmatrix} P^{-1} \alpha(A)$$

as desired.

Example 7.6   To demonstrate the use of Ackermann's formula, we now repeat the feedback gain computation first considered in Example 7.4 for the state equation in controller canonical form specified by

$$A_{\text{CCF}} = \begin{bmatrix} 0 & 1 & 0 \\ 0 & 0 & 1 \\ -18 & -15 & -2 \end{bmatrix} \qquad B_{\text{CCF}} = \begin{bmatrix} 0 \\ 0 \\ 1 \end{bmatrix}$$

for which, by inspection,

$$P_{\text{CCF}}^{-1} = \begin{bmatrix} 15 & 2 & 1 \\ 2 & 1 & 0 \\ 1 & 0 & 0 \end{bmatrix}$$

For the desired closed-loop characteristic polynomial

$$\alpha(s) = s^3 + \alpha_2 s^2 + \alpha_1 s + \alpha_0 = s^3 + 16s^2 + 39.55s + 53.26$$

we compute, with the aid of MATLAB,

$$\alpha(A_{\text{CCF}}) = A_{\text{CCF}}^3 + 16 A_{\text{CCF}}^3 + 39.55 A_{\text{CCF}} + 53.26 I$$

$$= \begin{bmatrix} 35.26 & 24.55 & 14.00 \\ -252.00 & -174.74 & -3.45 \\ 62.10 & -200.25 & -167.84 \end{bmatrix}$$

Ackermann's formula then gives

$$K = \begin{bmatrix} 0 & 0 & 1 \end{bmatrix} P_{\text{CCF}}^{-1} \alpha(A_{\text{CCF}})$$

$$= \begin{bmatrix} 1 & 0 & 0 \end{bmatrix} \begin{bmatrix} 35.26 & 24.55 & 14.00 \\ -252.00 & -174.74 & -3.45 \\ 62.10 & -200.25 & -167.84 \end{bmatrix}$$

$$= \begin{bmatrix} 35.26 & 24.55 & 14.00 \end{bmatrix}$$

This result agrees with K_{CCF} as computed in Example 7.4. This is so because for controllable single-input state equations, the feedback gain vector is uniquely determined by the desired closed-loop eigenvalues. □

Multiple-Input, Multiple-Output Eigenvalue Placement

One approach to addressing eigenvalue placement for controllable multiple-input state equations is to first derive a multiple-input version of our controller canonical form. This is considerably more complicated in the multiple-input case, but once this has been accomplished, the subsequent construction of a state feedback gain matrix that achieves desired closed-loop eigenvalues follows in a straightforward fashion. We do not pursue the details here and instead refer the interested reader to Chapters 13 and 14 in Rugh (1996) and Chapters 6 and 7 in Kailath (1980). We are reassured by the fact that for computational purposes, the MATLAB **place** function can accommodate multiple-input state equations (see Section 7.6).

7.4 STABILIZABILITY

We have seen that for a controllable state equation, arbitrary closed-loop eigenvalue placement can be achieved using state feedback. Since the freedom to assign closed-loop eigenvalues implies that we can asymptotically stabilize the closed-loop system via state feedback, we can say that controllability is a *sufficient* condition for asymptotic stabilization via state feedback. In situations where the plant is not controllable, it is natural to ask whether or not asymptotic stabilization via state feedback is still possible. We answer this question by introducing the concept of *stabilizability*. It turns out that we have already developed the key ingredients for the requisite analysis in our previous investigations of controllability and state feedback. To motivate the ensuing discussion, we consider the following example. The state equation

$$
\dot{x}(t) = \begin{bmatrix} 1 & 0 & 0 \\ 1 & -1 & 1 \\ 0 & 0 & -2 \end{bmatrix} x(t) + \begin{bmatrix} 1 \\ 1 \\ 0 \end{bmatrix} u(t)
$$

$$
y(t) = \begin{bmatrix} 1 & 0 & 0 \end{bmatrix} x(t)
$$

is already in the standard form for an uncontrollable state equation, as can be seen from the partitioning

$$
\begin{bmatrix} A_{11} & A_{12} \\ 0 & A_{22} \end{bmatrix} = \begin{bmatrix} 1 & 0 & 0 \\ 1 & -1 & 1 \\ \hline 0 & 0 & -2 \end{bmatrix} \qquad \begin{bmatrix} B_1 \\ 0 \end{bmatrix} = \begin{bmatrix} 1 \\ 1 \\ \hline 0 \end{bmatrix}
$$

in which the pair (A_{11}, B_1) specifies a controllable two-dimensional sub-system. In terms of a state feedback gain vector $K = \begin{bmatrix} k_1 & k_2 & k_3 \end{bmatrix}$, we have

$$
A - BK = \begin{bmatrix} 1 & 0 & 0 \\ 1 & -1 & 1 \\ 0 & 0 & -2 \end{bmatrix} - \begin{bmatrix} 1 \\ 1 \\ 0 \end{bmatrix} \begin{bmatrix} k_1 & k_2 & | & k_3 \end{bmatrix}
$$

$$
= \begin{bmatrix} \begin{bmatrix} 1 & 0 \\ 1 & -1 \end{bmatrix} - \begin{bmatrix} 1 \\ 1 \end{bmatrix} \begin{bmatrix} k_1 & k_2 \end{bmatrix} & \begin{bmatrix} 0 \\ 1 \end{bmatrix} - \begin{bmatrix} 1 \\ 1 \end{bmatrix} k_3 \\ 0 \quad 0 & -2 \end{bmatrix}
$$

The gains k_1 and k_2 can be chosen to arbitrarily locate the eigenvalues of the controllable subsystem. Because of the upper block triangular structure, the three eigenvalues of $A - BK$ are the two freely assigned by proper choice of k_1 and k_2 together with the third at -2 contributed by the uncontrollable subsystem. Thus we conclude that even though the state equation is not controllable, and hence *arbitrary* eigenvalue placement via state feedback is not achievable, it is still possible to construct a state feedback gain vector such that $A - BK$ specifies an asymptotically stable closed-loop state equation. This is a direct result of the fact that the uncontrollable subsystem, which cannot be influenced by state feedback, is itself asymptotically stable. We also see that the gain k_3 has no influence on the eigenvalues of $A - BK$. We now formalize this notion in the following definition.

Definition 7.2 *The linear state equation (7.1) [or the pair (A, B), for short] is stabilizable if there exists a state feedback gain matrix K for which all eigenvalues of $A - BK$ have strictly negative real part.*

Note that as a consequence of this definition and our prior eigenvalue placement results, we see that controllability implies stabilizability. On the other hand, the preceding example indicates that the converse does not hold. A state equation can be stabilizable but not controllable. We therefore regard stabilizability as a weaker condition than controllability.

The preceding example provides hints for general characterizations of stabilizability. If the pair (A, B) is controllable, then it is stabilizable as noted above and so we consider the case in which the pair (A, B) is not controllable. We know that there exists a coordinate transformation $x(t) = Tz(t)$ such that the transformed state equation has

$$
\hat{A} = \begin{bmatrix} A_{11} & A_{12} \\ 0 & A_{22} \end{bmatrix} \qquad \hat{B} = \begin{bmatrix} B_1 \\ 0 \end{bmatrix}
$$

in which (A_{11}, B_1) is a controllable pair. With $\hat{K} = \begin{bmatrix} K_1 & K_2 \end{bmatrix}$ a conformably partitioned state feedback gain matrix, we have

$$\hat{A} - \hat{B}\hat{K} = \begin{bmatrix} A_{11} - B_1 K_1 & A_{12} - B_1 K_2 \\ 0 & A_{22} \end{bmatrix}$$

whose eigenvalues are those of $A_{11} - B_1 K_1$, along with those of the A_{22}. Because (A_{11}, B_1) is a controllable pair, K_1 can be chosen such that the eigenvalues of $A_{11} - B_1 K_1$ have negative real part. However, the uncontrollable subsystem associated with A_{22} is uninfluenced by state feedback, and the eigenvalues of A_{22} will be among the eigenvalues of $\hat{A} - \hat{B}\hat{K}$ for any $\hat{K}$. Consequently, in order for every eigenvalue of $\hat{A} - \hat{B}\hat{K}$ to have negative real parts, the eigenvalues of A_{22} must have negative real parts to begin with. We therefore see that in the case where the pair (A, B) is not controllable, stabilizability requires that the uncontrollable subsystem be asymptotically stable. We also observe that K_2 plays no role in this analysis. Our next step is to adapt our previously derived Popov-Belevitch-Hautus tests for controllability in order to provide algebraic criteria for stabilizability that do not require a particular coordinate transformation.

Theorem 7.3 *The following statements are equivalent:*

1. *The pair (A, B) is stabilizable.*
2. *There exists no left eigenvector of A associated with an eigenvalue having nonnegative real part that is orthogonal to the columns of B;*
3. *The matrix $\begin{bmatrix} \lambda I - A & B \end{bmatrix}$ has full row-rank for all complex λ with nonnegative real part.*

Proof. To establish the equivalence of these statements, it is enough to prove the chain of implications: 1 implies 2, 2 implies 3, and 3 implies 1.

For the first implication, we prove the contrapositive: If there does exists a left eigenvector of A associated with an eigenvalue having nonnegative real part that is orthogonal to the columns of B, then the pair (A, B) is not stabilizable. Suppose that for $\lambda \in \sigma(A)$ with $\text{Re}(\lambda) \geq 0$ there exists $w \in \mathbf{C}^n$ that satisfies

$$w \neq 0 \qquad w^* A = \lambda w^* \qquad w^* B = 0$$

Then, for any state feedback gain matrix K, we have

$$w^*(A - BK) = w^* A - (w^* B)K$$
$$= \lambda w^*$$

which indicates that λ is also an eigenvalue of $A - BK$ with non-negative real part. Thus, by definition, the pair (A, B) is not stabilizable.

We next prove the contrapositive of the second implication: If $\begin{bmatrix} \lambda I - A & B \end{bmatrix}$ has less than full-row rank for some complex λ with nonnegative real part, then there does exists a left eigenvector of A associated with an eigenvalue having nonnegative real part that is orthogonal to the columns of B. Suppose that $\lambda \in \mathbb{C}$ with $\text{Re}(\lambda) \geq 0$ is such that the matrix $\begin{bmatrix} \lambda I - A & B \end{bmatrix}$ has linearly dependent rows. Consequently, there exists a nontrivial linear combination of the rows of $\begin{bmatrix} \lambda I - A & B \end{bmatrix}$ yielding a $1 \times (n + m)$ zero vector. This is equivalent to the existence of a nonzero vector $w \in \mathbb{C}^n$ that satisfies

$$ w^* \begin{bmatrix} \lambda I - A & B \end{bmatrix} = \begin{bmatrix} 0 & 0 \end{bmatrix} $$

which can be reorganized as

$$ w^* A = \lambda w^* \quad w^* B = 0 $$

Thus λ with $\text{Re}(\lambda) \geq 0$ is an eigenvalue of A with associated left eigenvector w that is orthogonal to the columns of B.

Finally, we prove the contrapositive of the third implication: If the pair (A, B) is not stabilizable, then there exists $\lambda \in \mathbb{C}$ with $\text{Re}(\lambda) \geq 0$ for which the matrix $\begin{bmatrix} \lambda I - A & B \end{bmatrix}$ has less than full-row rank. If the pair (A, B) is not stabilizable, then it is not controllable, and a change of coordinates yields the transformed state equation with

$$ \hat{A} = \begin{bmatrix} A_{11} & A_{12} \\ 0 & A_{22} \end{bmatrix} \qquad \hat{B} = \begin{bmatrix} B_1 \\ 0 \end{bmatrix} $$

in which the pair (A_{11}, B_1) is controllable. If it is not possible to find a stabilizing feedback gain matrix, then it must be the case that A_{22} has an eigenvalue $\lambda \in \mathbb{C}$ with $\text{Re}(\lambda) \geq 0$. By definition, $\lambda I - A_{22}$ is singular and therefore has linearly dependent rows, implying the same for

$$ \begin{bmatrix} \lambda I - \hat{A} & \hat{B} \end{bmatrix} = \begin{bmatrix} \lambda I - A_{11} & -A_{12} & B_1 \\ 0 & \lambda I - A_{22} & 0 \end{bmatrix} $$

Finally, the identity

$$ \begin{bmatrix} \lambda I - A & B \end{bmatrix} = T \begin{bmatrix} \lambda I - \hat{A} & \hat{B} \end{bmatrix} \begin{bmatrix} T^{-1} & 0 \\ 0 & I \end{bmatrix} $$

together with the fact that matrix rank is unaffected by pre- and postmultiplication by nonsingular matrices, implies that $\begin{bmatrix} \lambda I - A & B \end{bmatrix}$ has less than full row rank for this $\lambda \in \mathbb{C}$ with $\text{Re}(\lambda) \geq 0$. $\qquad \square$

Example 7.7 Recall the uncontrollable state equation introduced in Example 3.3 and revisited in Examples 3.5 and 3.6:

$$
\begin{bmatrix} \dot{x}_1(t) \\ \dot{x}_2(t) \\ \dot{x}_3(t) \end{bmatrix} = \begin{bmatrix} 0 & 1 & 0 \\ 0 & 0 & 1 \\ -6 & -11 & -6 \end{bmatrix} \begin{bmatrix} x_1(t) \\ x_2(t) \\ x_3(t) \end{bmatrix} + \begin{bmatrix} 0 \\ 1 \\ -3 \end{bmatrix} u(t)
$$

We saw in Example 3.5 that the coordinate transformation $x(t) = Tz(t)$ with

$$
T = \begin{bmatrix} 0 & 1 & 0 \\ 1 & -3 & 0 \\ -3 & 7 & 1 \end{bmatrix}
$$

yields the transformed state equation with coefficient matrices

$$
\hat{A} = T^{-1}AT \qquad \hat{B} = T^{-1}B
$$

$$
= \left[\begin{array}{cc|c} 0 & -2 & 1 \\ 1 & -3 & 0 \\ \hline 0 & 0 & -3 \end{array} \right] \qquad = \begin{bmatrix} 1 \\ 0 \\ \hline 0 \end{bmatrix}
$$

which is in the standard form for an uncontrollable state equation. Moreover, the uncontrollable subsystem is one-dimensional with $A_{22} = -3$. Hence this state equation is stabilizable. Arbitrarily choosing two closed-loop eigenvalues to be $-2 + j2$ and $-2 - j2$, the Bass-Gura formula applied to the controllable pair (A_{11}, B_1) extracted from the pair $(\hat{A}, \hat{B})$ yields

$$
K_1 = \begin{bmatrix} 1 & 3 \end{bmatrix}
$$

for which

$$
A_{11} - B_1 K_1 = \begin{bmatrix} 0 & -2 \\ 1 & -3 \end{bmatrix} - \begin{bmatrix} 1 \\ 0 \end{bmatrix} \begin{bmatrix} 1 & 3 \end{bmatrix} = \begin{bmatrix} -1 & -5 \\ 1 & -3 \end{bmatrix}
$$

has these specified eigenvalues. Therefore

$$
\hat{K} = \begin{bmatrix} 1 & 3 & 0 \end{bmatrix}
$$

is such that $\hat{A} - \hat{B}\hat{K}$ has the three eigenvalues $-2 + j2, -2 - j2$, and -3. Finally, the feedback gain vector

$$
K = \hat{K}T^{-1} = \begin{bmatrix} 1 & 3 & 0 \end{bmatrix} \begin{bmatrix} 3 & 1 & 0 \\ 1 & 0 & 0 \\ 2 & 3 & 1 \end{bmatrix} = \begin{bmatrix} 6 & 1 & 0 \end{bmatrix}
$$

yields $A - BK$ also having closed-loop eigenvalues $-2 + j2, -2 - j2$, and -3.

We also saw in Example 3.6 that $w_3 = \begin{bmatrix} 2 & 3 & 1 \end{bmatrix}^T$ is a left eigenvector of A associated with the eigenvalue $\lambda_3 = -3 < 0$ for which $w_3^T B = 0$. We can make the stronger assertion that the only left eigenvectors of A orthogonal to B are nonzero scalar multiples of w_3 and therefore are also associated with λ_3. We conclude from Theorem 7.3 that since there exists no left eigenvector of A associated with an eigenvalue having nonnegative real part that is orthogonal to B, the state equation is stabilizable. We also see that $\lambda_3 = -3$ is the only value of λ for which $\begin{bmatrix} \lambda I - A & B \end{bmatrix}$ has less than rank 3, so $\begin{bmatrix} \lambda I - A & B \end{bmatrix}$ has full-row rank 3 for all λ with nonnegative real part, and again, stabilizability follows from Theorem 7.3.

$\square$

7.5 STEADY-STATE TRACKING

We have studied in detail the extent to which state feedback control laws can influence the transient response characteristics by virtue of our complete freedom to specify closed-loop eigenvalues for a controllable state equation and our understanding of how eigenvalue locations affect the transient response. In this section we address a steady-state tracking requirement for step reference inputs. The first approach we consider involves a modification of our state feedback control law to include a gain that multiplies the reference input. This idea has been implemented already, albeit informally, in Example 7.2. Our second approach revisits in our state-space framework a familiar concept from classical control: Zero steady-state tracking error for step reference inputs can be achieved for so-called type 1 systems that include integral action on the tracking error. We refer to such control systems as *servomechanisms*.

Input Gain

As noted earlier, we now consider state feedback control laws of the form

$$u(t) = -Kx(t) + Gr(t)$$

in which the reference input is now multiplied by a gain G to be chosen so that for a step reference input $r(t) = R, t \geq 0$, the output of the closed-loop state equation

$$\dot{x}(t) = (A - BK)x(t) + BGr(t)$$
$$y(t) = Cx(t)$$

satisfies

$$y_{ss} \triangleq \lim_{t \to \infty} y(t) = R$$

We note at the outset that this steady-state tracking requirement makes sense only when the output $y(t)$ and the reference input $r(t)$ have the same dimension p in the multiple-input, multiple-output case. This, in turn, dictates that the gain matrix G has dimension $m \times p$. In the single-input, single-output case, $y(t), r(t)$, and $u(t)$ are scalar signals, and G is a scalar gain. The ensuing analysis is cast in the multi-input, multi-output case for which we will encounter the restriction $m \geq p$; i.e., the open-loop state equation has at least as many inputs as outputs. Clearly, the single-input, single-output case meets this requirement. We also assume from this point on that the feedback gain K has been chosen to yield an asymptotically stable closed-loop state equation.

From a frequency-domain viewpoint, the steady-state tracking objective requires that the closed-loop transfer function

$$H_{CL}(s) = C(sI - A + BK)^{-1}BG$$

have what we refer to as identity dc gain; that is, $H_{CL}(0) = I$ ($p \times p$). For then, with $R(s) = R(1/s)$ we may apply the final-value theorem (because the closed-loop state equation is asymptotically stable) to obtain

$$y_{ss} = \lim_{t \to \infty} y(t) = \lim_{s \to 0} sY(s) = \lim_{s \to 0} s H_{CL}(s) R \frac{1}{s} = H_{CL}(0)R$$

and so $y_{ss} = R$ for any constant vector $R \in \mathbb{R}^p$. The closed-loop dc gain is given by

$$H_{CL}(0) = -C(A - BK)^{-1}BG$$

in which $A - BK$ is guaranteed to be nonsingular because it is assumed to have strictly negative real-part eigenvalues. The question now is whether the gain matrix G can be chosen to yield $H_{CL}(0) = I$. We first consider the case $m = p$; i.e., the open-loop state equation has as many inputs as outputs. In this case, the dc gain is the product of the $p \times p$ matrix $-C(A - BK)^{-1}B$ and the $p \times p$ input gain matrix G. If the first factor is nonsingular, we can achieve identity dc gain by setting

$$G = -[C(A - BK)^{-1}B]^{-1}$$

If $-C(A - BK)^{-1}B$ is singular, the closed-loop state equation is said to have a *transmission zero* at $s = 0$. In the single-input, single-output case,

$-C(A - BK)^{-1}B$ is a scalar quantity and is therefore singular when and only when it is zero, meaning that the scalar transfer function $C(sI - A + BK)^{-1}B$ has a zero at $s = 0$ in the familiar sense.

For the case $m \geq p$, the factor $-C(A - BK)^{-1}B$ has dimension $p \times m$ and therefore has more columns than rows. If this matrix has full-row rank p then we can take the $m \times p$ input gain matrix G to be the *Moore-Penrose pseudoinverse* given by

$$G = -[C(A - BK)^{-1}B]^T \left[C(A - BK)^{-1}B[C(A - BK)^{-1}B]^T\right]^{-1}$$

which the reader should verify yields $H_{CL}(0) = I$.

We can arrive at the same conclusion from a time-domain viewpoint. For the constant reference input $r(t) = R, t \geq 0$, steady state corresponds to an equilibrium condition for the closed-loop state equation involving an equilibrium state denoted by x_{ss} that satisfies

$$0 = (A - BK)x_{ss} + BG\,R$$

which can be solved to give $x_{ss} = -(A - BK)^{-1}BG\,R$. The steady-state output then is obtained from

$$y_{ss} = Cx_{ss} = -C(A - BK)^{-1}BG\,R = H_{CL}(0)R$$

from which the same formula for the input gain results.

The preceding analysis can be generalized to address other steady-state objectives involving a closed-loop dc gain other than an identity matrix. For the $m = p$ case, let K_{dc} denote the desired closed-loop dc gain. We need only adjust the input gain formula according to

$$G = -[C(A - BK)^{-1}B]^{-1}K_{dc}$$

to achieve the desired result.

Example 7.8   We modify the state feedback control law computed for the state equation of Example 7.4 to include an input gain chosen so that the open-loop and closed-loop unit step responses reach the same steady-state value. The open-loop state equation is specified by the following coefficient matrices which although are in controller canonical form, we omit the corresponding subscripts

$$A = \begin{bmatrix} 0 & 1 & 0 \\ 0 & 0 & 1 \\ -18 & -15 & -2 \end{bmatrix} \quad B = \begin{bmatrix} 0 \\ 0 \\ 1 \end{bmatrix} \quad C = \begin{bmatrix} 1 & 0 & 0 \end{bmatrix}$$

The open-loop transfer function is, by inspection,

$$H(s) = \frac{1}{s^3 + 2s^2 + 15s + 18}$$

from which the open-loop DC gain is $H(0) = 1/18 = 0.056$.

In Example 7.4, the state feedback gain was computed to be

$$K = \begin{bmatrix} 35.26 & 24.55 & 14.00 \end{bmatrix}$$

to yield closed-loop eigenvalues at $\lambda_{1,2} = -1.33 \pm j1.49, -13.33$ in order to achieve a percent overshoot of 6 percent and a settling time of 3 s for the closed-loop unit step response. Without an input gain correction, the closed-loop transfer function is

$$C(sI - A + BK)^{-1}B = \frac{1}{s^3 + 16s^2 + 39.55s + 53.26}$$

which again is determined by inspection from the closed-loop state equation, which is also in controller canonical form. From this we compute $-C(A - BK)^{-1}B = 1/53.26 = 0.0188$. Clearly, unity closed-loop dc gain is achieved for $G = 53.26$, and the closed-loop dc gain will match the open-loop dc gain for

$$G = (53.26)\frac{1}{18} = 2.96$$

This input gain is used to produce the closed-loop unit step response plotted in of Figure 7.11. The closed-loop unit step response exhibits the improved transient response for which the feedback gain vector K was constructed, as well as a steady-state value that matches the open-loop unit step response by virtue of our input gain selection. □

Servomechanism Design

Using an input gain to influence closed-loop steady-state performance requires accurate knowledge of the open-loop state equation's coefficient matrices in order to invert $-C(A - BK)^{-1}B$. In practice, model uncertainty, parameter variations, or intentional approximations can result in deviations between the nominal coefficient matrices used for the input gain computation and the actual system description. This has the potential to significantly alter the actual steady-state behavior.

Here we present a servomechanism design methodology (Ogata, 2002) that combines the classical approach of adding an integral-error term to

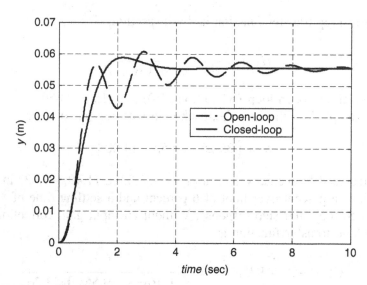

FIGURE 7.11   Open-loop versus closed-loop unit step response for Example 7.8.

obtain a type I system that yields zero steady-state tracking error for step reference inputs with our methods for state feedback design that deliver closed-loop stability and desirable transient response characteristics. This approach is *robust* with respect to uncertainty in the open-loop state equation in that the steady-state tracking performance is preserved as long as closed-loop stability is maintained. In this section we focus on the single-input, single-output case and impose the following additional assumptions:

Assumptions

1. The open-loop state equation, i.e., the pair (A, B), is controllable.
2. The open-loop state equation has no pole/eigenvalue at $s = 0$.
3. The open-loop state equation has no zero at $s = 0$.

Our control law will be of the form

$$\dot{\xi}(t) = r(t) - y(t)$$
$$u(t) = -Kx(t) + k_I\xi(t)$$

in which $r(t)$ is the step reference input to be tracked by the output $y(t)$. By setting the time derivative $\dot{\xi}(t)$ to equal the tracking error $r(t) - y(t)$, we see that $\xi(t)$ represents the integral of the tracking error. Taking

Laplace transforms for zero initial condition $\xi(0^-) = 0$ gives

$$s\xi(s) = R(s) - Y(s) = E(s)$$

$$\xi(s) = \frac{E(s)}{s}$$

which indicates that the integral error term introduces an open-loop pole at $s = 0$. Assumption 2 is in place so that the transfer function associated with the open-loop state equation does not itself contribute a pole at $s = 0$, in which case the new state variable $\xi(t)$ is not required. Assumption 3 prevents the pole at $s = 0$ introduced by the control law from being canceled by a zero at $s = 0$. We thus are guaranteed that the integral error term in the control law yields a type I system. The remainder of the control law can be written as

$$u(t) = -\begin{bmatrix} K & -k_I \end{bmatrix} \begin{bmatrix} x(t) \\ \xi(t) \end{bmatrix}$$

which we interpret as a state feedback law involving the *augmented* $(n + 1)$-dimensional state vector consisting of the open-loop state vector $x(t)$ together with the integrator state variable $\xi(t)$. The associated closed-loop system block diagram is shown in Figure 7.12, where "A, B Plant" refers to the open-loop state differential equation in Equation (7.1).

The interconnection of the feedback control law and the open-loop plant yields the $(n + 1)$-dimensional closed-loop state equation:

$$\begin{bmatrix} \dot{x}(t) \\ \dot{\xi}(t) \end{bmatrix} = \begin{bmatrix} A - BK & Bk_I \\ -C & 0 \end{bmatrix} \begin{bmatrix} x(t) \\ \xi(t) \end{bmatrix} + \begin{bmatrix} 0 \\ 1 \end{bmatrix} r(t)$$

$$y(t) = \begin{bmatrix} C & 0 \end{bmatrix} \begin{bmatrix} x(t) \\ \xi(t) \end{bmatrix}$$

Closed-loop stability and transient response performance are governed by the $(n + 1)$ eigenvalues of the $(n + 1) \times (n + 1)$ closed-loop system

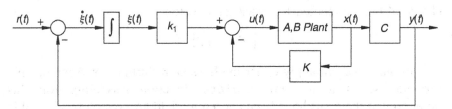

FIGURE 7.12 Closed-loop system block diagram.

dynamics matrix

$$\begin{bmatrix} A - BK & Bk_{\mathrm{I}} \\ -C & 0 \end{bmatrix} = \begin{bmatrix} A & 0 \\ -C & 0 \end{bmatrix} - \begin{bmatrix} B \\ 0 \end{bmatrix} \begin{bmatrix} K & -k_{\mathrm{I}} \end{bmatrix}$$

In order to arbitrarily place these closed-loop eigenvalues we require

$$\left(\begin{bmatrix} A & 0 \\ -C & 0 \end{bmatrix}, \begin{bmatrix} B \\ 0 \end{bmatrix} \right)$$

to be a controllable pair, which we have not yet explicitly assumed. By assumption 2, the open-loop system dynamics matrix A is nonsingular, and hence $|A| \neq 0$ because a zero eigenvalue has been disallowed. Next, assumption 3 implies that

$$H(0) = C(sI - A)^{-1}B|_{s=0} = -CA^{-1}B \neq 0$$

For nonsingular A, the block determinant formula

$$\begin{vmatrix} A & B \\ -C & 0 \end{vmatrix} = |CA^{-1}B| \, |A|$$

$$= CA^{-1}B|A|$$

shows that as a consequence of assumptions 2 and 3,

$$\begin{bmatrix} A & B \\ -C & 0 \end{bmatrix}$$

is a nonsingular $(n + 1) \times (n + 1)$ matrix. Analytical Exercise 7.1 asserts that controllability of the open-loop pair (A, B) as per assumption 1, along with nonsingularity of the preceding matrix, are necessary and sufficient for

$$\left(\begin{bmatrix} A & 0 \\ -C & 0 \end{bmatrix}, \begin{bmatrix} B \\ 0 \end{bmatrix} \right)$$

to be a controllable pair. Consequently, under assumptions 1, 2, and 3, the $(n + 1)$ closed-loop eigenvalues can be placed arbitrarily by appropriate choice of the augmented feedback gain vector

$$\begin{bmatrix} K & -k_{\mathrm{I}} \end{bmatrix}$$

For instance, we can apply either the Bass-Gura formula or Ackermann's formula, as well as the MATLAB **place** function. Assuming from this point on that the gains have been chosen to yield an asymptotically stable closed-loop state equation, we must show next that for step reference input

$r(t) = R, t \geq 0$, we have

$$y_{ss} = \lim_{t \to \infty} y(t) = R$$

This involves checking that the closed-loop state equation has an equilibrium condition involving constant x_{ss}, ξ_{ss}, u_{ss}, along with $y_{ss} = R$. Using the relationship $u_{ss} = -Kx_{ss} + k_I\xi_{ss}$ leads to

$$\begin{bmatrix} 0 \\ 0 \end{bmatrix} = \begin{bmatrix} A - BK & Bk_I \\ -C & 0 \end{bmatrix} \begin{bmatrix} x_{ss} \\ \xi_{ss} \end{bmatrix} + \begin{bmatrix} 0 \\ 1 \end{bmatrix} R$$

$$= \begin{bmatrix} A & 0 \\ -C & 0 \end{bmatrix} \begin{bmatrix} x_{ss} \\ \xi_{ss} \end{bmatrix} + \begin{bmatrix} B \\ 0 \end{bmatrix} u_{ss} + \begin{bmatrix} 0 \\ 1 \end{bmatrix} R$$

$$= \begin{bmatrix} A & B \\ -C & 0 \end{bmatrix} \begin{bmatrix} x_{ss} \\ u_{ss} \end{bmatrix} + \begin{bmatrix} 0 \\ 1 \end{bmatrix} R$$

With A nonsingular and $H(0) = -CA^{-1}B \neq 0$, a block matrix inverse formula gives

$$\begin{bmatrix} A & B \\ -C & 0 \end{bmatrix}^{-1} = \begin{bmatrix} A^{-1} - A^{-1}B(CA^{-1}B)^{-1}CA^{-1} & -A^{-1}B(CA^{-1}B)^{-1} \\ (CA^{-1}B)^{-1}CA^{-1} & (CA^{-1}B)^{-1} \end{bmatrix}$$

which allows us to solve for $[x_{ss}^T, u_{ss}]^T$ to obtain

$$\begin{bmatrix} x_{ss} \\ u_{ss} \end{bmatrix} = -\begin{bmatrix} A & B \\ -C & 0 \end{bmatrix}^{-1} \begin{bmatrix} 0 \\ R \end{bmatrix}$$

$$= \begin{bmatrix} A^{-1}B(CA^{-1}B)^{-1}R \\ -(CA^{-1}B)^{-1}R \end{bmatrix}$$

in which, from before, $-(CA^{-1}B)^{-1} = H^{-1}(0)$, the inverse of the open-loop dc gain. We may now solve for ξ_{ss}, that is,

$$\xi_{ss} = \frac{1}{k_I}[u_{ss} + Kx_{ss}]$$

$$= \frac{1}{k_I}\left[1 - KA^{-1}B\right]H^{-1}(0)R$$

We observe that $k_I \neq 0$, for otherwise we would have a closed-loop eigenvalue at $\lambda = 0$.

Finally,

$$y_{ss} = Cx_{ss}$$
$$= C[A^{-1}B(CA^{-1}B)^{-1}R]$$
$$= (CA^{-1}B)(CA^{-1}B)^{-1}R$$
$$= R$$

We therefore have achieved the desired steady-state tracking objective. All that is required is that the augmented feedback gain vector $\begin{bmatrix} K & -k_{\mathrm{I}} \end{bmatrix}$ stabilize the closed-loop state equation, thereby ensuring an asymptotically stable equilibrium for any constant reference input. This does not require perfect knowledge of the open-loop state equation. At this equilibrium, $\xi(t) = \xi_{ss}$ implies that $0 = \dot{\xi}(t) = R - y_{ss}$, so $y_{ss} = R$ is assured.

Example 7.9   We design a type I servomechanism for the state equation of Example 7.4 so that closed-loop unit step response reaches a steady-state value of 1 corresponding to zero steady-state error between the reference input $r(t)$ and the system output $y(t)$. We begin by checking that the three assumptions are satisfied. The state equation is in controller canonical form and is therefore controllable. We see from the associated transfer function given in Example 7.8 that there is neither a pole nor a zero at $s = 0$.

We may then proceed with the construction of a state feedback gain vector for the controllable pair

$$\begin{bmatrix} A & 0 \\ -C & 0 \end{bmatrix} = \left[\begin{array}{ccc|c} 0 & 1 & 0 & 0 \\ 0 & 0 & 1 & 0 \\ -18 & -15 & -2 & 0 \\ \hline -1 & 0 & 0 & 0 \end{array} \right] \qquad \begin{bmatrix} B \\ 0 \end{bmatrix} = \begin{bmatrix} 0 \\ 0 \\ 1 \\ 0 \end{bmatrix}$$

We select eigenvalues based on the ITAE criterion presented in Section 7.2 using an undamped natural frequency of $\omega_n = 2$ rad/s. This yields the desired fourth-order characteristic polynomial

$$s^4 + 2.1\,\omega_n s^3 + 3.4\,\omega_n^2 s^2 + 2.7\,\omega_n^3 s + \omega_n^4$$
$$= s^4 + 4.2s^3 + 13.6s^2 + 21.6s + 16$$

and associated eigenvalues

$$\lambda_{1,2} = -0.848 \pm j2.53 \qquad \lambda_{3,4} = -1.25 \pm j0.828$$

We observe that, although we started with a state equation in controller canonical form, the four-dimensional controllable pair is not in controller canonical form, so we must use our general purpose formulas to compute the state feedback gain vector. Either the Bass-Gura formula or Ackermann's formula yields

$$\begin{bmatrix} K & -k_I \end{bmatrix} = \begin{bmatrix} 3.6 & -1.4 & 2.2 \mid -16 \end{bmatrix}$$

and the associated control law is given by

$$\dot{\xi}(t) = r(t) - y(t)$$

$$u(t) = -\begin{bmatrix} 3.6 & -1.4 & 2.2 \end{bmatrix} \begin{bmatrix} x_1(t) \\ x_2(t) \\ x_3(t) \end{bmatrix} + 16\xi(t)$$

The resulting closed-loop state equation is

$$\begin{bmatrix} \dot{x}_1(t) \\ \dot{x}_2(t) \\ \dot{x}_3(t) \\ \dot{\xi}(t) \end{bmatrix} = \left[\begin{array}{ccc|c} 0 & 1 & 0 & 0 \\ 0 & 0 & 1 & 0 \\ -21.6 & -13.6 & -4.2 & 16 \\ \hline -1 & 0 & 0 & 0 \end{array} \right] \begin{bmatrix} x_1(t) \\ x_2(t) \\ x_3(t) \\ \xi(t) \end{bmatrix} + \begin{bmatrix} 0 \\ 0 \\ 0 \\ \hline 1 \end{bmatrix} r(t)$$

$$y(t) = \begin{bmatrix} 1 & 0 & 0 & \mid & 0 \end{bmatrix} \begin{bmatrix} x_1(t) \\ x_2(t) \\ x_3(t) \\ \xi(t) \end{bmatrix}$$

To illustrate the robustness properties of type 1 servomechanisms, we consider a perturbation to the original system dynamics matrix that yields

$$\tilde{A} = \begin{bmatrix} 0 & 1 & 0 \\ 0 & 0 & 1 \\ -16 & -16 & -1 \end{bmatrix}$$

having eigenvalues $-1, \pm j4$, indicating a marginally stable system. Using the same control law designed based on the nominal system dynamics matrix now yields the closed-loop system dynamics matrix

$$\begin{bmatrix} \tilde{A} - BK & Bk_I \\ -C & 0 \end{bmatrix} = \left[\begin{array}{ccc|c} 0 & 1 & 0 & 0 \\ 0 & 0 & 1 & 0 \\ -19.6 & -14.6 & -3.2 & 16 \\ \hline -1 & 0 & 0 & 0 \end{array} \right]$$

with eigenvalues

$$\lambda_{1,2} = -0.782 \pm j3.15 \quad \lambda_{3,4} = -0.818 \pm j0.922$$

Thus, despite the perturbation, the closed-loop system remains asymptotically stable. The closed-loop unit step responses for both the nominal and perturbed cases are shown in Figure 7.13. We see that the perturbation has an effect on the transient response but, because closed-loop asymptotic stability has been maintained, zero steady-state error is still achieved as forced by the integral-error term in the control law.

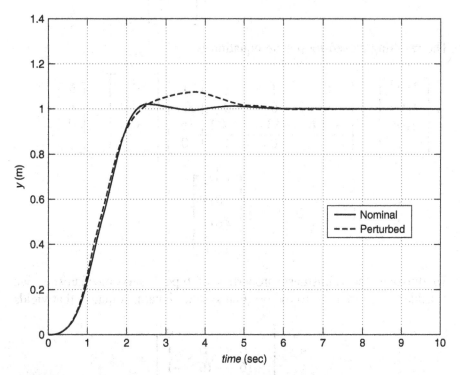

FIGURE 7.13 Nominal and perturbed closed-loop unit step responses for Example 7.9.

7.6 MATLAB FOR STATE FEEDBACK CONTROL LAW DESIGN

MATLAB for Shaping the Dynamic Response

The MATLAB functions that are useful for dynamic shaping are discussed in the MATLAB sections of Chapters 1 and 2. With either dynamic shaping method (dominant/augmented eigenvalues or ITAE eigenvalues), a useful MATLAB capability is provided in conjunction with the step function:

```
figure;
DesSys = tf(numDes,denDes);
step(DesSys);
```

where **denDes** contains the $n + 1$ coefficients of the nth-order desired closed-loop characteristic polynomial (found based on dominant-plus-augmented eigenvalues or ITAE eigenvalues), and **numDes** is the constant desired behavior numerator, chosen to normalize the steady-state unit step response value to 1.0. After executing the step function with the desired system, one can right-click in the figure window, as described earlier, to display the performance measures on the plot (rise time, peak time, percent overshoot, and settling time). MATLAB calculates these values numerically from the response data by applying the definition for each performance measure; i.e, they should be accurate even for non-second-order systems.

MATLAB for Control Law Design and Evaluation

The following MATLAB functions are useful for design of linear state feedback control laws:

K=place(A,B,DesEig) Solve for state feedback gain matrix **K** to place the desired eigenvalues **DesEig** of the closed-loop system dynamics matrix $A-BK$.

K=acker(A,B,DesEig) Solve for state feedback gain matrix **K** to place the desired eigenvalues **DesEig** of the closed-loop system dynamics matrix $A - BK$ using Ackermann's formula (for single-input, single-output systems only).

conv Multiply polynomial factors to obtain a polynomial product.

Continuing MATLAB Example

Shaping the Dynamic Response For the Continuing MATLAB Example (rotational mechanical system), we compute two desired eigenvalues for control law design to improve the performance relative to the open-loop responses of Figure 2.2. We use a desired percent overshoot and settling time of 3 percent and 0.7 s, respectively, to find the desired second-order control law eigenvalues. The following MATLAB code performs the dynamic shaping calculations for the Continuing MATLAB Example:

```
%-----------------------------------------------------------
%   Chapter 7.  Dynamic Shaping
%-----------------------------------------------------------

PO = 3;  ts = 0.7;              % Specify percent
                                % overshoot and settling
                                % time
term = pi^2 + log(PO/100)^2;
zeta = log(PO/100)/sqrt(term)   % Damping ratio from PO
wn   = 4/(zeta*ts)              % Natural frequency from
                                % settling time and zeta
num2  = wn^2;                   % Generic desired
                                % second-order system
den2  = [1 2*zeta*wn wn^2]
DesEig2 = roots(den2)           % Desired control law
                                % eigenvalues
Des2 = tf(num2,den2);           % Create desired system
                                % from num2 and den2

figure;
td = [0:0.01:1.5];
step(Des2,td);                  % Right-click to get
                                % performance measures
```

This m-file generates the following results. It also generates the desired closed-loop response shown in Figure 7.14, with the performance specifications displayed via right-clicking. We see that the 3 percent overshoot is achieved exactly as in theory, whereas the settling time is close to the desired 0.7 s value.

```
zeta =
    0.7448

wn =
    7.6722

den2 =
    1.0000    11.4286    58.8627
```

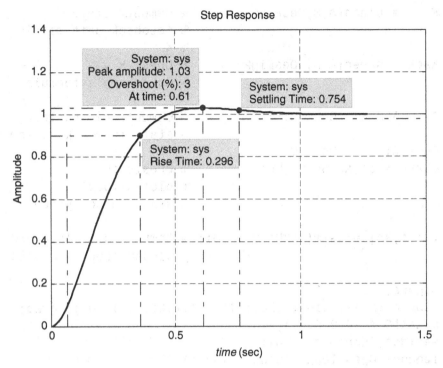

FIGURE 7.14   Desired second-order closed-loop response.

```
DesEig2 =
  -5.7143 + 5.1195i
  -5.7143 - 5.1195i
```

Control Law Design and Evaluation   Next, we design a state feed-back control law for the Continuing MATLAB Example. That is, we calculate the control law gain matrix K given A, B, and the desired control law eigenvalues developed in the previous subsection. The following MAT-LAB code segment performs this control law design for the continuing example:

```
%-------------------------------------------------------
%   Chapter 7.  Design of Linear State Feedback Control
% Laws
%-------------------------------------------------------
```

```
K     = place(A,B,DesEig2)        % Compute state
                                  % feedback gain matrix
                                  % K
Kack = acker(A,B, DesEig2);       % Check K via
                                  % Ackermann's formula

Ac = A-B*K;   Bc = B;             % Compute closed-loop
                                  % state feedback system
Cc = C;        Dc = D;
JbkRc = ss(Ac,Bc,Cc,Dc);         % Create the
                                  % closed-loop
                                  % state-space system

[Yc,t,Xc] = lsim(JbkRc,U,t,X0);  % Compare open-loop and
                                  % closed-loop responses

figure;
subplot(211), plot(t,Xo(:,1),'r',t,Xc(:,1),'g'); grid;
axis([0 4 -0.2 0.5]);
set(gca,'FontSize',18);
legend('Open-loop','Closed-loop');
ylabel(' \itx _1')
subplot(212), plot(t,Xo(:,2),'r',t,Xc(:,2),'g'); grid;
axis([0 4 -2 1]);
set(gca,'FontSize',18);
xlabel('\ittime (sec)'); ylabel('\itx_2');
```

This m-file, combined with the previous chapter m-files, yields the following output, plus the comparison of open- versus closed-loop state responses shown in Figure 7.15.

```
K =
18.86        7.43
```

Figure 7.15 shows that the simulated closed-loop system performs better than the open-loop system in terms of reduced overshoot and faster rise and settling times. There is less vibration with the closed-loop responses, and the steady-state zero values are obtained sooner than in the open-loop case.

Closed-loop system simulation also can be performed using MATLAB'S Simulink. Again, a detailed treatment of Simulink is beyond the scope of

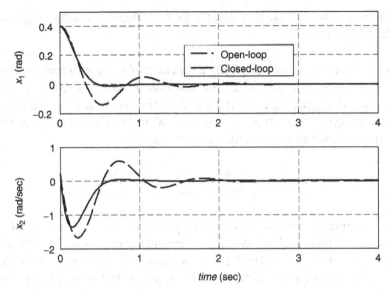

FIGURE 7.15 Open- versus closed-loop state responses for the Continuing MAT-LAB Example.

this book. The reader is encouraged to explore this powerful MATLAB toolbox on her or his own.

7.7 CONTINUING EXAMPLES: SHAPING DYNAMIC RESPONSE AND CONTROL LAW DESIGN

Continuing Example 1: Two-Mass Translational Mechanical System

Shaping the Dynamic Response For control law design, we need to specify desired eigenvalues that would improve the open-loop system performance. There are infinite possibilities; in this example we will use the two methods discussed in Section 7.2 to be applied later in control law design for Continuing Example 1.

We will use a dominant second-order system for desired eigenvalues to achieve a 5 percent overshoot and a 2 s settling time. These two eigenvalues must be augmented with two additional nondominant eigenvalues (we need 4 desired eigenvalues for our fourth-order system): real, negative, and at least 10 times farther to the left in the complex plane. We will use this specification of eigenvalues for multiple-input, multiple-output case *a*.

We also will use a fourth-order ITAE approach for desired eigenvalues, with the same undamped natural frequency ω_n as case *a* for easy comparison. We will use this specification of eigenvalues for single-input, single-output case *b*.

Case a Percent overshoot is only a function of ξ; substituting the desired 5 percent overshoot yields the dimensionless damping ratio $\xi = 0.69$. With this value, plus the desired 2 s settling time specification, we then find the undamped natural frequency $\omega_n = 2.90$ rad/s. This yields the desired dominant second-order transfer function, namely,

$$H(s) = \frac{\omega_n^2}{s^2 + 2\xi \omega_n s + \omega_n^2} = \frac{8.40}{s^2 + 4s + 8.40}$$

whose dominant, complex conjugate poles are $s_{1,2} = -2 \pm 2.10i$. The response of this desired dominant second-order system is shown in Figure 7.16 with the second-order performance measures. This figure was produced using the MATLAB step function with the preceding desired second-order numerator and denominator and then right-clicking to add the performance measures. As seen in Figure 7.16, we have obtained the

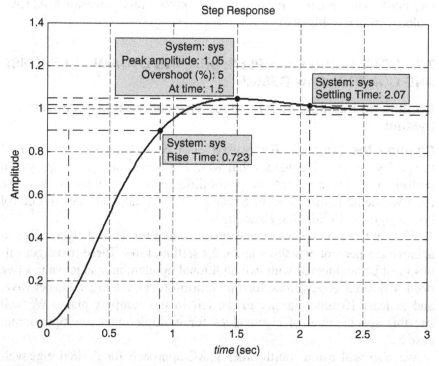

FIGURE 7.16 Dominant second-order response with performance characteristics.

desired 5 percent overshoot (at a peak time of 1.5 s) and a settling time of 2.07 s (2 s was specified). The 10 to 90 percent rise time is 0.723 s. We augment the dominant second-order eigenvalues to fit our fourth-order system as follows:

$$\lambda_{1,2} = -2 \pm 2.10i \qquad \lambda_{3,4} = -20, -21$$

Note that we do not specify repeated eigenvalues for $\lambda_{3,4}$ to avoid any numerical issues with repeated roots. The transfer function for the fourth-order desired behavior that mimics second-order behavior is (normalized for a steady-state value of 1)

$$H_4(s) = \frac{3528}{s^4 + 45s^3 + 592s^2 + 2024s + 3528}$$

We will wait to plot the step response of this augmented fourth-order transfer function until the following subsection, where we compare all responses on one graph.

Case b. For a fourth-order system, the optimal ITAE characteristic polynomial is

$$\alpha_{ITAE4}(s) = s^4 + 2.1 \, \omega_n s^3 + 3.4 \, \omega_n^2 s^2 + 2.7 \, \omega_n^3 s + \omega_n^4$$

In this example we will use the same natural frequency from above, that is, $\omega_n = 2.90$ rad/s yielding

$$\alpha_{ITAE4}(s) = s^4 + 6.09s^3 + 28.56s^2 + 65.72s + 70.54$$

For this fourth-order desired characteristic polynomial, the four desired eigenvalues are

$$\lambda_{1,2} = -1.23 \pm 3.66i \qquad \lambda_{3,4} = -1.81 \pm 1.20i$$

Figure 7.17 plots the fourth-order ITAE desired response, along with the dominant second-order and augmented fourth-order desired responses from case *a*. All are normalized to a steady-state value of 1.0 for easy comparison.

From Figure 7.17 we see that the augmented fourth-order response (solid) mimics the dominant second-order response (dashed) closely, as desired. The augmented fourth-order response lags the dominant second-order response, but it matches the required 5 percent overshoot and 2 s settling time well. The fourth-order ITAE response (dotted) did not

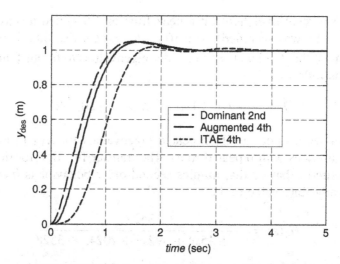

FIGURE 7.17 Dynamic shaping example results.

involve a percent overshoot or settling time specification, but we used the same natural frequency as in the dominant second-order response for comparison purposes. The ITAE response lags even further and demonstrates a fourth-order wiggle not present in the augmented fourth-order response.

Control Law Design and Evaluation   For both cases of Continuing Example 1, we calculate a state feedback gain matrix K to move the closed-loop system eigenvalues as desired and hence achieve the control objectives. In both cases, we simulate the closed-loop behavior and compare the open-loop system responses with those of the closed-loop system.

For case a (multiple-input, multiple-output), we design the control law based on the desired eigenvalues developed earlier and we simulate the closed-loop system response given the same conditions as the open-loop simulation of Chapter 2: zero initial state and step inputs of magnitudes 20 and 10 N, respectively, for $u_1(t)$ and $u_2(t)$.

For case b [input $u_2(t)$ and output $y_1(t)$], we design the control law based on the desired eigenvalues developed earlier and we simulate the closed-loop system response given the same conditions as the open-loop simulation of Chapter 2: zero input $u_2(t)$ and initial state $x(0) = [0.1, 0, 0.2, 0]^T$.

Case a.   The four desired eigenvalues corresponding to a desired 5% overshoot and 2 sec settling time were given earlier for case a : $\lambda_{1,2} =$

$-2 \pm 2.10i$ and $\lambda_{3,4} = -20, -21$. By using the MATLAB function **place** we find the following 2×4 state feedback gain matrix K:

$$K = \begin{bmatrix} 606 & 858 & 637 & 47 \\ -3616 & -166 & 759 & 446 \end{bmatrix}$$

On checking, the eigenvalues of $A - BK$ are indeed those specified in the **place** command. The reference input $r(t)$ we take to be the same as the open-loop input $u(t)$, i.e., step inputs of magnitudes 20 and 10 N, respectively, for $r_1(t)$ and $r_2(t)$. Simulating the closed-loop system response and comparing it with the open-loop system response yields Figure 7.18.

The first thing evident in Figure 7.18 is that the closed-loop control law attenuates the output (see Section 7.4). This is so because the control law in effect adds "virtual" springs in addition to the "real" springs in the system model; a stiffer set of springs will cause the output attenuation seen in Figure 7.18. Before we can discuss the performance of the control law, we must ensure that the steady-state closed-loop response match that of the open-loop system. This can be done in two ways: First, the output attenuation correction factors (simple term-by-term input gains) are 14.3 and 3.99 for outputs $y_1(t)$ and $y_2(t)$, respectively. The

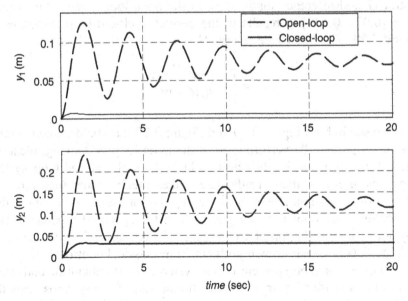

FIGURE 7.18 Open- versus closed-loop responses for case a, with output attenuation.

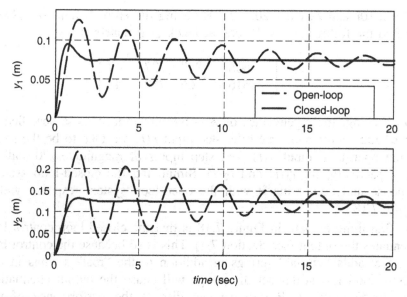

FIGURE 7.19 Open- versus closed-loop responses for case *a* corrected via scalar input gains.

corrected closed-loop responses following this approach are shown in Figure 7.19. Second, using the modified control law with attenuation matrix G and reference input equal to the open-loop steady state values ($R = [0.075, 0.125]^T$), results in the corrected closed-loop responses of Figure 7.20. The input gain matrix G is

$$G = \begin{bmatrix} 1206 & 437 \\ -3816 & 959 \end{bmatrix}$$

Now we see in both Figure 7.19 and Figure 7.20 that the designed control law has improved the performance of the open-loop system significantly. The settling time of 2 s has been achieved, so the closed-loop system responses approach their steady-state values much sooner than those of the open-loop system. However, in the top plot of Figure 7.19 we see that the percent overshoot is much greater than the specified 5 percent. This was not visible in the top plot of Figure 7.18 because of the vertical axis scaling. This is a well-known problem from classical controls—there are zeros (numerator roots) present in this system. The dominant second-order eigenvalue specification method does not account for any zeros; thus the results are skewed in the presence of zeros. In classical controls, a way to handle this is preshaping the input via filters.

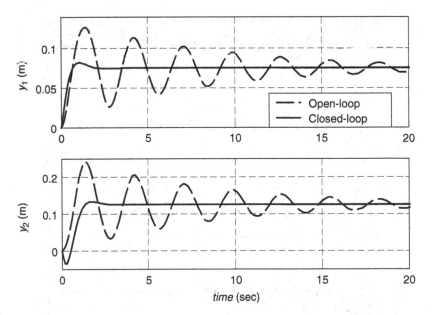

FIGURE 7.20 Open- versus closed-loop responses for case *a* corrected via input gain matrix.

Note that this overshoot problem is less severe when using the more general *G*-method as shown in Figure 7.20. However, for the output $y_2(t)$, there is a noticeable undershoot before it attains the desired value of 0.125.

Case b. The four desired closed-loop eigenvalues derived from the fourth-order ITAE characteristic polynomial are $\lambda_{1,2} = -1.23 \pm 3.66i$ and $\lambda_{3,4} = -1.81 \pm 1.20i$. Again using the MATLAB function **place** , we find the following 1×4 state feedback gain vector K:

$$K = \begin{bmatrix} -145 & -61 & 9 & 97 \end{bmatrix}$$

Since case *b* is a single-input, single-output situation, we can check this result using Ackermann's formula (MATLAB function **acker**); the results are identical. A quick check confirms that the eigenvalues of $A - BK$ are indeed those specified at the outset. For zero reference input and initial state $x(0) = [0.1, 0, 0.2, 0]^T$, the closed-loop state variable responses are compared with the open-loop state variable responses in Figure 7.21.

Since the closed-loop state variables all tend to zero in steady state as the open-loop states do, there is no output attenuation issue. We see in

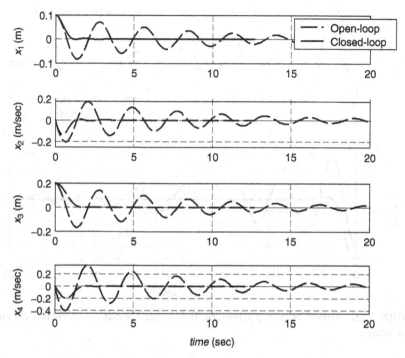

FIGURE 7.21 Open- versus closed-loop responses for case *b*.

Figure 7.21 that the designed control law has improved the performance of the open-loop system significantly. The closed-loop system responses meet their zero steady-state values much sooner than those of the open-loop system. The displacements do not overshoot significantly, and the velocity responses are both better than their open-loop counterparts.

Continuing Example 2: Rotational Electromechanical System

Shaping the Dynamic Response For control law design, we must specify desired eigenvalues to improve the open-loop system performance. In this example we will use a desired dominant first-order system, to be applied later in control law design for Continuing Example 2. We will use a dominant first-order system with time constant $\tau = \frac{1}{4}$ sec. We then augment the dominant eigenvalue associated with this time constant with two additional nondominant eigenvalues (we need three desired eigenvalues for our third-order system): real, negative, and larger amplitude so that their effect does not change the dominant behavior very much.

The relationship between the desired dominant eigenvalue a and first-order time constant τ is $e^{at} = e^{-t/\tau}$; therefore, $a = -1/\tau = -4$. This yields the desired dominant first-order transfer function:

$$H(s) = \frac{-a}{s-a} = \frac{4}{s+4}$$

Let us augment the dominant first-order eigenvalue to fit our third-order system desired eigenvalues requirement as follows: We choose additional eigenvalues approximately three times farther to the left to yield

$$\lambda_{1,2,3} = -4, -12, -13$$

Note that we do not specify repeated eigenvalues for $\lambda_{2,3}$ to avoid any numerical issues with repeated roots. The transfer function for the third-order desired behavior that mimics first-order behavior is (normalizing for a steady-state value of 1):

$$H_3(s) = \frac{624}{s^3 + 29s^2 + 256s + 624}$$

Figure 7.22 plots the augmented third-order desired response, along with the dominant first-order system from which it was derived. Both are normalized to a steady-state value of 1.0 for easy comparison. From Figure 7.22 we see that the augmented third-order response (solid) mimics the dominant first-order response (dashed) fairly closely. We see in the dashed curve that after three time constants (at $t = 0.75$ s), the dominant first-order response has achieved 95 percent of the steady-state value of 1.0. The augmented third-order response lags the dominant first-order

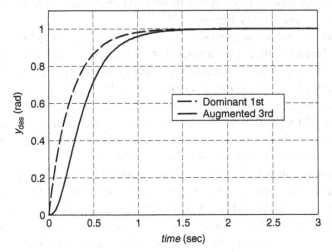

FIGURE 7.22 Dynamic shaping for Continuing Example 2.

response; we can make this arbitrarily close to the dashed curve by moving the augmented eigenvalues even farther to the left. However, this may lead to large numerical values for control gains, which is generally to be avoided. This could potentially require large actuator amplitudes, perhaps exceeding physical limits.

Control Law Design and Evaluation

For Continuing Example 2, we compute a state feedback gain matrix K to move the closed-loop system eigenvalues where desired and hence achieve the control objectives. Then we simulate the closed-loop behavior and compare the open-loop system responses with those of the closed-loop system.

We design the control law based on the desired eigenvalues developed in the preceding subsection, and we simulate the closed-loop system response given the same conditions as the open-loop simulation of Chapter 2 [zero initial state and a unit step input in voltage $v(t)$].

The three desired eigenvalues for this case, $\lambda_{1,2,3} = -4, -12, -13$, were specified by a dominant first-order eigenvalue augmented by two more real, negative eigenvalues farther to the left by a factor of three. By using MATLAB functions **place** or **acker** we find the following 1×3 state feedback gain matrix K:

$$K = \begin{bmatrix} 312 & 127 & 13 \end{bmatrix}$$

The eigenvalues of $A - BK$ are indeed those that were specified, as the reader may verify. Simulating the closed-loop system response for zero initial state and unit step reference input and comparing with the open-loop unit step response yields Figure 7.23.

There is no output attenuation issue because the open-loop response increases linearly after the transient response has died out; this is as expected because there is no torsional spring in the system model. However, we could use the input gain matrix method to achieve any desired steady-state angular displacement for the closed-loop system.

The closed-loop angular displacement $x_1(t)$ in Figure 7.23 (top plot) was artificially scaled to achieve a steady-state value of 0.5 rad. Comparing the open- and closed-loop angular displacement, we see that the state feedback control law has effectively added a virtual spring whereby we can servo to commanded angles rather than having the shaft angle increase linearly without bound as in the open-loop case. In Figure 7.23, the closed-loop angular velocity and acceleration both experience a transient and then tend to zero steady-state values. The open-loop values are the same as those plotted in Figure 2.5.

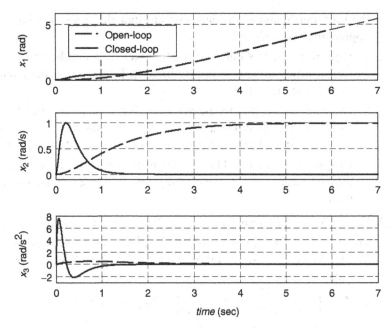

FIGURE 7.23 Open- versus closed-loop responses for Continuing Example 2.

For this example, the closed-loop system dynamics matrix is

$$A - BK = \begin{bmatrix} 0 & 1 & 0 \\ 0 & 0 & 1 \\ -624 & -256 & -29 \end{bmatrix}$$

Now the (3,1) element of $A - BK$ is no longer 0 as it was for the open-loop A matrix; this nonzero term represents the virtual spring provided by the control law, allowing constant commanded shaft angles to be achieved.

The coefficients of the closed-loop characteristic polynomial can be seen in the third row of $A - BK$, in ascending order of powers of s, with negative signs. The closed-loop system is asymptotically stable, changed from the marginally stable open-loop system because all three eigenvalues are now negative real numbers.

7.8 HOMEWORK EXERCISES

Numerical Exercises

NE7.1 For the following cases, determine acceptable closed-loop system eigenvalues to achieve the required behavior. In each case, plot the unit step response to demonstrate that the desired behavior is approximately satisfied.

a. Determine acceptable eigenvalues for a second-, third-, and fourth-order system to approximate a first-order system with a time constant of 0.5 s.

b. Determine acceptable eigenvalues for a second-, third-, and fourth- order system to approximate a second-order system with a percent overshoot of 6 percent and a settling time of 4 s.

c. Co-plot the desired ITAE responses for second-, third-, and fourth- order systems assuming $\omega_n = 5$ rad/s. Discuss your results.

NE7.2 For each (A, B) pair below, use the Bass-Gura formula to calculate the state feedback gain vector K to place the given eigenvalues of the closed-loop system dynamics matrix $A - BK$. Check your results.

a. $A = \begin{bmatrix} -1 & 0 \\ 0 & -4 \end{bmatrix}$ $B = \begin{bmatrix} 1 \\ 1 \end{bmatrix}$ $\lambda_{1,2} = -2 \pm 3i$

b. $A = \begin{bmatrix} 0 & 1 \\ -6 & -8 \end{bmatrix}$ $B = \begin{bmatrix} 0 \\ 1 \end{bmatrix}$ $\lambda_{1,2} = -4, -5$

c. $A = \begin{bmatrix} 0 & 1 \\ -6 & 0 \end{bmatrix}$ $B = \begin{bmatrix} 0 \\ 1 \end{bmatrix}$ $\lambda_{1,2} = -4, -5$

d. $A = \begin{bmatrix} 0 & 8 \\ 1 & 10 \end{bmatrix}$ $B = \begin{bmatrix} 1 \\ 0 \end{bmatrix}$ $\lambda_{1,2} = -1 \pm i$

NE7.3 Repeat **NE 7.2** using Ackermann's Formula.

Analytical Exercises

AE7.1 Show that the $(n + 1)$-dimensional single-input, single-output state equation

$$\begin{bmatrix} \dot{x}(t) \\ \dot{\xi}(t) \end{bmatrix} = \begin{bmatrix} A & 0 \\ C & 0 \end{bmatrix} \begin{bmatrix} x(t) \\ \xi(t) \end{bmatrix} + \begin{bmatrix} B \\ 0 \end{bmatrix} u(t)$$

with output $\xi(t)$ is controllable if and only if the n-dimensional single-input, single-output state equation represented by (A, B, C) is controllable and the $(n + 1) \times (n + 1)$ matrix

$$\begin{bmatrix} A & B \\ C & 0 \end{bmatrix}$$

is nonsingular. What is the interpretation of the state variable $\xi(t)$?

AE7.2 Suppose that the pair (A, B) is controllable, and for any finite $t_f >$ 0, consider the state feedback gain matrix $K = -B^T W^{-1}(0, t_f)$. Show that the closed-loop state equation

$$\dot{x}(t) = (A - BK)x(t)$$

is asymptotically stable. Hint: Recall AE6.5.

AE7.3 Suppose that given the symmetric positive definite matrix Q, the *Riccati matrix equation*

$$A^T P + PA - PBB^T P + Q = 0$$

has a symmetric positive-definite solution P. Show that the state feedback law $u(t) = -K\,x(t)$ with $K = -B^T P$ yields the closed-loop state equation

$$\dot{x}(t) = (A - BK)x(t)$$

that is asymptotically stable, thereby implying that the pair (A, B) is stabilizable.

AE7.4 For the linear state equation (7.1) and state feedback law (7.4), show that the open- and closed-loop transfer functions are related by

$$H_{CL}(s) = H(s)[I + K(sI - A)^{-1}B]^{-1}G$$

This indicates that the same closed-loop input-output behavior can be achieved using a dynamic precompensator instead of state feedback. Hint: Use AE1.8.

Continuing MATLAB Exercises

CME7.1 For the system given in CME1.1:

    a. Determine the desired eigenvalues for a generic second-order system to obtain 1 percent overshoot and a 1-s settling time. Plot the unit step response for this desired behavior and discuss your results.

    b. Design a state feedback control law to achieve the eigenvalue placement of part a. Compare open- and closed-loop responses to a unit step input, assuming zero initial conditions. Normalize the closed-loop output level to match that of the open-loop system for easy comparison. Were your control law design goals met?

c. Design a state feedback servomechanism to achieve the eigenvalue placement of part a and a steady-state output value of 1. Compare open- and closed-loop responses to a unit step input. In this case, normalize the open-loop output level to 1 for easy comparison. Were your control law design goals met?

CME7.2 For the system given in CME1.2:

a. Determine the desired eigenvalues for a first-order system with a time constant of 1 s, augmented by two additional real, negative eigenvalues (the first exactly 10 times the dominant eigenvalue, the third one less than the second). Plot the unit step response for this desired behavior, and discuss your results.

b. Design a state feedback control law to achieve the eigenvalue placement of part a. Compare open- and closed-loop responses to a unit step input. Normalize the closed-loop output level to match that of the open-loop system for easy comparison. Were your control law design goals met?

c. Design a state feedback servomechanism to achieve the eigenvalue placement of part a and a steady-state output value of 1. Compare open- and closed-loop responses to a unit step input. In this case, normalize the open-loop output level to 1 for easy comparison. Were your control law design goals met?

CME7.3 For the system given in CME1.3:

a. Determine the desired eigenvalues for a fourth–order ITAE system with $\omega_n = 2$ rad/s. Plot the unit step response for this desired behavior, and discuss your results.

b. Design a state feedback control law to achieve the eigenvalue placement of part a. Compare open- and closed-loop responses to a unit step input. Normalize the closed-loop output level to match that of the open-loop system for easy comparison. Were your control law design goals met?

c. Design a state feedback servomechanism to achieve the eigenvalue placement of part a and a steady-state output value of 1. Compare open- and closed-loop responses to a unit step input. In this case normalize, the open-loop output level to 1 for easy comparison. Were your control law design goals met?

CME7.4 For the system given in CME1.4:

a) Determine the desired eigenvalues for a generic second-order system to obtain 2 percent overshoot and a 2-s settling time. Plot the unit step response for this desired behavior and discuss your results.

b) Design a state feedback control law to achieve the eigenvalue placement of part a. Compare open- and closed-loop responses to a unit step input. Normalize the closed-loop output level to match that of the open-loop system for easy comparison. Were your control law design goals met?

c) Design a state feedback servomechanism to achieve the eigenvalue placement of part a and a steady-state output value of 1. Compare open- and closed-loop responses to a unit step input. In this case, normalize the open-loop output level to 1 for easy comparison. Were your control law design goals met?

Continuing Exercises

CE7.1a Given control law design criteria of 3 percent overshoot and a 3-s settling time: (i) Calculate the undamped natural frequency and damping ratio required for a standard generic second-order system to achieve this (assuming a unit step input). What are the associated system eigenvalues? Plot the unit step response for the result and demonstrate how well the design criteria are met (normalize your output to ensure the final value is 1.0). Display the resulting rise time, peak time, settling time, and percent overshoot on your graph. (ii) Augment these desired second-order system eigenvalues for control law design in CE1; since this is a sixth-order system you, will need four additional eigenvalues (real, negative, approximately 10 times farther to the left). Choose the first additional eigenvalue to be exactly ten times the real part of the dominant second-order system eigenvalues. For the remaining three, successively subtract one from the first additional eigenvalue (to avoid repeated eigenvalues). (iii) Also compute the optimal ITAE sixth-order coefficients and eigenvalues using an undamped natural frequency twice that from the dominant second-order approach. Plot both the sixth-order ITAE and the augmented sixth-order desired step responses with the dominant second-order step response of part (i) (normalize to ensure steady-state values of 1.0); compare and discuss.

CE7.1b For the desired closed-loop eigenvalues from CE7.1a, design state feedback control laws (i.e., calculate K) for all three cases from CE2.1. For cases (i) and (ii), use the augmented sixth-order eigenvalues based on dominant second-order behavior; for case (iii), use the sixth-order ITAE eigenvalues. In each case, evaluate your results: Plot and compare the simulated open- versus closed-loop output responses for the same input cases as in CE2.1a; use output attenuation correction so that the closed-loop steady-state values match the open-loop steady-state values for easy comparisons.

CE7.2a Since this is a fourth-order system, we will need four desired eigenvalues for future control law design in CE2. Use a fourth-order ITAE approach with undamped natural frequency $\omega_n = 3$ rad/s to generate the four desired eigenvalues. Plot the desired system impulse response.

CE7.2b For the desired closed-loop eigenvalues from CE7.2a, design state feedback control laws (i.e., calculate K) for all three cases from CE2.2. In each case, evaluate your results: Plot and compare the simulated open- versus closed-loop output responses for the same input cases as in CE2.2a. Be sure to scale the vertical axes so that the closed-loop responses are clearly visible.

CE7.3a Use a dominant first-order system with time constant $\tau = 0.5$ s. What is the associated desired eigenvalue? Augment this desired first-order eigenvalue for future control law design in CE3; since this is a third-order system, you will need two additional eigenvalues (real, negative, approximately 10 times farther to the left). Choose the first additional eigenvalue to be exactly 10 times the dominant first-order eigenvalue. For the remaining eigenvalue, subtract one from the first additional eigenvalue (to avoid repeated eigenvalues). Plot this augmented third-order desired step response versus the dominant first-order step response (normalize to ensure steady-state values of 1.0); compare and discuss.

CE7.3b For the desired closed-loop eigenvalues from CE7.3a, design state feedback control laws (i.e., calculate K) for both cases from CE2.3. In each case, evaluate your results: Plot and compare the simulated open- versus closed-loop output responses for the same input cases as in CE2.3a [for case (ii), use output attenuation correction so that the closed-loop steady-state values match the open-loop steady-state values for easy comparison].

CE7.4a Based on your knowledge of the CE4 system and the methods of this chapter, calculate a good set of desired closed-loop eigenvalues to be used as input to the control law design problem. Plot the responses using appropriate input signals, and discuss your results.

CE7.4b For your desired closed-loop eigenvalues from CE7.4a, design a state feedback control law (i.e., calculate K). Evaluate your results: Plot and compare the simulated open- versus closed-loop output responses for a unit impulse input and zero initial conditions as in CE2.4a.

CE7.5 In Chapter 5 we found that the CE1.5 system was not minimal. However, since the open-loop system is controllable, for this problem, use the fourth-order system for control law design, i.e., do not use the minimal second-order system. In this way you can preserve the physical interpretation of the four state variables.

CE7.5a Based on your knowledge of the CE5 system and the methods of this chapter, compute a good set of desired closed-loop eigenvalues to be used as input to the control law design problem. Plot the responses using appropriate input signals, and discuss your results.

CE7.5b For your desired closed-loop eigenvalues from CE7.5a, design a state feedback control law (i.e., calculate K). Evaluate your results: Plot and compare the simulated open- versus closed-loop output responses for the initial conditions given in CE2.5a.

8

OBSERVERS AND OBSERVER-BASED COMPENSATORS

For the linear time-invariant state equation

$$\begin{aligned}\dot{x}(t) &= Ax(t) + Bu(t) \\ y(t) &= Cx(t)\end{aligned} \qquad x(0) = x_0 \qquad (8.1)$$

we know from Chapter 7 that if the pair (A, B) is controllable then, a state feedback control law can be constructed to arbitrarily locate the closed-loop eigenvalues. While this is an extremely powerful result, there is one serious drawback. As we noted in Chapter 4, it is unreasonable to expect in real-world applications that every state variable is measurable, which jeopardizes are ability to implement a state feedback control law. This fact motivated our pursuit of an estimate of the state vector derived from measurements of the input and output and led us to the concept of observability. In particular, we established the fundamental result that the initial state can be uniquely determined (and therefore, the state trajectory can be reconstructed) from input and output measurements when and only when the state equation is observable. Unfortunately, the scheme proposed for computing the initial state is not directly amenable to real-time control.

In this chapter we present the *linear state observer*, also known as the *Luenberger observer*, named after D. G. Luenberger, who did much of the

Linear State-Space Control Systems, by Robert L. Williams II and Douglas A. Lawrence
Copyright © 2007 John Wiley & Sons, Inc.

groundbreaking work on state observation in the 1960s. As we will see, an observer is itself a dynamic system that for an observable state equation can be designed to provide an asymptotically convergent estimate of the state. We then pursue related topics, including the system property of *detectability*, which is related to observability in the same way that stabilizability is related to controllability, and the construction of observers of reduced dimension. Following this, we resolve the dilemma raised earlier by showing that state feedback control laws can be implemented using the observer-generated state estimate in place of the actual state. The result is an observer-based compensator that has very special properties. Finally, we conclude the chapter with MATLAB demonstrations of these concepts using our Continuing MATLAB Example and Continuing Examples 1 and 2.

8.1 OBSERVERS

For the n-dimensional linear state equation (8.1), we define a linear state observer to also be an n-dimensional linear state equation that accepts $u(t)$ and $y(t)$ as inputs and whose state represents the estimate of $x(t)$. The observer assumes the form

$$
\begin{aligned}
\dot{\hat{x}}(t) &= A\hat{x}(t) + Bu(t) + L[y(t) - \hat{y}(t)] \qquad \hat{x}(0) = \hat{x}_0 \qquad (8.2) \\
\hat{y}(t) &= C\hat{x}(t)
\end{aligned}
$$

which looks like a copy of the state equation (8.1) driven by an error term $y(t) - \hat{y}(t)$ that enters the dynamics through an $n \times p$ *observer gain matrix* L. This error term is intended to drive the state estimate $\hat{x}(t)$ to the actual state $x(t)$ over time. To further explore this key convergence issue, we define the *estimation error* $\tilde{x}(t) = x(t) - \hat{x}(t)$ in terms of which we derive the *error dynamics*

$$
\begin{aligned}
\dot{\tilde{x}}(t) &= \dot{x}(t) - \dot{\hat{x}}(t) \\
&= [Ax(t) + Bu(t)] - \{A\hat{x}(t) + Bu(t) + L(y(t) - \hat{y}(t))\} \\
&= Ax(t) - A\hat{x}(t) - L[Cx(t) - C\hat{x}(t)] \\
&= A\tilde{x}(t) - LC\tilde{x}(t) \\
&= (A - LC)\tilde{x}(t) \qquad\qquad\qquad\qquad\qquad (8.3)
\end{aligned}
$$

for which the initial state is given by $\tilde{x}(0) = x(0) - \hat{x}(0) = x_0 - \hat{x}_0$. Note that since the error dynamics specify a homogeneous linear state equation, if we could initialize the observer by $\hat{x}(0) = \hat{x}_0 = x_0$ to yield zero initial error $\tilde{x}(0) = 0$, then $\tilde{x}(t) \equiv 0$ and thus $\hat{x}(t) = x(t)$ for all

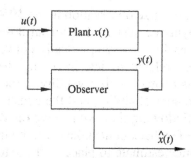

FIGURE 8.1 High-level observer block diagram.

$t \geq 0$. However, since the initial state of Equation (8.1), $x(0) = x_0$, is presumably unknown, we cannot achieve perfect estimation, so we set our sights on generating an asymptotically convergent estimate starting with any initial error. This corresponds to the requirement that the error dynamics (8.3) be asymptotically stable. We know from Chapter 6 that the error dynamics will be asymptotically stable if and only if the matrix $A - LC$ that governs the error dynamics has strictly negative real-part eigenvalues. Clearly, this eigenvalue condition depends on the observer gain vector L.

A high-level observer diagram is shown in Figure 8.1. The true state $x(t)$ of the Plant is not known in general. The observer inputs are system input $u(t)$ and output (via sensors) $y(t)$, and the observer output is the state estimate $\hat{x}(t)$. The observer state equation (8.2) is shown graphically in the block diagram of Figure 8.2. This figure provides details for Figure 8.1.

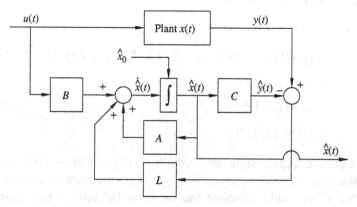

FIGURE 8.2 Detailed observer block diagram.

Observer Error Eigenvalue Placement

Since $A - LC$ and $(A - LC)^T = A^T - C^T L^T$ have identical eigenvalues, we see that on making the following associations

$$A \leftrightarrow A^T \quad B \leftrightarrow C^T$$
$$K \leftrightarrow L^T$$

the problem of choosing the observer gain vector L to place the eigenvalues of $A - LC$ at desirable locations is equivalent to the fundamental problem considered in Chapter 6: choosing the "state feedback gain matrix" L^T to locate the eigenvalues of $A^T - C^T L^T$. Not surprisingly, the duality between these problems allows us to leverage the heavy lifting of Chapter 6 to yield a streamlined treatment of the problem at hand.

Theorem 8.1 *For any symmetric set of n complex numbers* $\{\mu_1, \mu_2, \dots, \mu_n\}$, *there exists an observer gain matrix* L *such that* $\sigma(A - LC) = \{\mu_1, \mu_2, \dots, \mu_n\}$ *if and only if the pair* (A, C) *is observable.*

Proof. By duality, the pair (A, C) is observable if and only if the pair (A^T, C^T) is controllable. By Theorem 7.1, there exists an L^T such that $\sigma(A^T - C^T L^T) = \{\mu_1, \mu_2, \dots, \mu_n\}$ if and only if the pair (A^T, C^T) is controllable. These relationships are illustrated in the following diagram, in which the double arrows represent the equivalences reflected by our "if and only if" statements and the eigenvalue relationship $\sigma(A - LC) = \sigma(A^T - C^T L^T)$.

(A, C) is observable	$\Leftrightarrow$	(A^T, C^T) is controllable
		$\Updownarrow$
Arbitrary eigenvalue placement for $A - LC$	$\Leftrightarrow$	Arbitrary eigenvalue placement for $A^T - C^T L^T$

Consequently, we conclude that there exists an observer gain matrix L such that $\sigma(A - LC) = \{\mu_1, \mu_2, \dots, \mu_n\}$ if and only if the pair (A, C) is observable. $\square$

We also may exploit the duality between these two eigenvalue placement problems to derive observer gain formulas for the single-output case by appropriately modifying the feedback gain formulas from Chapter 6. We again refer the interested reader to Rugh (1996) and Kailath (1980) for observer gain constructions in the multiple-output case. The Chapter 6 discussion relating transient response characteristics to eigenvalue locations also applies to the problem of shaping the estimation error response. We note here that it is generally desirable to have an observer error response that is much faster (typically an order of magnitude) than the desired closed-loop transient response. We will elaborate on this point in Section 8.4.

Observer Gain Formula for Observer Canonical Form

The observer canonical form involves the observable pair (A_{OCF}, C_{OCF}) specified by

$$A_{OCF} = \begin{bmatrix} 0 & 0 & \cdots & 0 & -a_0 \\ 1 & 0 & \cdots & 0 & -a_1 \\ 0 & 1 & \cdots & 0 & -a_2 \\ \vdots & \vdots & \ddots & \vdots & \vdots \\ 0 & 0 & \cdots & 1 & -a_{n-1} \end{bmatrix} \quad C_{OCF} = [0 \quad 0 \quad \cdots \quad 0 \quad 1]$$

We know that the characteristic polynomial of A_{OCF} is also written down, by inspection, as

$$|sI - A_{OCF}| = s^n + a_{n-1}s^{n-2} + \cdots + a_2 s^2 + a_1 s + a_0$$

and as in Chapter 6, we represent our desired eigenvalues $\{\mu_1, \mu_2, \ldots, \mu_n\}$ by the associated characteristic polynomial

$$\alpha(s) = (s - \mu_1)(s - \mu_2) \cdots (s - \mu_n)$$
$$= s^n + \alpha_{n-1}s^{n-1} + \cdots + \alpha_2 s^2 + \alpha_1 s + \alpha_0$$

Thus, on setting

$$L_{OCF} = \begin{bmatrix} \alpha_0 - a_0 \\ \alpha_1 - a_1 \\ \alpha_2 - a_2 \\ \vdots \\ \alpha_{n-1} - a_{n-1} \end{bmatrix}$$

we obtain

$$A_{OCF} - L_{OCF} C_{OCF} = \begin{bmatrix} 0 & 0 & \cdots & 0 & -a_0 \\ 1 & 0 & \cdots & 0 & -a_1 \\ 0 & 1 & \cdots & 0 & -a_2 \\ \vdots & \vdots & \ddots & \vdots & \vdots \\ 0 & 0 & \cdots & 1 & -a_{n-1} \end{bmatrix}$$

$$-\begin{bmatrix} \alpha_0 - a_0 \\ \alpha_1 - a_1 \\ \alpha_2 - a_2 \\ \vdots \\ \alpha_{n-1} - a_{n-1} \end{bmatrix} \begin{bmatrix} 0 & 0 & \cdots & 0 & 1 \end{bmatrix}$$

$$= \begin{bmatrix} 0 & 0 & \cdots & 0 & -\alpha_0 \\ 1 & 0 & \cdots & 0 & -\alpha_1 \\ 0 & 1 & \cdots & 0 & -\alpha_2 \\ \vdots & \vdots & \ddots & \vdots & \vdots \\ 0 & 0 & \cdots & 1 & -\alpha_{n-1} \end{bmatrix}$$

having the characteristic polynomial

$$|sI - A_{\text{OCF}} + L_{\text{OCF}} C_{\text{OCF}}| = s^n + \alpha_{n-1} s^{n-1} + \cdots + \alpha_2 s^2 + \alpha_1 s + \alpha_0$$

as desired.

Bass-Gura Formula

Given the observable pair (A, C), we know from Chapter 6 that corresponding to the controllable pair (A^T, C^T),

$$L^T = \begin{bmatrix} (\alpha_0 - a_0) & (\alpha_1 - a_1) & (\alpha_2 - a_2) & \cdots & (\alpha_{n-1} - a_{n-1}) \end{bmatrix}$$

$$\times \begin{bmatrix} P_{(A^T, C^T)} \begin{bmatrix} a_1 & a_2 & \cdots & a_{n-1} & 1 \\ a_2 & a_3 & \cdots & 1 & 0 \\ \vdots & \vdots & \ddots & \vdots & \vdots \\ a_{n-1} & 1 & \cdots & 0 & 0 \\ 1 & 0 & \cdots & 0 & 0 \end{bmatrix} \end{bmatrix}^{-1}$$

is such that

$$|sI - (A^T - C^T L^T)| = s^n + \alpha_{n-1} s^{n-1} + \cdots + \alpha_2 s^2 + \alpha_1 s + \alpha_0$$

Thus, by simply taking the transpose of this expression and using

$$Q_{(A,C)} = P^T_{(A^T, C^T)}$$

along with the fact that

$$Q_{OCF}^{-1} = \begin{bmatrix} a_1 & a_2 & \cdots & a_{n-1} & 1 \\ a_2 & a_3 & \cdots & 1 & 0 \\ \vdots & \vdots & \ddots & \vdots & \vdots \\ a_{n-1} & 1 & \cdots & 0 & 0 \\ 1 & 0 & \cdots & 0 & 0 \end{bmatrix} = P_{CCF}^{-1}$$

is symmetric, we obtain the Bass-Gura formula for the observer gain vector, that is,

$$L = \begin{bmatrix} \begin{bmatrix} a_1 & a_2 & \cdots & a_{n-1} & 1 \\ a_2 & a_3 & \cdots & 1 & 0 \\ \vdots & \vdots & \ddots & \vdots & \vdots \\ a_{n-1} & 1 & \cdots & 0 & 0 \\ 1 & 0 & \cdots & 0 & 0 \end{bmatrix} Q_{(A,C)} \end{bmatrix}^{-1} \begin{bmatrix} \alpha_0 - a_0 \\ \alpha_1 - a_1 \\ \alpha_2 - a_2 \\ \vdots \\ \alpha_{n-1} - a_{n-1} \end{bmatrix}$$

Example 8.1 We construct an asymptotic state observer for the three-dimensional state equation

$$\begin{bmatrix} \dot{x}_1(t) \\ \dot{x}_2(t) \\ \dot{x}_3(t) \end{bmatrix} = \begin{bmatrix} 0 & 1 & 0 \\ 0 & 0 & 1 \\ -4 & -4 & -1 \end{bmatrix} \begin{bmatrix} x_1(t) \\ x_2(t) \\ x_3(t) \end{bmatrix} + \begin{bmatrix} 0 \\ 0 \\ 1 \end{bmatrix} u(t)$$

$$y(t) = \begin{bmatrix} 1 & 0 & 0 \end{bmatrix} \begin{bmatrix} x_1(t) \\ x_2(t) \\ x_3(t) \end{bmatrix}$$

Since this state equation is in controller canonical form, we obtain the open-loop characteristic polynomial by inspection, from which we determine the polynomial coefficients and eigenvalues as follows:

$$a(s) = s^3 + s^2 + 4s + 4$$
$$= (s+1)(s-j2)(s+j2) \qquad a_2 = 1 \quad a_1 = 4 \quad a_0 = 4$$
$$\lambda_{1,2,3} = -1, \pm j2$$

We choose the following eigenvalues for the observer error dynamics:

$$\{\mu_1, \mu_2, \mu_3\} = \{-1, -2, -3\}$$

so that the desired characteristic polynomial and associated polynomial coefficients are

$$\begin{aligned}\alpha(s) &= (s+1)(s+2)(s+3) \\ &= s^3 + 6s^2 + 11s + 6\end{aligned} \qquad \alpha_2 = 6 \quad \alpha_1 = 11 \quad \alpha_0 = 6$$

We check observability of the pair (A, C) using the observability matrix

$$Q = \begin{bmatrix} C \\ CA \\ CA^2 \end{bmatrix} = \begin{bmatrix} 1 & 0 & 0 \\ 0 & 1 & 0 \\ 0 & 0 & 1 \end{bmatrix}$$

which is clearly nonsingular, so this state equation is observable. We next proceed directly to the observer gain vector computation using the Bass-Gura formula:

$$\begin{aligned} L &= \left[\begin{bmatrix} 4 & 1 & 1 \\ 1 & 1 & 0 \\ 1 & 0 & 0 \end{bmatrix} \begin{bmatrix} 1 & 0 & 0 \\ 0 & 1 & 0 \\ 0 & 0 & 1 \end{bmatrix} \right]^{-1} \begin{bmatrix} 6-4 \\ 11-4 \\ 6-1 \end{bmatrix} \\ &= \begin{bmatrix} 0 & 0 & 1 \\ 0 & 1 & -1 \\ 1 & -1 & -3 \end{bmatrix} \begin{bmatrix} 2 \\ 7 \\ 5 \end{bmatrix} \\ &= \begin{bmatrix} 5 \\ 2 \\ -20 \end{bmatrix} \end{aligned}$$

With the aid of MATLAB, we see that

$$A - LC = \begin{bmatrix} -5 & 1 & 0 \\ -2 & 0 & 1 \\ 16 & -4 & -1 \end{bmatrix}$$

has eigenvalues $\{-1, -2, -3\}$ and characteristic polynomial $\alpha(s) = s^3 + 6s^2 + 11s + 6$, so the desired result has been achieved. The asymptotic state observer is given by

$$\begin{bmatrix} \dot{\hat{x}}_1(t) \\ \dot{\hat{x}}_2(t) \\ \dot{\hat{x}}_3(t) \end{bmatrix} = \begin{bmatrix} -5 & 1 & 0 \\ -2 & 0 & 1 \\ 16 & -4 & -1 \end{bmatrix} \begin{bmatrix} \hat{x}_1(t) \\ \hat{x}_2(t) \\ \hat{x}_3(t) \end{bmatrix} + \begin{bmatrix} 0 \\ 0 \\ 1 \end{bmatrix} u(t) + \begin{bmatrix} 5 \\ 2 \\ -20 \end{bmatrix} y(t)$$

We investigate the observer performance for the following initial conditions and input signal:

$$x_0 = \begin{bmatrix} 1 \\ -1 \\ 1 \end{bmatrix} \quad \hat{x}_0 = \begin{bmatrix} 0 \\ 0 \\ 0 \end{bmatrix} \quad u(t) = \sin(t) \quad t \geq 0$$

Since in practice the initial state x_0 is unknown, the initial state for the observer typically is set to $\hat{x}_0 = 0$ because in a probabilistic sense this represents the expected value of an initial state that is a zero mean random variable. Figure 8.3 compares the state response $x(t)$ with the observer response $\hat{x}(t)$. Even though the original state equation is not asymptotically stable and is driven by a sinusoidal input signal, $\hat{x}(t)$ asymptotically converges to $x(t)$, as dictated by the asymptotically stable observer error dynamics.   □

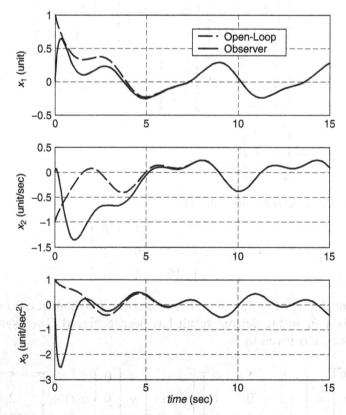

FIGURE 8.3 Open-loop state response and observer estimates for Example 8.1.

Example 8.2   In this example we design an observer for the state equation appearing in Example 7.4, which we repeat here for convenience. Thus

$$\begin{bmatrix} \dot{x}_1(t) \\ \dot{x}_2(t) \\ \dot{x}_3(t) \end{bmatrix} = \begin{bmatrix} 0 & 1 & 0 \\ 0 & 0 & 1 \\ -18 & -15 & -2 \end{bmatrix} \begin{bmatrix} x_1(t) \\ x_2(t) \\ x_3(t) \end{bmatrix} + \begin{bmatrix} 0 \\ 0 \\ 1 \end{bmatrix} u(t)$$

$$y(t) = [1 \quad 0 \quad 0] \begin{bmatrix} x_1(t) \\ x_2(t) \\ x_3(t) \end{bmatrix}$$

which, as we recall, is in controller canonical form. Therefore, we see by inspection that the characteristic polynomial and polynomial coefficients are

$$a(s) = s^3 + 2s^2 + 15s + 18 \quad a_2 = 2 \quad a_1 = 15 \quad a_0 = 18$$

In Example 7.4 we specified desired closed-loop eigenvalues to be $\mu_{1,2,3} = -1.33 \pm j1.49, -13.3$ to achieve a percent overshoot of 6 percent and a settling time of 3 s for the closed-loop step response. We select eigenvalues for the observer error dynamics by scaling the preceding eigenvalues by 10 to give $\mu_{1,2,3} = -13.3 \pm j14.9, -133.3$. The desired characteristic polynomial for the observer error dynamics then becomes

$$\alpha(s) = s^3 + 160s^2 + 3955s + 53260$$

$$\alpha_2 = 160 \quad \alpha_1 = 3955 \quad \alpha_0 = 53260$$

Note that when comparing this characteristic polynomial for the observer error dynamics with $s^3 + 16s^2 + 39.55s + 53.26$ associated with the eigenvalues achieved by state feedback in Example 7.4, we see that $160/16 = 10$, $3955/39.55 = 100$, and $53260/53.26 = 1000$ as a result of the factor of 10 relating the respective eigenvalues.

We next check observability of the pair (A, C) via

$$Q = \begin{bmatrix} C \\ CA \\ CA^2 \end{bmatrix} = \begin{bmatrix} 1 & 0 & 0 \\ 0 & 1 & 0 \\ 0 & 0 & 1 \end{bmatrix}$$

which is obviously nonsingular, so the state equation is observable. We then proceed to compute the observer gain vector using the Bass-Gura formula, that is,

$$
L = \left[\begin{bmatrix} 15 & 2 & 1 \\ 2 & 1 & 0 \\ 1 & 0 & 0 \end{bmatrix} \begin{bmatrix} 1 & 0 & 0 \\ 0 & 1 & 0 \\ 0 & 0 & 1 \end{bmatrix} \right]^{-1} \begin{bmatrix} (53260 - 18) \\ (3955 - 15) \\ (160 - 2) \end{bmatrix}
$$

$$
= \begin{bmatrix} 0 & 0 & 1 \\ 0 & 1 & -2 \\ 1 & -2 & -11 \end{bmatrix} \begin{bmatrix} 53242 \\ 3940 \\ 158 \end{bmatrix}
$$

$$
= \begin{bmatrix} 158 \\ 3624 \\ 43624 \end{bmatrix}
$$

We conclude the example by verifying that

$$
A - LC = \begin{bmatrix} 0 & 1 & 0 \\ 0 & 0 & 1 \\ -18 & -15 & -2 \end{bmatrix} - \begin{bmatrix} 158 \\ 3624 \\ 43624 \end{bmatrix} \begin{bmatrix} 1 & 0 & 0 \end{bmatrix}
$$

$$
= \begin{bmatrix} -158 & 1 & 0 \\ -3624 & 0 & 1 \\ -43642 & -15 & -2 \end{bmatrix}
$$

has eigenvalues $\lambda_{1,2,3} = -13.3 \pm j14.9, -133.3$ as required. We observe that the "fast" observer eigenvalues we specified resulted in large observer gains that may be impractical. Therefore, we may need to trade off the rapid convergence of the observer error with practical limitations on achievable observer gains. □

Ackermann's Formula

Given the observable pair (A, C), again, our strategy is to first apply our Chapter 6 result to the controllable pair (A^T, C^T) to write

$$
L^T = [0 \quad 0 \quad \cdots \quad 0 \quad 1] P_{(A^T, C^T)}^{-1} \alpha(A^T)
$$

Using

$$
\alpha(A^T) = (A^T)^n + \alpha_{n-1}(A^T)^{n-1} + \cdots + \alpha_2(A^T)^2 + \alpha_1 A^T + \alpha_0 I
$$

$$
= (A^n)^T + \alpha_{n-1}(A^{n-1})^T + \cdots + \alpha_2(A^2)^T + \alpha_1 A^T + \alpha_0 I
$$

$$
= [\alpha(A)]^T
$$

we obtain $[\alpha(A^T)]^T = \alpha(A)$, which along with

$$[P^{-1}_{(A^T,C^T)}]^T = [P^T_{(A^T,C^T)}]^{-1} = Q^{-1}_{(A,C)}$$

gives

$$L = \alpha(A)Q^{-1}_{(A,C)} \begin{bmatrix} 0 \\ 0 \\ \vdots \\ 0 \\ 1 \end{bmatrix}$$

Example 8.3 We demonstrate the use of Ackermann's formula for the state equation given in Example 8.1. We have previously computed

$$Q = \begin{bmatrix} C \\ CA \\ CA^2 \end{bmatrix} = \begin{bmatrix} 1 & 0 & 0 \\ 0 & 1 & 0 \\ 0 & 0 & 1 \end{bmatrix}$$

In terms of the desired characteristic polynomial $\alpha(s) = s^3 + 6s^2 + 11s + 6$, we construct

$$\alpha(A) = A^3 + 6A^2 + 11A + 6I$$

$$= \begin{bmatrix} -4 & -4 & -1 \\ 4 & 0 & -3 \\ 12 & 16 & 3 \end{bmatrix} + 6\begin{bmatrix} 0 & 0 & 1 \\ -4 & -4 & -1 \\ 4 & 0 & -2 \end{bmatrix}$$

$$+ 11\begin{bmatrix} 0 & 1 & 0 \\ 0 & 0 & 1 \\ -4 & -4 & -1 \end{bmatrix} + 6\begin{bmatrix} 1 & 0 & 0 \\ 0 & 1 & 0 \\ 0 & 0 & 1 \end{bmatrix}$$

$$= \begin{bmatrix} 2 & 7 & 5 \\ -20 & -18 & 2 \\ -8 & -28 & -20 \end{bmatrix}$$

so that

$$L = \begin{bmatrix} 2 & 7 & 5 \\ -20 & -18 & 2 \\ -8 & -28 & -20 \end{bmatrix}\begin{bmatrix} 1 & 0 & 0 \\ 0 & 1 & 0 \\ 0 & 0 & 1 \end{bmatrix}^{-1}\begin{bmatrix} 0 \\ 0 \\ 1 \end{bmatrix} = \begin{bmatrix} 5 \\ 2 \\ -20 \end{bmatrix}$$

which agrees with the previously-derived result. $\square$

8.2 DETECTABILITY

We have seen that for an observable state equation, a state observer can be constructed for which the observer error dynamics have arbitrarily assignable eigenvalues by proper choice of the observer gain matrix. Since the freedom to assign these eigenvalues implies that we can asymptotically stabilize the observer error dynamics, we can say that observability is a *sufficient* condition for asymptotic stabilization of the observer error dynamics via the observer gain matrix. In situations where the state equation is not observable, we investigate whether or not asymptotic stabilization of the observer error dynamics is still possible. To do so, we introduce the concept of *detectability*. We will see that a duality relationship exists between detectability and stabilizability that is similar to the duality relationship between observability and controllability that we established in Chapter 4. To get started, we return to the example that we considered in Section 7.4 to introduce stabilizability. The state equation

$$
\dot{x}(t) = \begin{bmatrix} 1 & 0 & 0 \\ 1 & -1 & 1 \\ 0 & 0 & -2 \end{bmatrix} x(t) + \begin{bmatrix} 1 \\ 1 \\ 0 \end{bmatrix} u(t)
$$

$$
y(t) = [1 \quad 0 \quad 0]x(t)
$$

is already in the standard form for an unobservable state equation, as can be seen from the partitioning

$$
\begin{bmatrix} A_{11} & 0 \\ A_{21} & A_{22} \end{bmatrix} = \begin{bmatrix} 1 & 0 & 0 \\ 1 & -1 & 1 \\ 0 & 0 & -2 \end{bmatrix} \quad [C_1 \quad 0] = [1| \quad 0 \quad 0]
$$

in which the pair (A_{11}, C_1) specifies an observable one-dimensional subsystem. In terms of an observer gain vector $L = [l_1 \quad l_2 \quad l_3]^T$ we have

$$
A - LC = \begin{bmatrix} 1 & 0 & 0 \\ 1 & -1 & 1 \\ 0 & 0 & -2 \end{bmatrix} - \begin{bmatrix} l_1 \\ l_2 \\ l_3 \end{bmatrix} [1| \quad 0 \quad 0]
$$

$$
= \begin{bmatrix} 1+l_1 & 0 & 0 \\ \begin{bmatrix} 1 \\ 0 \end{bmatrix} - \begin{bmatrix} l_2 \\ l_3 \end{bmatrix} & -1 & 1 \\ & 0 & -2 \end{bmatrix} \quad (1)
$$

in which the observer gain l_1 can be chosen easily to place the lone eigen-value of the observable subsystem. Because of the lower block triangular structure, the three eigenvalues of $A - LC$ are $1 + l_1$, along with those of the unobservable subsystem, located at -1 and -2. Thus we conclude that even though the state equation is not observable, it is still possible to construct an observer gain vector specifying an observer that yields asymptotically stable observer error dynamics governed by the eigenvalues of $A - LC$. We emphasize that asymptotically stable observer error dynamics are achievable directly because the unobservable subsystem, which cannot be influenced by the observer gain vector, is asymptotically stable to begin with. We also see that the observer gains l_2 and l_3 have no influence on the eigenvalues of $A - LC$. We now formalize these notions in the following definition:

Definition 8.2   *The linear state equation* (8.1) *[or the pair* (A, C), *for short] is detectable if there exists an observer gain matrix L for which all eigenvalues of* $A - LC$ *have strictly negative real part.*

We see from this definition and our prior eigenvalue placement results that observability implies detectability. On the other hand, the preceding example indicates that the converse does not hold. A state equation can be detectable but not observable. We therefore regard detectability as a weaker condition than observability, just as stabilizability is a weaker condition than controllability.

The analysis in the preceding example can be generalized as follows: If the pair (A, C) is observable, then it is detectable, as we have already noted, so we consider the case in which the pair (A, C) is not observable. We know that there exists a coordinate transformation $x(t) = Tz(t)$ such that the transformed state equation has

$$\hat{A} = \begin{bmatrix} A_{11} & 0 \\ A_{21} & A_{22} \end{bmatrix} \qquad \hat{C} = [\, C_1 \quad 0\,]$$

in which the pair (A_{11}, C_1) is observable. With $\hat{L} = [\, L_1^T \quad L_2^T \,]^T$ a conformably partitioned observer gain matrix, we have

$$\hat{A} - \hat{L}\hat{C} = \begin{bmatrix} A_{11} - L_1 C_1 & 0 \\ A_{21} - L_2 C_1 & A_{22} \end{bmatrix}$$

whose eigenvalues are those of $A_{11} - L_1 C_1$, along with those of the A_{22}. Because (A_{11}, C_1) is an observable pair, L_1 can be chosen such that the eigenvalues of $A_{11} - L_1 C_1$ have strictly negative real part. However, the unobservable subsystem associated with A_{22} is uninfluenced by the observer gain matrix, and the eigenvalues of A_{22} will be among the eigenvalues of $\hat{A} - \hat{L}\hat{C}$ for any $\hat{L}$. Consequently, in order for every eigenvalue of $\hat{A} - \hat{L}\hat{C}$ to have a strictly negative real part, the eigenvalues of A_{22} must have strictly negative real parts. We therefore see that in the case where the pair (A, C) is not observable, detectability requires that the unobservable subsystem be asymptotically stable. We also note that L_2 plays no role in this analysis.

With the aim of deriving Popov-Belevitch-Hautus tests that provide algebraic criteria for detectability comparable with those presented previously for stabilizability, we first establish that a formal duality relationship exists between stabilizability and detectability analogous to that relating controllability and observability. Specifically, we can make the following statements:

The pair (A, B) is stabilizable if and only if the pair (A^T, B^T) is detectable.

The pair (A, C) is detectable if and only if the pair (A^T, C^T) is stabilizable.

For the first statement, if the pair (A, B) is stabilizable, then, by definition, there exists a state feedback gain matrix K for which all eigenvalues of $A - BK$ have strictly negative real parts. Since the matrix transpose operation does not affect eigenvalues, it follows that all eigenvalues of $(A - BK)^T = A^T - K^T B^T$ also have strictly negative real parts. By interpreting K^T as an observer gain matrix, we conclude that the pair (A^T, B^T) is detectable. The converse can be established by reversing the preceding steps, and the second statement can be argued in an analogous fashion. The Popov-Belevitch-Hautus tests for stabilizability now can be dualized to yield the following result for detectability:

Theorem 8.3 *The following statements are equivalent:*

1. *The pair (A, C) is detectable.*

2. *There exists no right eigenvector of A associated with an eigenvalue having nonnegative real part that is orthogonal to the rows of C.*

3. *The matrix $\begin{bmatrix} C \\ \lambda I - A \end{bmatrix}$ has full-column rank for all complex λ with nonnegative real parts.*

Proof. A proof of the theorem can be assembled from the following relationships:

- The pair (A, C) is detectable if and only if the pair (A^T, C^T) is stabilizable.
- There exists a right eigenvector of A associated with an eigenvalue having nonnegative real part that is orthogonal to the rows of C if and only if there exists a left eigenvector of A^T associated with an eigenvalue having nonnegative real part that is orthogonal to the columns of C^T,
- rank $\begin{bmatrix} C \\ \lambda I - A \end{bmatrix} = \text{rank}\,[\,\lambda I - A^T \quad C^T\,]$ for all complex λ.

The details are left to the reader.   □

Example 8.4   Recall the unobservable state equation introduced in Example 4.3 and revisited in Examples 4.7 and 4.8, that is,

$$
\begin{bmatrix} \dot{x}_1(t) \\ \dot{x}_2(t) \\ \dot{x}_3(t) \end{bmatrix} = \begin{bmatrix} 0 & 0 & -6 \\ 1 & 0 & -11 \\ 0 & 1 & -6 \end{bmatrix} \begin{bmatrix} x_1(t) \\ x_2(t) \\ x_3(t) \end{bmatrix}
$$

$$
y(t) = [\,0 \quad 1 \quad -3\,] \begin{bmatrix} x_1(t) \\ x_2(t) \\ x_3(t) \end{bmatrix}
$$

It is straightforward to check that the coordinate transformation $x(t) = T\,z(t)$ with

$$
T = \begin{bmatrix} 3 & 1 & 2 \\ 1 & 0 & 3 \\ 0 & 0 & 1 \end{bmatrix}
$$

yields the transformed state equation given by coefficient matrices

$$
\hat{A} = T^{-1}AT
$$
$$
= \left[\begin{array}{cc|c} 0 & 1 & 1 \\ -2 & -3 & 0 \\ \hline 1 & 0 & -3 \end{array}\right] \qquad \begin{array}{l} \hat{C} = CT \\ = [1 \quad 0 \ | \ 0] \end{array}
$$

which, based on the preceding partitioning, is in the standard form for an unobservable state equation. Moreover, the unobservable subsystem is one-dimensional with $A_{22} = -3$. Hence this state equation is detectable. Arbitrarily choosing two eigenvalues of the observer error dynamics to be $-2 + j2, -2 - j2$, the Bass-Gura formula applied to the observable pair (A_{11}, C_1) identified earlier yields

$$
L_1 = \begin{bmatrix} 1 \\ 3 \end{bmatrix}
$$

for which

$$A_{11} - L_1 C_1 = \begin{bmatrix} 0 & 1 \\ -2 & -3 \end{bmatrix} - \begin{bmatrix} 1 \\ 3 \end{bmatrix} [1 \quad 0] = \begin{bmatrix} -1 & 1 \\ -5 & -3 \end{bmatrix}$$

has these specified eigenvalues. Therefore,

$$\hat{L} = \begin{bmatrix} 1 \\ 3 \\ 0 \end{bmatrix}$$

is such that $\hat{A} - \hat{L}\hat{C}$ has the three eigenvalues $-2 + j2, -2 - j2$, and -3. Finally, the observer gain vector

$$L = T\hat{L} = \begin{bmatrix} 3 & 1 & 2 \\ 1 & 0 & 3 \\ 0 & 0 & 1 \end{bmatrix} \begin{bmatrix} 1 \\ 3 \\ 0 \end{bmatrix} = \begin{bmatrix} 6 \\ 1 \\ 0 \end{bmatrix}$$

yields $A - LC$ also having eigenvalues $-2 + j2, -2 - j2$, and -3.

We also saw in Example 4.7 that $v_3 = [2 \quad 3 \quad 1]^T$ is a right eigenvector of A associated with the eigenvalue $\lambda_3 = -3 < 0$, for which $Cv_3 = 0$. We can make the stronger assertion that the only right eigenvectors of A orthogonal to C are nonzero scalar multiples of v_3 and therefore are also associated with λ_3. We conclude from Theorem 8.3 that since there exists no right eigenvector of A associated with an eigenvalue having nonnegative real part that is orthogonal to the rows of C, the state equation is detectable. We also see that $\lambda_3 = -3$ is the only value of λ for which rank $\begin{bmatrix} C \\ \lambda I - A \end{bmatrix}$ has less than rank 3, so rank $\begin{bmatrix} C \\ \lambda I - A \end{bmatrix}$ has full-column rank 3 for all λ with nonnegative real parts, and again detectability follows from Theorem 8.3. $\qquad\square$

8.3 REDUCED-ORDER OBSERVERS

The observer design presented in Section 8.1 has dimension equal to that of the original state equation (8.1) and therefore is referred to as a *full-order observer*. We now show that having access to p independent output measurements provides information about the state vector that can be used in the estimation process to allow a reduction in the observer dimension. The development in this subsection follows the presentation in Friedland (1986).

We first suppose that the state vector can be decomposed into

$$x(t) = \begin{bmatrix} x_1(t) \\ x_2(t) \end{bmatrix}$$

in terms of which the output is $y(t) = x_1(t)$. This reflects the real-world situation in which a subset of the system's state variables is sensed directly and therefore need not be estimated. We partition the state equation accordingly to yield

$$\begin{bmatrix} \dot{x}_1(t) \\ \dot{x}_2(t) \end{bmatrix} = \begin{bmatrix} A_{11} & A_{12} \\ A_{21} & A_{22} \end{bmatrix} \begin{bmatrix} x_1(t) \\ x_2(t) \end{bmatrix} + \begin{bmatrix} B_1 \\ B_2 \end{bmatrix} u(t)$$

$$y(t) = [I \quad 0] \begin{bmatrix} x_1(t) \\ x_2(t) \end{bmatrix}$$

That $x_1(t)$ is determined directly from $y(t)$ suggests the perfect estimate $\hat{x}_1(t) = y(t) = x_1(t)$, and it remains to produce an estimate of $x_2(t)$. For this, we consider an estimate of the form

$$\hat{x}_2(t) = Ly(t) + w(t)$$

in which $w(t)$ is generated by the following $(n - p)$-dimensional state equation having $y(t)$ and $u(t)$ as inputs, that is,

$$\dot{w}(t) = Jw(t) + Ny(t) + Mu(t)$$

As in the full-order observer case, we consider the estimation error

$$\tilde{x}(t) = \begin{bmatrix} \tilde{x}_1(t) \\ \tilde{x}_2(t) \end{bmatrix} = \begin{bmatrix} x_1(t) - \hat{x}_1(t) \\ x_2(t) - \hat{x}_2(t) \end{bmatrix}$$

Since $\hat{x}_1(t) \equiv x_1(t)$, we have $\tilde{x}_1(t) \equiv 0$, and we need only consider the dynamics governing $\tilde{x}_2(t)$, that is,

$$\dot{\tilde{x}}_2(t) = \dot{x}_2(t) - \dot{\hat{x}}_2(t)$$
$$= [A_{21}x_1(t) + A_{22}x_2(t) + B_2u(t)] - [L\dot{y}(t) + \dot{w}(t)]$$
$$= [A_{21}x_1(t) + A_{22}x_2(t) + B_2u(t)] - [L(A_{11}x_1(t) + A_{12}x_2(t)$$
$$+ B_1u(t)) + Jw(t) + Ny(t) + Mu(t)]$$

By substituting

$$w(t) = \hat{x}_2(t) - Lx_1(t)$$
$$= x_2(t) - \tilde{x}_2(t) - Lx_1(t)$$

we obtain after some algebra

$$\dot{\tilde{x}}_2(t) = J\tilde{x}_2(t) + (A_{21} + JL - N - LA_{11})x_1(t)$$
$$+ (A_{22} - J - LA_{12})x_2(t)$$
$$+ (B_2 - M - LB_1)u(t)$$

To obtain homogeneous error dynamics, we set

$$J = A_{22} - LA_{12}$$
$$N = A_{21} + JL - LA_{11}$$
$$M = B_2 - LB_1$$

to cancel the $x_2(t)$, $x_1(t)$, and $u(t)$ terms, respectively, which results in

$$\dot{\tilde{x}}_2(t) = J\tilde{x}_2(t) \qquad \tilde{x}_2(0) = x_2(0) - \hat{x}_2(0)$$

To ensure that $\tilde{x}_2(t)$ tends to zero as time tends to infinity for all $\tilde{x}_2(0)$, we require that

$$J = A_{22} - LA_{12}$$

has strictly negative real-part eigenvalues. The following result is of interest in this regard:

Proposition 8.4   *The pair (A_{22}, A_{12}) is observable if and only if the pair*

$$\left(\begin{bmatrix} A_{11} & A_{12} \\ A_{21} & A_{22} \end{bmatrix}, [I \quad 0] \right)$$

is observable.

The proof is left as an analytical exercise for the reader. We offer the hint that the Popov-Belevitch-Hautus rank test for observability makes for an economical development.

To summarize the preceding discussion, for the case in which $y(t) = x_1(t)$, a reduced-order observer of dimension $n - p$ is specified by

$$\dot{w}(t) = (A_{22} - LA_{12})w(t) + (A_{21} + JL - LA_{11})y(t)$$
$$+ (B_2 - LB_1)u(t)$$
$$\begin{bmatrix} \hat{x}_1(t) \\ \hat{x}_2(t) \end{bmatrix} = \begin{bmatrix} y(t) \\ Ly(t) + w(t) \end{bmatrix}$$

in which the $(n - p) \times p$ observer gain matrix L is constructed so that the $n - p$ eigenvalues of $J = A_{22} - LA_{12}$ have strictly negative real parts.

We now proceed to the general case beginning with the state equation (8.1), which we assume to be observable; in addition, we also assume that C has full row rank. The latter assumption guarantees that the output $y(t)$ provides p independent measurements of the state vector. This assumption comes with no real loss of generality because measurements corresponding to certain rows in C that are linearly dependent on the remaining rows can be discarded. As a consequence of this rank assumption, it is possible to find an $(n - p) \times n$ matrix E such that

$$\begin{bmatrix} C \\ E \end{bmatrix}$$

is $n \times n$ and nonsingular. Setting

$$T^{-1} = \begin{bmatrix} C \\ E \end{bmatrix}, \qquad T = [P \quad Q]$$

we apply the coordinate transformation

$$z(t) = \begin{bmatrix} z_1(t) \\ z_2(t) \end{bmatrix} = T^{-1}x(t) = \begin{bmatrix} Cx(t) \\ Ex(t) \end{bmatrix}$$

which yields $z_1(t) = y(t)$. Letting

$$T^{-1}AT = \begin{bmatrix} C \\ E \end{bmatrix} A[P \quad Q] = \begin{bmatrix} CAP & CAQ \\ EAP & EAQ \end{bmatrix}$$

$$= \begin{bmatrix} A_{11} & A_{12} \\ A_{21} & A_{22} \end{bmatrix}$$

$$T^{-1}B = \begin{bmatrix} C \\ E \end{bmatrix} B$$

$$= \begin{bmatrix} B_1 \\ B_2 \end{bmatrix}$$

$$CT = C[P \quad Q] = [CP \quad CQ]$$

$$= [I \quad 0]$$

yields the transformed state equation that corresponds to the special case we first considered:

$$\begin{bmatrix} \dot{z}_1(t) \\ \dot{z}_2(t) \end{bmatrix} = \begin{bmatrix} A_{11} & A_{12} \\ A_{21} & A_{22} \end{bmatrix} \begin{bmatrix} z_1(t) \\ z_2(t) \end{bmatrix} + \begin{bmatrix} B_1 \\ B_2 \end{bmatrix} u(t)$$

$$y(t) = [I \quad 0] \begin{bmatrix} z_1(t) \\ z_2(t) \end{bmatrix}$$

Since coordinate transformations do not affect observability, by virtue of Proposition 8.4, the following statements are equivalent:

- The pair (A, C) is observable.
- The pair $\left(\begin{bmatrix} A_{11} & A_{12} \\ A_{21} & A_{22} \end{bmatrix}, [I \quad 0] \right)$ is observable.
- The pair (A_{22}, A_{12}) is observable.

All that is left for us to do is to relate the previously derived reduced-order observer description to the original state equation. As before, we choose the observer gain matrix L to locate the $n - p$ eigenvalues of $J = A_{22} - LA_{12}$ at desired locations. This is possible under the assumption that the pair (A, C) is observable because this, as indicated earlier, is equivalent to observability of the pair (A_{22}, A_{12}). Following this, we set

$$
\begin{aligned}
J &= A_{22} - LA_{12} \\
&= (EAQ) - L(CAQ) \\
&= (E - LC)AQ \\
N &= A_{21} + JL - LA_{11} \\
&= (EAP) + ((E - LC)AQ)L - L(CAP) \\
&= (E - LC)A(P + QL) \\
M &= B_2 - LB_1 \\
&= (E - LC)B
\end{aligned}
$$

With these modified definitions, the reduced-order observer dynamics again are given by

$$
\dot{w}(t) = Jw(t) + Ny(t) + Mu(t)
$$

from which we construct the estimate

$$
\begin{aligned}
\hat{x}(t) &= T\hat{z}(t) \\
&= [P \quad Q] \begin{bmatrix} \hat{z}_1(t) \\ \hat{z}_2(t) \end{bmatrix} \\
&= [P \quad Q] \begin{bmatrix} y(t) \\ Ly(t) + w(t) \end{bmatrix} \\
&= Qw(t) + (P + QL)y(t)
\end{aligned}
$$

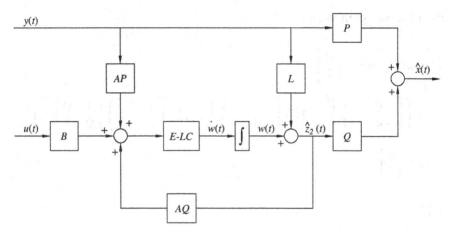

FIGURE 8.4 Reduced-order observer block diagram.

On rearranging these relationships slightly to yield

$$\dot{w}(t) = (E - LC)[AQ\hat{z}_2(t) + APy(t) + Bu(t)]$$
$$\hat{z}_2(t) = Ly(t) + w(t)$$
$$\hat{x}(t) = Py(t) + Q\hat{z}_2(t)$$

we see that the reduced-order observer can be represented by the block diagram of Figure 8.4.

Example 8.5 We design a reduced-order observer for the state equation of Example 8.1. The output equation already has the required form, so a state coordinate transformation is not required and we partition the remaining state equation coefficient matrices as

$$\begin{bmatrix} A_{11} & A_{12} \\ A_{21} & A_{22} \end{bmatrix} = \begin{bmatrix} 0 & 1 & 0 \\ 0 & 0 & 1 \\ -4 & -4 & -1 \end{bmatrix} \qquad \begin{bmatrix} B_1 \\ B_2 \end{bmatrix} = \begin{bmatrix} 0 \\ 0 \\ 1 \end{bmatrix}$$

The observable pair (A_{22}, A_{12}) has the observability matrix

$$Q_2 = \begin{bmatrix} A_{12} \\ A_{12}A_{22} \end{bmatrix} = \begin{bmatrix} 1 & 0 \\ 0 & 1 \end{bmatrix}$$

Upon selecting desired eigenvalues $\{-5, -5\}$ yielding the desired characteristic polynomial $\alpha(s) = s^2 + 10s + 25$, Ackermann's formula gives the observer gain vector

$$L = \alpha(A_{22})Q_2^{-1}\begin{bmatrix} 0 \\ 1 \end{bmatrix}$$

$$= \left(\begin{bmatrix} 0 & 1 \\ -4 & -1 \end{bmatrix}^2 + 10 \cdot \begin{bmatrix} 0 & 1 \\ -4 & -1 \end{bmatrix} + 25 \cdot \begin{bmatrix} 1 & 0 \\ 0 & 1 \end{bmatrix} \right) \begin{bmatrix} 1 & 0 \\ 0 & 1 \end{bmatrix}^{-1} \begin{bmatrix} 0 \\ 1 \end{bmatrix}$$

$$= \begin{bmatrix} 9 \\ 12 \end{bmatrix}$$

With

$$J = A_{22} - LA_{12}$$

$$= \begin{bmatrix} 0 & 1 \\ -4 & -1 \end{bmatrix} - \begin{bmatrix} 9 \\ 12 \end{bmatrix}\begin{bmatrix} 1 & 0 \end{bmatrix} = \begin{bmatrix} -9 & 1 \\ -16 & -1 \end{bmatrix}$$

$$N = A_{21} + JL - LA_{11}$$

$$= \begin{bmatrix} 0 \\ -4 \end{bmatrix} + \begin{bmatrix} -9 & 1 \\ -16 & -1 \end{bmatrix}\begin{bmatrix} 9 \\ 12 \end{bmatrix} - \begin{bmatrix} 9 \\ 12 \end{bmatrix}(0) = \begin{bmatrix} -69 \\ -160 \end{bmatrix}$$

$$M = B_2 - LB_1 = \begin{bmatrix} 0 \\ 1 \end{bmatrix} - \begin{bmatrix} 9 \\ 12 \end{bmatrix}(0) = \begin{bmatrix} 0 \\ 1 \end{bmatrix}$$

we construct the two-dimensional reduced-order observer

$$\begin{bmatrix} \dot{w}_1(t) \\ \dot{w}_2(t) \end{bmatrix} = \begin{bmatrix} -9 & 1 \\ -16 & -1 \end{bmatrix}\begin{bmatrix} w_1(t) \\ w_2(t) \end{bmatrix} + \begin{bmatrix} -69 \\ -160 \end{bmatrix} y(t) + \begin{bmatrix} 0 \\ 1 \end{bmatrix} u(t)$$

$$\begin{bmatrix} \hat{x}_1(t) \\ \hat{x}_2(t) \\ \hat{x}_3(t) \end{bmatrix} = \begin{bmatrix} y(t) \\ 9y(t) + w_1(t) \\ 12y(t) + w_2(t) \end{bmatrix}$$

The combined open-loop system and reduced-order observer is given by the five-dimensional state equation

$$\begin{bmatrix} \dot{x}_1(t) \\ \dot{x}_2(t) \\ \dot{x}_3(t) \\ \dot{w}_1(t) \\ \dot{w}_2(t) \end{bmatrix} = \left[\begin{array}{ccc|cc} 0 & 1 & 0 & 0 & 0 \\ 0 & 0 & 1 & 0 & 0 \\ -4 & -4 & -1 & 0 & 0 \\ \hline -69 & 0 & 0 & -9 & 1 \\ -160 & 0 & 0 & -16 & -1 \end{array} \right] \begin{bmatrix} x_1(t) \\ x_2(t) \\ x_3(t) \\ w_1(t) \\ w_2(t) \end{bmatrix} + \begin{bmatrix} 0 \\ 0 \\ 1 \\ 0 \\ 1 \end{bmatrix} u(t)$$

$$
\begin{bmatrix} \hat{x}_1(t) \\ \hat{x}_2(t) \\ \hat{x}_3(t) \end{bmatrix} = \begin{bmatrix} 1 & 0 & 0 & | & 0 & 0 \\ 9 & 0 & 0 & | & 1 & 0 \\ 12 & 0 & 0 & | & 0 & 1 \end{bmatrix} \begin{bmatrix} x_1(t) \\ x_2(t) \\ x_3(t) \\ w_1(t) \\ w_2(t) \end{bmatrix}
$$

We investigate the reduced-order observer performance for the following initial conditions and input signal:

$$
x_0 = \begin{bmatrix} 1 \\ -1 \\ 1 \end{bmatrix} \qquad w_0 = \begin{bmatrix} -9 \\ -12 \end{bmatrix} \qquad u(t) = \sin(t) \quad t \geq 0
$$

This corresponds to the same open-loop initial state and input signal as in Example 8.1. The reduced-order observer's initial state was chosen to yield $\hat{x}_2(0) = 9y(0) + w_1(0) = 0$ and $\hat{x}_3(0) = 12y(0) + w_2(0) = 0$. Figure 8.5 compares the state response $x(t)$ with the reduced-order observer estimate $\hat{x}(t)$. We observe that $\hat{x}_1(t)$ exactly matches $x_1(t)$ because we have direct access to $y(t) = x_1(t)$. For the remaining two state variables, we see that $\hat{x}_2(t)$ and $\hat{x}_3(t)$ asymptotically converge to $x_2(t)$ and $x_3(t)$, respectively, with a rate of convergence that is faster than that exhibited in Example 8.1. This is because the two eigenvalues that govern this convergence were chosen to be farther to the left in the complex plane compared with the three eigenvalues selected in Example 8.1.

8.4 OBSERVER-BASED COMPENSATORS AND THE SEPARATION PROPERTY

For the linear state equation (8.1), we know that

- Controllability of the pair (A, B) is necessary and sufficient for arbitrary eigenvalue placement by state feedback

$$
u(t) = -Kx(t) + r(t)
$$

- Observability of the pair (A, C) is necessary and sufficient for state estimation via

$$
\dot{\hat{x}}(t) = A\hat{x}(t) + Bu(t) + L[y(t) - C\hat{x}(t)]
$$

to yield observer error dynamics with arbitrarily placed eigenvalues.

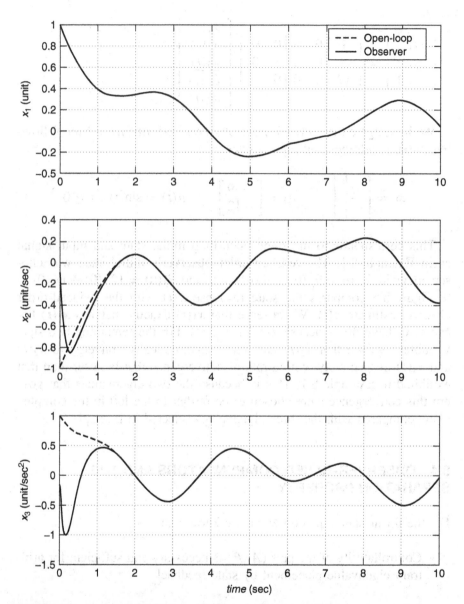

FIGURE 8.5 Open-loop state response and reduced-order observer estimates for Example 8.5.

We now investigate what happens if we combine these results by replacing the true state vector $x(t)$ in the state feedback control law with the state estimate $\hat{x}(t)$ produced by the observer to yield

$$u(t) = -K\hat{x}(t) + r(t)$$

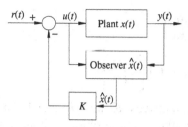

FIGURE 8.6   High-level observer-based compensator block diagram.

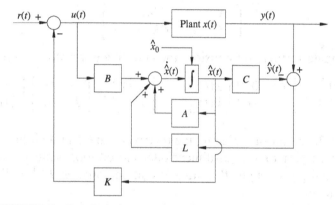

FIGURE 8.7   Detailed observer-based compensator block diagram.

This interconnection of a state feedback control law and a state observer results in a dynamic observer-based compensator given by the state equation

$$\dot{\hat{x}}(t) = A\hat{x}(t) + Bu(t) + L[y(t) - C\hat{x}(t)]$$
$$u(t) = -K\hat{x}(t) + r(t)$$

or

$$\dot{\hat{x}}(t) = (A - BK - LC)\hat{x}(t) + Ly(t) + Br(t)$$
$$u(t) = -K\hat{x}(t) + r(t) \tag{8.4}$$

which is a particular type of dynamic output feedback compensator. A high-level observer-based compensator block diagram is shown in Figure 8.6; details are shown in Figure 8.7.

The feedback interconnection of the open-loop state equation (8.1) and the observer-based compensator (8.4) yields the $2n$-dimensional closed-loop state equation

$$\begin{bmatrix} \dot{x}(t) \\ \dot{\hat{x}}(t) \end{bmatrix} = \begin{bmatrix} A & -BK \\ LC & A - BK - LC \end{bmatrix} \begin{bmatrix} x(t) \\ \hat{x}(t) \end{bmatrix} + \begin{bmatrix} B \\ B \end{bmatrix} r(t)$$

$$\begin{bmatrix} x(0) \\ \hat{x}(0) \end{bmatrix} = \begin{bmatrix} x_0 \\ \hat{x}_0 \end{bmatrix}$$

$$y(t) = [C \quad 0] \begin{bmatrix} x(t) \\ \hat{x}(t) \end{bmatrix} \tag{8.5}$$

Although the gain matrices K and L can be chosen to solve their respective eigenvalue placement problems, it is not yet clear how this relates to closed-loop stability, which involves the $2n$ eigenvalues of

$$\begin{bmatrix} A & -BK \\ LC & A - BK - LC \end{bmatrix}$$

To investigate further, we consider the coordinate transformation

$$\begin{bmatrix} x(t) \\ \tilde{x}(t) \end{bmatrix} = \begin{bmatrix} I & 0 \\ I & -I \end{bmatrix} \begin{bmatrix} x(t) \\ \hat{x}(t) \end{bmatrix} \qquad \begin{bmatrix} x(t) \\ \hat{x}(t) \end{bmatrix} = \begin{bmatrix} I & 0 \\ I & -I \end{bmatrix} \begin{bmatrix} x(t) \\ \tilde{x}(t) \end{bmatrix}$$

which has the effect of replacing the observer state $\hat{x}(t)$ with the observer error $\tilde{x}(t)$ as part of the $2n$-dimensional closed-loop state vector. Note that interestingly enough, this transformation satisfies $T^{-1} = T$. Direct calculations give

$$\begin{aligned} \dot{x}(t) &= Ax(t) - BK\hat{x}(t) + Br(t) \\ &= Ax(t) - BK[x(t) - \tilde{x}(t)] + Br(t) \\ &= (A - BK)x(t) - BK\tilde{x}(t) + Br(t) \end{aligned}$$

and

$$\begin{aligned} \dot{\tilde{x}}(t) &= \dot{x}(t) - \dot{\hat{x}}(t) \\ &= [Ax(t) - BK\hat{x}(t) + Br(t)] - [(A - BK - LC)\hat{x}(t) + LCx(t) \\ &\quad + Br(t)] \\ &= A[x(t) - \hat{x}(t)] - LC[x(t) - \hat{x}(t)] \\ &= (A - LC)\tilde{x}(t) \end{aligned}$$

so the closed-loop state equation becomes in the new coordinates

$$\begin{bmatrix} \dot{x}(t) \\ \dot{\tilde{x}}(t) \end{bmatrix} = \begin{bmatrix} A - BK & BK \\ 0 & A - LC \end{bmatrix} \begin{bmatrix} x(t) \\ \tilde{x}(t) \end{bmatrix} + \begin{bmatrix} B \\ 0 \end{bmatrix} r(t)$$

$$y(t) = [C \quad 0] \begin{bmatrix} x(t) \\ \tilde{x}(t) \end{bmatrix} \tag{8.6}$$

Because of the block triangular structure of the transformed closed-loop system dynamics matrix, we see that the $2n$ closed-loop eigenvalues are given by

$$\sigma\left(\begin{bmatrix} A - BK & BK \\ 0 & A - LC \end{bmatrix}\right) = \sigma(A - BK) \cup \sigma(A - LC)$$

Because state coordinate transformations do not affect eigenvalues, we conclude that

$$\sigma\left(\begin{bmatrix} A & -BK \\ LC & A - BK - LC \end{bmatrix}\right) = \sigma(A - BK) \cup \sigma(A - LC)$$

This indicates that the $2n$ closed-loop eigenvalues can be placed by separately and independently locating eigenvalues of $A - BK$ and $A - LC$ via choice of the gain matrices K and L, respectively. We refer to this as the *separation property* of observer-based compensators.

We generally require the observer error to converge faster than the desired closed-loop transient response dynamics. Therefore, we choose the eigenvalues of $A - LC$ to be about 10 times farther to the left in the complex plane than the (dominant) eigenvalues of $A - BK$.

Example 8.6   We return to the open-loop state equation in Example 8.1. The open-loop characteristic polynomial is $a(s) = s^3 + s^3 + 4s + 4$. The open-loop the eigenvalues $\lambda_{1,2,3} = -1, \pm j2$ indicate that this is a marginally stable state equation because of the nonrepeated zero-real-part eigenvalues. This state equation is not bounded-input, bounded output stable since the bounded input $u(t) = \sin(2t)$ creates a resonance with the imaginary eigenvalues and an unbounded zero-state response results. We therefore seek to design an observer-based compensator by combining the observer designed previously with a state feedback control law.

Given the following eigenvalues to be placed by state feedback, namely,

$$\mu_{1,2,3} = -\tfrac{1}{2} \pm j\tfrac{\sqrt{3}}{2}, -1$$

the desired characteristic polynomial is $\alpha(s) = s^3 + 2s^2 + 2s + 1$. Since the open-loop state equation is in controller canonical form we easily compute state feedback gain vector to be

$$
\begin{aligned}
K &= K_{\text{CCF}} \\
  &= [(\alpha_0 - a_0) \quad (\alpha_1 - a_1) \quad (\alpha_2 - a_2)] \\
  &= [(1 - 4) \quad (2 - 4) \quad (2 - 1)] \\
  &= [-3 \quad -2 \quad 1]
\end{aligned}
$$

On inspection of

$$A - BK = \begin{bmatrix} 0 & 1 & 0 \\ 0 & 0 & 1 \\ -1 & -2 & -2 \end{bmatrix}$$

we see from the companion form that the desired characteristic polynomial has been achieved. The previously specified observer (from Example 8.1), together with the state feedback gain vector computed independently earlier yields the observer-based compensator (8.4) [with $r(t) = 0$], that is,

$$\begin{bmatrix} \dot{\hat{x}}_1(t) \\ \dot{\hat{x}}_2(t) \\ \dot{\hat{x}}_3(t) \end{bmatrix} = \begin{bmatrix} -5 & 1 & 0 \\ -2 & 0 & 1 \\ 19 & -2 & -2 \end{bmatrix} \begin{bmatrix} \hat{x}_1(t) \\ \hat{x}_2(t) \\ \hat{x}_3(t) \end{bmatrix} + \begin{bmatrix} 5 \\ 2 \\ -20 \end{bmatrix} y(t)$$

$$u(t) = \begin{bmatrix} 3 & 2 & -1 \end{bmatrix} \begin{bmatrix} \hat{x}_1(t) \\ \hat{x}_2(t) \\ \hat{x}_3(t) \end{bmatrix}$$

yielding the six-dimensional homogeneous closed-loop state equation

$$\begin{bmatrix} \dot{x}_1(t) \\ \dot{x}_2(t) \\ \dot{x}_3(t) \\ \hline \dot{\hat{x}}_1(t) \\ \dot{\hat{x}}_2(t) \\ \dot{\hat{x}}_3(t) \end{bmatrix} = \left[\begin{array}{ccc|ccc} 0 & 1 & 0 & 0 & 0 & 0 \\ 0 & 0 & 1 & 0 & 0 & 0 \\ -4 & -4 & -1 & 3 & 2 & -1 \\ \hline 5 & 0 & 0 & -5 & 1 & 0 \\ 2 & 0 & 0 & -2 & 0 & 1 \\ -20 & 0 & 0 & 19 & -2 & -2 \end{array} \right] \begin{bmatrix} x_1(t) \\ x_2(t) \\ x_3(t) \\ \hline \hat{x}_1(t) \\ \hat{x}_2(t) \\ \hat{x}_3(t) \end{bmatrix}$$

$$y(t) = \begin{bmatrix} 1 & 0 & 0 & | & 0 & 0 & 0 \end{bmatrix} \begin{bmatrix} x_1(t) \\ x_2(t) \\ x_3(t) \\ \hline \hat{x}_1(t) \\ \hat{x}_2(t) \\ \hat{x}_3(t) \end{bmatrix}$$

We investigate the closed-loop performance for the following initial conditions and zero reference input:

$$x_0 = \begin{bmatrix} 1 \\ -1 \\ 1 \end{bmatrix} \qquad \hat{x}_0 = \begin{bmatrix} 0 \\ 0 \\ 0 \end{bmatrix} \qquad r(t) = 0 \quad t \geq 0$$

These are the same initial conditions as in Example 8.1, for which the zero-input response of the closed-loop state equation is shown in

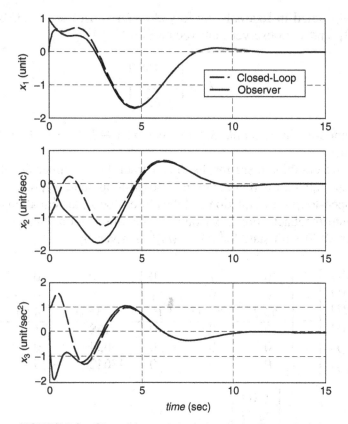

FIGURE 8.8 Closed-loop state response for Example 8.6.

Figure 8.8. Each state variable is plotted along with its estimate, and the responses show that the estimates converge to the actual state variables and that all six closed-loop state variables asymptotically tend to zero as time tends to infinity, as expected. □

Example 8.7 We return to the open-loop state equation

$$\begin{bmatrix} \dot{x}_1(t) \\ \dot{x}_2(t) \\ \dot{x}_3(t) \end{bmatrix} = \begin{bmatrix} 0 & 1 & 0 \\ 0 & 0 & 1 \\ -18 & -15 & -2 \end{bmatrix} \begin{bmatrix} x_1(t) \\ x_2(t) \\ x_3(t) \end{bmatrix} + \begin{bmatrix} 0 \\ 0 \\ 1 \end{bmatrix} u(t)$$

$$y(t) = [\,1 \quad 0 \quad 0\,] \begin{bmatrix} x_1(t) \\ x_2(t) \\ x_3(t) \end{bmatrix}$$

for which the state feedback gain vector

$$K = [\,35.26 \quad 24.55 \quad 14.00\,]$$

was constructed in Example 7.4 to yield $\sigma(A - BK) = \{-1.33 \pm j1.49,$ $-13.33\}$, and the observer gain vector

$$L = \begin{bmatrix} 158 \\ 3624 \\ 43624 \end{bmatrix}$$

was computed in Example 8.2 to give $\sigma(A - LC) = \{-13.3 \pm j14.9,$ $-133.3\}$.

Here we combine these results to construct the six-dimensional closed-loop state equation and compare the closed-loop state response with that of the open-loop state equation and the closed-loop state equation resulting from state feedback. For this example, the closed-loop state equation of the form (8.6) with state $[x^T(t), \tilde{x}^T(t)]^T$ is given by

$$\begin{bmatrix} \dot{x}_1(t) \\ \dot{x}_2(t) \\ \dot{x}_3(t) \\ \dot{\tilde{x}}_1(t) \\ \dot{\tilde{x}}_2(t) \\ \dot{\tilde{x}}_3(t) \end{bmatrix} = \left[\begin{array}{ccc|ccc} 0 & 1 & 0 & 0 & 0 & 0 \\ 0 & 0 & 1 & 0 & 0 & 0 \\ -53.26 & -39.55 & -16 & 35.26 & 24.55 & 14 \\ \hline 0 & 0 & 0 & -158 & 1 & 0 \\ 0 & 0 & 0 & -3624 & 0 & 1 \\ 0 & 0 & 0 & -43642 & -15 & -2 \end{array} \right]$$

$$\times \begin{bmatrix} x_1(t) \\ x_2(t) \\ x_3(t) \\ \hline \tilde{x}_1(t) \\ \tilde{x}_2(t) \\ \tilde{x}_3(t) \end{bmatrix}$$

$$y(t) = [1 \quad 0 \quad 0 \mid 0 \quad 0 \quad 0] \begin{bmatrix} x_1(t) \\ x_2(t) \\ x_3(t) \\ \tilde{x}_1(t) \\ \tilde{x}_2(t) \\ \tilde{x}_3(t) \end{bmatrix}$$

The unit step response plots for this example are presented in Figure 8.9 (for all three states). Shown are the open-loop response, plus the closed-loop responses for state feedback and observer-based compensation, all plotted together for each state variable. Note that we assumed an initial observer error of 0.0005 on the first state and zero initial observer error on the other two state variables. In simulation we cannot assume that the true

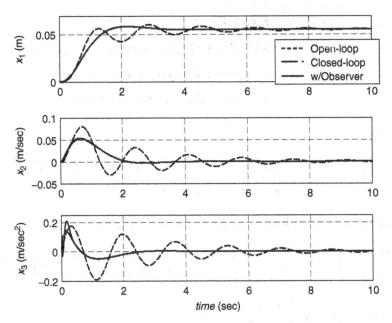

FIGURE 8.9 Open-loop, closed-loop with state feedback, and closed-loop with observer responses for Example 8.7.

initial state x_0 is known; otherwise the simulated closed-loop responses for state feedback and observer-based compensation will be identical, and we can not evaluate the observer performance.

In Figure 8.9, we see that the closed-loop responses for observer-based compensation converge rapidly to the closed-loop responses for the state feedback case. The reader should test the closed-loop observer-based state equation over a range of different initial errors.   □

Reduced-Order Observer-Based Compensators

We again consider the strategy of replacing the state $x(t)$ appearing in the state feedback control law with an estimate $\hat{x}(t)$, that is,

$$u(t) = -K\hat{x}(t) + r(t)$$

except that now $\hat{x}(t)$ is generated by a reduced-order observer. We show that the separation property also holds when the compensator is based on a reduced-order observer. Such a compensator of dimension $n - p$ is given by

$$\dot{w}(t) = Jw(t) + Ny(t) + Mu(t)$$

$$u(t) = -K\hat{x}(t) + r(t)$$
$$= -K[Py(t) + Q\hat{z}_2(t)] + r(t)$$
$$= -K\{Py(t) + Q[Ly(t) + w(t)]\} + r(t)$$
$$= -KQw(t) - K(P + QL)y(t) + r(t)$$

so that

$$\dot{w}(t) = (J - MKQ)w(t) + [-MK(P + QL) + N]y(t) + Mr(t)$$
$$u(t) = -KQw(t) - K(P + QL)y(t) + r(t)$$

and now the closed-loop state equation is

$$\begin{bmatrix} \dot{x}(t) \\ \dot{w}(t) \end{bmatrix} = \begin{bmatrix} A - BK(P + QL)C & -BKQ \\ -MK(P + QL)C + NC & J - MKQ \end{bmatrix} \begin{bmatrix} x(t) \\ w(t) \end{bmatrix} + \begin{bmatrix} B \\ M \end{bmatrix} r(t)$$

$$y(t) = \begin{bmatrix} C & 0 \end{bmatrix} \begin{bmatrix} x(t) \\ w(t) \end{bmatrix}$$

We use the following closed-loop coordinate transformation:

$$\begin{bmatrix} x(t) \\ \tilde{z}_2(t) \end{bmatrix} = \begin{bmatrix} I & 0 \\ E - LC & -I \end{bmatrix} \begin{bmatrix} x(t) \\ w(t) \end{bmatrix}$$

$$\begin{bmatrix} x(t) \\ w(t) \end{bmatrix} = \begin{bmatrix} I & 0 \\ E - LC & -I \end{bmatrix} \begin{bmatrix} x(t) \\ \tilde{z}_2(t) \end{bmatrix}$$

where, once again, the coordinate transformation matrix equals its inverse. Then the similarity transformation

$$\begin{bmatrix} I & 0 \\ E - LC & -I \end{bmatrix} \begin{bmatrix} A - BK(P + QL)C & -BKQ \\ -MK(P + QL)C + NC & J - MKQ \end{bmatrix}$$

$$\times \begin{bmatrix} I & 0 \\ E - LC & -I \end{bmatrix}$$

along with the identities

$$J = (E - LC)AQ$$
$$N = (E - LC)A(P + QL)$$
$$M = (E - LC)B$$

yields

$$\begin{bmatrix} I & 0 \\ E - LC & -I \end{bmatrix}$$

$$\times \begin{bmatrix} A - BK(P + QL)C & -BKQ \\ (E - LC)(A - BK)(P + QL)C & (E - LC)(A - BK)Q \end{bmatrix}$$

$$\begin{bmatrix} I & 0 \\ E - LC & -I \end{bmatrix}$$

$$= \begin{bmatrix} A - BK(P + QL)C & -BKQ \\ (E - LC)A(I - PC - QLC) & -(E - LC)AQ \end{bmatrix} \begin{bmatrix} I & 0 \\ E - LC & -I \end{bmatrix}$$

$$= \begin{bmatrix} A - BK(P + QL)C - BKQ(E - LC) & BKQ \\ (E - LC)A(I - PC - QLC) - (E - LC)AQ(E - LC) & (E - LC)AQ \end{bmatrix}$$

Using

$$[P \quad Q] \begin{bmatrix} C \\ E \end{bmatrix} = PC + QE = I$$

the upper left block simplifies to

$$A - BK(P + QL)C - BKQ(E - LC)$$
$$= A - BK[(P + QL)C + Q(E - LC)]$$
$$= A - BK[PC + QLC + QE - QLC]$$
$$= A - BK[PC + QE]$$
$$= A - BK$$

and the lower left block becomes

$$(E - LC)A(I - PC - QLC) - (E - LC)AQ(E - LC)$$
$$= (E - LC)[A(QE - QLC) - A(QE - QLC)]$$
$$= (E + LC)[0]$$
$$= 0$$

In addition, recalling that $M = (E - LC)B$, we have

$$\begin{bmatrix} I & 0 \\ E - LC & -I \end{bmatrix} \begin{bmatrix} B \\ M \end{bmatrix} = \begin{bmatrix} B \\ (E - LC)B - M \end{bmatrix} = \begin{bmatrix} B \\ 0 \end{bmatrix}$$

This, along with $J = (E - LC)AQ$, allows us to write the transformed closed-loop state equation as

$$\begin{bmatrix} \dot{x}(t) \\ \dot{\tilde{z}}_2(t) \end{bmatrix} = \begin{bmatrix} A - BK & BKQ \\ 0 & J \end{bmatrix} \begin{bmatrix} x(t) \\ \tilde{z}_2(t) \end{bmatrix} + \begin{bmatrix} B \\ 0 \end{bmatrix} r(t)$$

so we have

$$\sigma\left(\begin{bmatrix} A - BK & BKQ \\ 0 & J \end{bmatrix}\right) = \sigma(A - BK) \cup \sigma(J)$$

which indicates that the $2n - p$ closed-loop eigenvalues are the n eigenvalues of $A - BK$, which can be located arbitrarily by the feedback gain matrix K for (A, B) controllable, together with the $n - p$ eigenvalues of $J = (E - LC)AQ = A_{22} - LA_{12}$, which can be placed arbitrarily by the reduced-order observer gain matrix L for (A, C) observable.

Example 8.8   Here we design a compensator for the state equation in Example 8.1 that incorporates the reduced-order observer constructed in Example 8.5. We will make direct use of the previously-computed coefficient matrices:

$$J = \begin{bmatrix} -9 & 1 \\ -16 & -1 \end{bmatrix} \qquad N = \begin{bmatrix} -69 \\ -160 \end{bmatrix} \qquad M = \begin{bmatrix} 0 \\ 1 \end{bmatrix}$$

In addition, we take

$$E = \begin{bmatrix} 0 & 1 & 0 \\ 0 & 0 & 1 \end{bmatrix}$$

to give

$$\begin{bmatrix} C \\ E \end{bmatrix} = \begin{bmatrix} 1 & 0 & 0 \\ 0 & 1 & 0 \\ 0 & 0 & 1 \end{bmatrix}$$

from which it follows that

$$P = \begin{bmatrix} 1 \\ 0 \\ 0 \end{bmatrix} \qquad Q = \begin{bmatrix} 0 & 0 \\ 1 & 0 \\ 0 & 1 \end{bmatrix}$$

trivially satisfies

$$[P \quad Q]^{-1} = \begin{bmatrix} C \\ E \end{bmatrix}$$

We also directly incorporate the feedback gain vector

$$K = [-3 \quad -2 \quad 1]$$

computed in Example 8.6 to locate the eigenvalues of $A - BK$ at $-\frac{1}{2} \pm j\frac{\sqrt{3}}{2}, -1$. Everything is now in place to compute the coefficient matrices of the reduced-order observer-based compensator:

$J - MKQ$

$$= \begin{bmatrix} -9 & 1 \\ -16 & -1 \end{bmatrix} - \begin{bmatrix} 0 \\ 1 \end{bmatrix} [-3 \quad -2 \quad 1] \begin{bmatrix} 0 & 0 \\ 1 & 0 \\ 0 & 1 \end{bmatrix} = \begin{bmatrix} -9 & 1 \\ -14 & -2 \end{bmatrix}$$

$- MK(P + QL) + N$

$$= -\begin{bmatrix} 0 \\ 1 \end{bmatrix} [-3 \quad -2 \quad 1] \begin{bmatrix} 1 \\ 9 \\ 12 \end{bmatrix} + \begin{bmatrix} -69 \\ -160 \end{bmatrix} = \begin{bmatrix} -69 \\ -151 \end{bmatrix}$$

$$- KQ = -[-3 \quad -2 \quad 1] \begin{bmatrix} 0 & 0 \\ 1 & 0 \\ 0 & 1 \end{bmatrix} = [2 \quad -1]$$

$$- K(P + QL) = -[-3 \quad -2 \quad 1] \begin{bmatrix} 1 \\ 9 \\ 12 \end{bmatrix} = 9$$

which yields the compensator state equation

$$\begin{bmatrix} \dot{w}_1(t) \\ \dot{w}_2(t) \end{bmatrix} = \begin{bmatrix} -9 & 1 \\ -14 & -2 \end{bmatrix} \begin{bmatrix} w_1(t) \\ w_2(t) \end{bmatrix} + \begin{bmatrix} -69 \\ -151 \end{bmatrix} y(t) + \begin{bmatrix} 0 \\ 1 \end{bmatrix} r(t)$$

$$u(t) = [2 \quad -1] \begin{bmatrix} w_1(t) \\ w_2(t) \end{bmatrix} + 9y(t) + r(t)$$

The resulting five-dimensional closed-loop state equation is given by

$$\begin{bmatrix} \dot{x}_1(t) \\ \dot{x}_2(t) \\ \dot{x}_3(t) \\ \dot{w}_1(t) \\ \dot{w}_2(t) \end{bmatrix} = \begin{bmatrix} 0 & 1 & 0 & 0 & 0 \\ 0 & 0 & 1 & 0 & 0 \\ 5 & -4 & -1 & 2 & -1 \\ -69 & 0 & 0 & -9 & 1 \\ -151 & 0 & 0 & -14 & -2 \end{bmatrix} \begin{bmatrix} x_1(t) \\ x_2(t) \\ x_3(t) \\ w_1(t) \\ w_2(t) \end{bmatrix} + \begin{bmatrix} 0 \\ 0 \\ 1 \\ 0 \\ 1 \end{bmatrix} r(t)$$

$$y(t) = [1 \quad 0 \quad 0 \mid 0 \quad 0] \begin{bmatrix} x_1(t) \\ x_2(t) \\ x_3(t) \\ w_1(t) \\ w_2(t) \end{bmatrix}$$

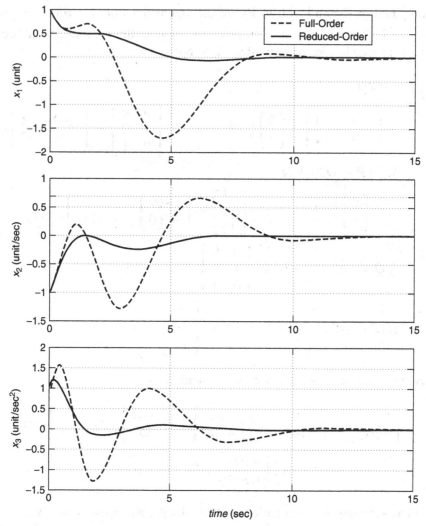

FIGURE 8.10 Closed-loop state responses for full-order and reduced-order observer-based compensators for Example 8.8.

The reader should verify that the closed-loop eigenvalues are indeed

$$\sigma(A - BK) \cup \sigma(J) = \{-\tfrac{1}{2} \pm j\tfrac{\sqrt{3}}{2}, -1\} \cup \{-5, -5\}$$

We investigate the closed-loop performance for the following initial conditions and zero reference input signal:

$$x_0 = \begin{bmatrix} 1 \\ -1 \\ 1 \end{bmatrix} \qquad w_0 = \begin{bmatrix} -9 \\ -12 \end{bmatrix} \qquad r(t) = 0 \qquad t \geq 0$$

Figure 8.10 compares the closed-loop state response $x(t)$ for the full-order observer-based compensator constructed in Example 8.6 with the closed-loop state response $x(t)$ for the reduced-order observer-based compensator constructed here. The difference between theses responses again can be attributed to the different eigenvalue specifications for the observer error dynamics for the full-order and reduced-order cases.

8.5 STEADY-STATE TRACKING WITH OBSERVER-BASED COMPENSATORS

Input Gain with Observers

We begin by observing that the structure of the closed-loop state equation (8.6) with state $[x^T(t), \tilde{x}^T(t)]$ easily allows us to compute the closed-loop transfer function

$$H_{\text{CL}}(s) = [C \quad 0] \begin{bmatrix} sI - A + BK & -BK \\ 0 & sI - A + LC \end{bmatrix}^{-1} \begin{bmatrix} B \\ 0 \end{bmatrix}$$

$$= [C \quad 0]$$

$$\times \begin{bmatrix} (sI - A + BK)^{-1} & (sI - A + BK)^{-1}BK(sI - A + LC)^{-1} \\ 0 & (sI - A + LC)^{-1} \end{bmatrix}$$

$$\times \begin{bmatrix} B \\ 0 \end{bmatrix}$$

$$= C(sI - A + BK)^{-1}B$$

which is exactly the same closed-loop transfer function that results from a direct application of state feedback. Of course, the same conclusion holds for the closed-loop state equation (8.5) with state $[x^T(t), \hat{x}^T(t)]$ because the two state equations are related by a state coordinate transformation that does not affect the transfer function. One interpretation of this outcome is the following: Transfer functions in general characterize input-output behavior for *zero initial state*. Initializing $[x^T(0), \tilde{x}^T(0)] = [0, 0]$ necessarily yields $\tilde{x}(t) \equiv 0$ for all $t \geq 0$, so the observer produces a perfect estimate of $x(t)$, and thus we should expect the closed-loop input-output behavior with an observer-based compensator to exactly match closed-loop input-output behavior with state feedback.

As a consequence, the input gain required to achieve identity closed-loop dc gain is identical to that derived in Chapter 7:

$$G = -[C(A - BK)^{-1}B]^{-1}$$

Observer-Based Servomechanism Design

In Section 7.5 we presented a servomechanism design methodology featuring integral error compensation together with state feedback that yielded asymptotic tracking of reference step inputs. This design was robust in the sense that accurate knowledge of the open-loop state equation coefficient matrices was not required to achieve zero steady-state error. This section extends that work by incorporating an observer to provide an estimate of the state feedback for direct substitution into the state feedback portion of the compensator. As with the servomechanism design based on state feedback presented in Section 7.5, this section follows the development in Ogata (2002).

For the open-loop state equation (8.1), we have the same objectives as in Section 7.5; i.e., we must achieve asymptotic stability for the closed-loop state equation and asymptotic tracking of step reference inputs with zero steady-state error. We adopt essentially the same assumptions as in Section 7.5 but also add observability of the open-loop state equation to assumption 1.

Assumptions

1. The open-loop state equation (8.1) is controllable and observable.
2. The open-loop state equation has no pole/eigenvalue at $s = 0$.
3. The open-loop state equation has no zero at $s = 0$.

Our observer-based compensator is of the form

$$\dot{\xi}(t) = r(t) - y(t)$$
$$\dot{\hat{x}}(t) = (A - LC)\hat{x}(t) + Bu(t) + Ly(t)$$
$$u(t) = -K\hat{x}(t) + k_I\xi(t)$$

which we can repackage as the $(n + 1)$-dimensional state equation

$$\begin{bmatrix} \dot{\xi}(t) \\ \dot{\hat{x}}(t) \end{bmatrix} = \begin{bmatrix} 0 & 0 \\ Bk_I & A - BK - LC \end{bmatrix} \begin{bmatrix} \xi(t) \\ \hat{x}(t) \end{bmatrix} + \begin{bmatrix} 1 \\ 0 \end{bmatrix} r(t) + \begin{bmatrix} -1 \\ L \end{bmatrix} y(t)$$

$$u(t) = \begin{bmatrix} k_I & -K \end{bmatrix} \begin{bmatrix} \xi(t) \\ \hat{x}(t) \end{bmatrix}$$

This, along with the open-loop state equation (8.1), yields the $(2n + 1)$-dimensional closed-loop state equation

$$\begin{bmatrix} \dot{x}(t) \\ \dot{\xi}(t) \\ \dot{\hat{x}}(t) \end{bmatrix} = \begin{bmatrix} A & Bk_I & -BK \\ -C & 0 & 0 \\ LC & Bk_I & A - BK - LC \end{bmatrix} \begin{bmatrix} x(t) \\ \xi(t) \\ \hat{x}(t) \end{bmatrix} + \begin{bmatrix} 0 \\ 1 \\ 0 \end{bmatrix} r(t)$$

$$y(t) = [C \quad 0 \quad 0] \begin{bmatrix} x(t) \\ \xi(t) \\ \hat{x}(t) \end{bmatrix}$$

Our first design objective is to achieve closed-loop asymptotic stability, for which we require that the $(2n + 1)$ eigenvalues of the $(2n + 1) \times (2n + 1)$ closed-loop system dynamics matrix have strictly negative real parts. It is not obvious how to proceed with the closed-loop state equation as written above, so we once again replace the observer state $\hat{x}(t)$ with the observer error $\tilde{x}(t) = x(t) - \hat{x}(t)$. As always, the observer error dynamics are given by

$$\dot{\tilde{x}}(t) = (A - LC)\tilde{x}(t)$$

and this leads to

$$\begin{bmatrix} \dot{x}(t) \\ \dot{\xi}(t) \\ \dot{\tilde{x}}(t) \end{bmatrix} = \left[\begin{array}{cc|c} A - BK & Bk_I & BK \\ -C & 0 & 0 \\ \hline 0 & 0 & A - LC \end{array} \right] \begin{bmatrix} x(t) \\ \xi(t) \\ \tilde{x}(t) \end{bmatrix} + \begin{bmatrix} 0 \\ 1 \\ 0 \end{bmatrix} r(t)$$

$$y(t) = [C \quad 0 \quad 0] \begin{bmatrix} x(t) \\ \xi(t) \\ \tilde{x}(t) \end{bmatrix}$$

Again, the block triangular structure of this closed-loop system dynamics matrix indicates that the $(2n + 1)$ closed-loop eigenvalues are the $(n + 1)$ eigenvalues of

$$\begin{bmatrix} A - BK & Bk_I \\ -C & 0 \end{bmatrix}$$

which we know from Chapter 7 that under our assumptions can be freely assigned by proper choice of the augmented state feedback gain vector $[K - k_I]$ together with n eigenvalues of $(A - LC)$, which, by virtue of our added observability assumption, can be arbitrarily located by the observer gain vector L.

To check that the steady-state tracking requirement has been met, we characterize the closed-loop equilibrium condition resulting from the constant reference input $r(t) = R, t \geq 0$. Using the relationship

$$u_{ss} = -K(x_{ss} - \tilde{x}_{ss}) + k_I \xi_{ss}$$
$$= -K \, x_{ss} + K \tilde{x}_{ss} + k_I \xi_{ss}$$

we see that

$$
\begin{bmatrix} 0 \\ 0 \\ 0 \end{bmatrix} = \left[\begin{array}{cc|c} A - BK & Bk_I & BK \\ -C & 0 & 0 \\ \hline 0 & 0 & A - LC \end{array} \right] \begin{bmatrix} x_{ss} \\ \xi_{ss} \\ \tilde{x}_{ss} \end{bmatrix} + \begin{bmatrix} 0 \\ 1 \\ 0 \end{bmatrix} R
$$

$$
= \left[\begin{array}{cc|c} A & B & 0 \\ -C & 0 & 0 \\ \hline 0 & 0 & A - LC \end{array} \right] \begin{bmatrix} x_{ss} \\ u_{ss} \\ \tilde{x}_{ss} \end{bmatrix} + \begin{bmatrix} 0 \\ 1 \\ 0 \end{bmatrix} R
$$

which corresponds to the decoupled equations

$$
(A - LC)\tilde{x}_{ss} = 0 \qquad \begin{bmatrix} 0 \\ 0 \end{bmatrix} = \begin{bmatrix} A & B \\ -C & 0 \end{bmatrix} \begin{bmatrix} x_{ss} \\ u_{ss} \end{bmatrix} + \begin{bmatrix} 0 \\ 1 \end{bmatrix} R
$$

the first of which, since $A - LC$ is necessarily nonsingular, forces $\tilde{x}_{ss} = 0$. The second equation we observe is exactly has encountered in Chapter 7 from which we conclude that

$$
\begin{bmatrix} x_{ss} \\ u_{ss} \end{bmatrix} = - \begin{bmatrix} A & B \\ -C & 0 \end{bmatrix}^{-1} \begin{bmatrix} 0 \\ R \end{bmatrix}
$$

$$
= \begin{bmatrix} A^{-1}B(CA^{-1}B)^{-1}R \\ -(CA^{-1}B)^{-1}R \end{bmatrix}
$$

$$
\xi_{ss} = \frac{1}{k_I}[u_{ss} + Kx_{ss}]
$$

$$
= \frac{1}{k_I}[1 - KA^{-1}B]H^{-1}(0)R
$$

and

$$
y_{ss} = Cx_{ss}
$$

$$
= C(A^{-1}B(CA^{-1}B)^{-1}R)
$$

$$
= (CA^{-1}B)(CA^{-1}B)^{-1}R
$$

$$
= R
$$

as desired.

The observer-based servomechanism block diagram is shown in Figure 8.11. Note that this block diagram is very similar to Figure 7.12, except that rather than directly applying the state feedback gain to the state $x(t)$, the state estimate $\hat{x}(t)$ produced by the *Observer* block is used

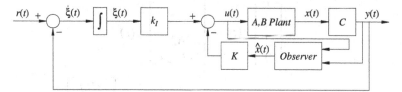

FIGURE 8.11   Observer-based servomechanism block diagram.

instead. The *Observer* block represents

$$\dot{\hat{x}}(t) = (A - LC)\hat{x}(t) + Bu(t) + Ly(t)$$

which, as shown in the block diagram, has $u(t)$ and $y(t)$ as inputs.

Example 8.9   We construct an observer-based type I servomechanism for the state equation of Example 8.2 by incorporating the observer designed in Example 8.2 into the state-feedback-based type 1 servomechanism designed in Example 7.9. In addition to the assumptions verified in Example 7.9, the state equation is observable as demonstrated in Example 8.2. We recall from Example 7.9 that the augmented state feedback gain vector

$$[\,K \quad -k_{\mathrm{I}}\,] = [\,3.6 \quad -1.4 \quad 2.2| \quad -16\,]$$

yields

$$\begin{bmatrix} A - BK & Bk_{\mathrm{I}} \\ -C & 0 \end{bmatrix} = \left[\begin{array}{ccc|c} 0 & 1 & 0 & 0 \\ 0 & 0 & 1 & 0 \\ -21.6 & -13.6 & -4.2 & 16 \\ \hline -1 & 0 & 0 & 0 \end{array}\right]$$

having eigenvalues at $-0.848 \pm j2.53$ and $-1.25 \pm j0.828$. In addition, the observer gain vector

$$L = \begin{bmatrix} 158 \\ 3624 \\ 43624 \end{bmatrix}$$

locates the eigenvalues of $A - LC$ at $-13.3 \pm j14.9, -133.3$, which are much farther to the left in the complex plane. Everything is in place to specify the four-dimensional observer-based compensator:

$$\begin{bmatrix} \dot{\xi}(t) \\ \dot{\hat{x}}_1(t) \\ \dot{\hat{x}}_2(t) \\ \dot{\hat{x}}_3(t) \end{bmatrix} = \left[\begin{array}{c|ccc} 0 & 0 & 0 & 0 \\ \hline 0 & -158 & 1 & 0 \\ 0 & -3624 & 0 & 1 \\ 16 & -43645.6 & -13.6 & -4.2 \end{array}\right] \begin{bmatrix} \xi(t) \\ \hat{x}_1(t) \\ \hat{x}_2(t) \\ \hat{x}_3(t) \end{bmatrix}$$

$$+ \begin{bmatrix} \frac{1}{0} \\ 0 \\ 0 \end{bmatrix} r(t) + \begin{bmatrix} \frac{-1}{158} \\ 158 \\ 3624 \\ 43624 \end{bmatrix} y(t)$$

$$u(t) \;=\; \begin{bmatrix} 16 & | -3.6 & 1.4 & -2.2 \end{bmatrix} \begin{bmatrix} \dot{\xi}(t) \\ \hat{x}_1(t) \\ \hat{x}_2(t) \\ \hat{x}_3(t) \end{bmatrix}$$

The resulting seven-dimensional closed-loop state equation is:

$$\begin{bmatrix} \dot{x}_1(t) \\ \dot{x}_2(t) \\ \dot{x}_3(t) \\ \dot{\xi}(t) \\ \dot{\hat{x}}_1(t) \\ \dot{\hat{x}}_2(t) \\ \dot{\hat{x}}_3(t) \end{bmatrix} = \begin{bmatrix} 0 & 1 & 0 & 0 & 0 & 0 & 0 \\ 0 & 0 & 1 & 0 & 0 & 0 & 0 \\ -18 & -15 & -2 & 16 & -3.6 & 1.4 & -2.2 \\ -1 & 0 & 0 & 0 & 0 & 0 & 0 \\ 158 & 0 & 0 & 0 & -158 & 1 & 0 \\ 3624 & 0 & 0 & 0 & -3624 & 0 & 1 \\ 43624 & 0 & 0 & 16 & -43645.6 & -13.6 & -4.2 \end{bmatrix}$$

$$\times \begin{bmatrix} x_1(t) \\ x_2(t) \\ x_3(t) \\ \xi(t) \\ \hat{x}_1(t) \\ \hat{x}_2(t) \\ \hat{x}_3(t) \end{bmatrix} + \begin{bmatrix} 0 \\ 0 \\ 0 \\ 1 \\ 0 \\ 0 \\ 0 \end{bmatrix} r(t)$$

$$y(t) = \begin{bmatrix} 1 & 0 & 0 & |0| & 0 & 0 & 0 \end{bmatrix} \begin{bmatrix} x_1(t) \\ x_2(t) \\ x_3(t) \\ \xi(t) \\ \hat{x}_1(t) \\ \hat{x}_2(t) \\ \hat{x}_3(t) \end{bmatrix}$$

The closed-loop unit step response for zero initial conditions is shown in Figure 8.12. Note that because in this case $\hat{x}_0 = x_0 = 0$, the observer state $\hat{x}(t)$ response exactly matches $x(t)$, so the output response exactly matches the nominal output response shown in Figure 7.13 for the state-feedback-based type 1 servomechanism design of Example 7.9. To see the effect of the observer, the closed-loop response for the following initial

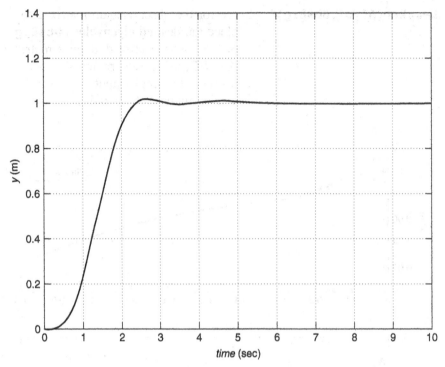

FIGURE 8.12 Closed-loop unit step response for Example 8.9.

conditions and zero reference input

$$x_0 = \begin{bmatrix} 0.1 \\ 0 \\ 0 \end{bmatrix} \qquad \hat{x}_0 = \begin{bmatrix} 0 \\ 0 \\ 0 \end{bmatrix} \qquad r(t) = 0 \quad t \geq 0$$

is plotted in Figure 8.13. The initial error $\tilde{x}_1(0) = x_1(0) - \hat{x}_1(0) = 0.1$ yields an error response for all three state variables that rapidly decays to zero (note the different time scale compared with Figure 8.12).

8.6   MATLAB FOR OBSERVER DESIGN

The following MATLAB functions are useful for observer design:

`L=place(A',C',ObsEig)'`   Solve for the observer gain matrix **L** to place the desired eigenvalues **ObsEig** of the observer error dynamics matrix $A - LC$.

`L=acker(A',C',ObsEig)'`   Solve for the observer gain matrix **L** to place the desired eigenvalues **ObsEig** of the observer error dynamics matrix $A - LC$, using Ackermann's formula, for single-input, single-output systems only.

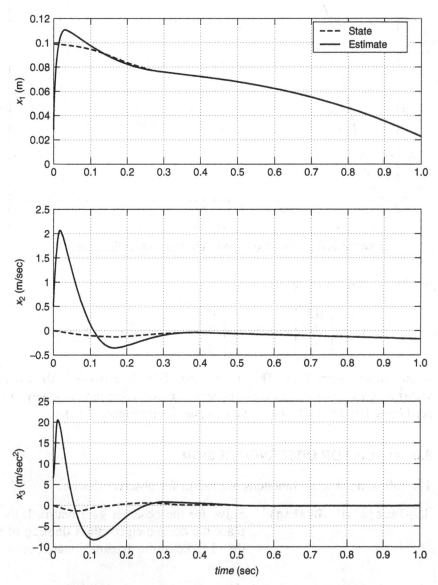

FIGURE 8.13   Observer error performance for Example 8.9.

For the Continuing MATLAB Example (rotational mechanical system), we now design an observer, i.e., calculate the observer gain matrix *L* given *A*, *C*, and reasonable observer eigenvalues to accompany the closed-loop eigenvalues of the Chapter 7 Continuing MATLAB Example. The following MATLAB code performs this observer design for the continuing example:

```
%------------------------------------------------------------
%   Chapter 8.  Design and Simulation of Linear
% Observers for State Feedback
%------------------------------------------------------------

% Select desired observer eigenvalues; ten times
% control law eigenvalues
ObsEig2 = 10*DesEig2;

L    = place(A',C', ObsEig2)';  % Compute observer gain
                                % matrix L
Lack = acker(A',C', ObsEig2)';  % Check L via
                                % Ackermann's formula

Ahat = A-L*C;           % Compute the closed-loop observer
                        % estimation error matrix
eig(Ahat);              % Check to ensure desired
   eigenvalues are in there

% Compute and simulate closed-loop system with control
% law and observer
Xr0 = [0.4;0.2;0.10;0];     % Define vector of
                            % initial conditions
Ar = [(A-B*K) B*K;zeros(size(A)) (A-L*C)];
Br = [B;zeros(size(B))];
Cr = [C zeros(size(C))];
Dr = D;
JbkRr = ss(Ar,Br,Cr,Dr);    % Create the closed-loop
                            % system with observer
r = [zeros(size(t))];       % Define zero reference
                            % input to go with t
[Yr,t,Xr] = lsim(JbkRr,r,t,Xr0);

% Compare Open, Closed, and Control Law/Observer
% responses
```

```
figure;
plot(t,Yo,'r',t,Yc,'g',t,Yr,'b'); grid;
axis([0 4 -0.2 0.5]);
set(gca,'FontSize',18);
legend('Open-loop','Closed-loop','w/ Observer');
xlabel('\ittime (sec)'); ylabel('\ity');

figure;                    % Plot observer errors
plot(t,Xr(:,3),'r',t,Xr(:,4),'g'); grid;
axis([0 0.2 -3.5 0.2]);
set(gca,'FontSize',18);
legend('Obs error 1','Obs error 2');
xlabel('\ittime (sec)'); ylabel('\ite');
```

This m-file, combined with the previous chapter m-files, yields the following output, plus the output response plot of Figure 8.14 and the observer error plot of Figure 8.15:

```
ObsEig2 =
-57.1429 +51.1954i
-57.1429 -51.1954i

L =
 110.29
5405.13

Ahat =
   -110.3      1.0
  -5445.1     -4.0

ans =
-57.1429 +51.1954i
-57.1429 -51.1954i
```

In Figure 8.15 we use e to represent the observer error $e(t) = \tilde{x}(t) = x(t) - \hat{x}(t)$. In the simulation, we started the observer with an error of 0.1 rad in the shaft angle θ estimate (and zero error in $\dot{\theta}$ estimate). In Figure 8.15 we see that the observer error for the shaft angle starts from the prescribed initial value and quickly goes to zero (the time scale of Figure 8.15 is greatly expanded compared with that of Figure 8.14). The observer velocity error goes to zero soon after, but with a large initial negative peak, even though the initial error was zero. However, this effect is not seen in Figure 8.14, where the closed-loop system response with observer

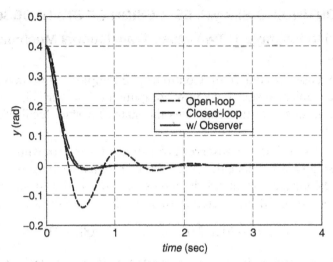

FIGURE 8.14 Open-loop, closed-loop with state feedback, and closed-loop with observer output response for the Continuing MATLAB Example.

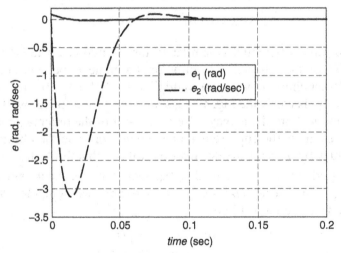

FIGURE 8.15 Observer error state responses for the Continuing MATLAB Example.

(solid) slightly lags the closed-loop system response with state feedback (dashed) but then quickly matches (around 1 s). Since the observer error eigenvalues were chosen to be 10 times greater than the closed-loop eigenvalues placed via state feedback, the observer error transient response goes to zero much faster than the closed-loop system with state feedback. The dashed and dotted responses in Figure 8.14 are identical to those of Figure 7.14 (top).

8.7   CONTINUING EXAMPLES: DESIGN OF STATE OBSERVERS

Continuing Example 1: Two-Mass Translational Mechanical System

For both cases a and b of Continuing Example 1 (two-mass translational mechanical system), we now compute an observer gain matrix L to construct an observer that estimates the states for state feedback control.

Case a. $A - LC$ is the observer error dynamics matrix whose eigenvalues we assign via L. Since the observer error dynamics must be faster than the closed-loop state feedback dynamics, we choose four desired observer eigenvalues to be 10 times greater than the four control law eigenvalues:

$$\lambda_{1,2} = -20 \pm 21i \qquad \lambda_{3,4} = -200, -210$$

The characteristic polynomial associated with these desired observer eigenvalues is

$$\alpha_{\text{OBS}}(s) = s^4 + 450s^3 + 5.924 \times 10^4 s^2 + 2.0244 \times 10^6 s$$
$$+ 3.5276 \times 10^7$$

Note that this polynomial is similar to the desired characteristic polynomial $\alpha(s)$ of the Chapter 7 Continuing Example 1, but the coefficients of the descending s powers are multiplied by $10^0, 10^1, 10^2, 10^3$, and 10^4, respectively, because the eigenvalues were each multiplied by 10. Therefore, numerical problems may arise as a result of the orders-of-magnitude differences in coefficients. Hence one should not scale the observer error eigenvalues by a factor greater than the rule-of-thumb of 10.

Taking advantage of the duality between state feedback and observer gain design, and using MATLAB function **place**, we computed the 4×2 observer gain matrix L to be

$$L = \begin{bmatrix} 195 & 1073 \\ -978 & 213836 \\ 2 & 254 \\ -101 & 11863 \end{bmatrix}$$

Because of the numerical issue pointed out earlier, the elements of L vary greatly in magnitude. The output response plots for the observer-based closed-loop system are shown in Figure 8.16 for case a.

We initialized the observer with an error of 0.5 and 1 mm for $y_1(t)$ and $y_2(t)$, respectively (and zero error for both velocity estimates). In Figure 8.16 we see that the response of the closed-loop system with

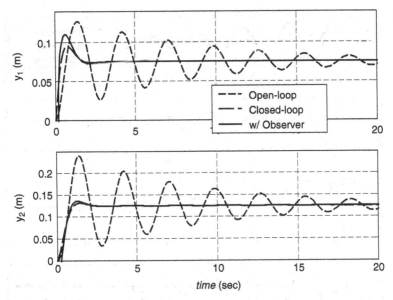

FIGURE 8.16 Open-loop, closed-loop with state feedback, and closed-loop with observer output response for Continuing Example 1, case a.

observer-based compensator (solid) initially overshoots that of the closed-loop system with state feedback (dashed) but then quickly matches (around 3 s). Since the observer error eigenvalues were chosen to be those placed by state feedback scaled by 10, the observer error transient response goes to zero faster than the transient response of the closed-loop system with state feedback. The open-loop and closed-loop responses in Figure 8.16 are identical to those of Figure 7.19.

Case b. The observer eigenvalues are chosen to be 10 times greater than the desired closed-loop eigenvalues used in Chapter 7, that is,

$$\lambda_{1,2} = -12.3 \pm 36.6i \qquad \lambda_{3,4} = -18.1 \pm 12.0i$$

Taking advantage of the duality between state feedback and observer gain design, and using MATLAB function **place**, we found the 4×1 observer gain matrix L to be

$$L = \begin{bmatrix} 60 \\ 2756 \\ 5970 \\ 135495 \end{bmatrix}$$

Again, we see that the terms of L vary greatly in magnitude as a result of the fact that the desired observer eigenvalues are ten times

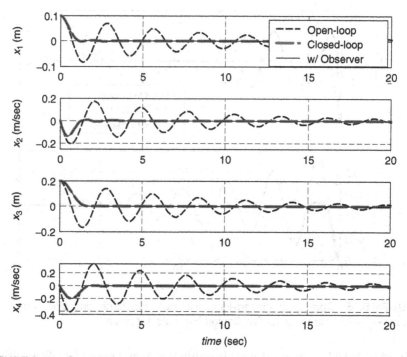

FIGURE 8.17   Open-loop, closed-loop with state feedback, and closed-loop with observer plant state variable responses for Continuing Example 1, case *b*.

greater than those placed by state feedback in Chapter 7. The plant state variable responses for the observer-based closed-loop system are shown in Figure 8.17 for case b.

We again initialized the observer with an error of 0.5 and 1 mm for $y_1(t)$ and $y_2(t)$, respectively (and zero error for both velocity estimates). In Figure 8.17 we see that in case *b* the observer-based closed-loop system (solid) matches the response of the closed-loop system with state feedback (dashed) very well on this time scale. There are observer errors, but they approach zero before 1 s. Because of the eigenvalue scaling, the observer error transient response goes to zero faster than that of the closed-loop system with state feedback. The dotted and dashed (the dashed curves are almost perfectly masked by the solid curves) responses in Figure 8.17 are identical to those of Figure 7.21.

Continuing Example 2: Rotational Electromechanical System

For Continuing Example 2 (rotational electromechanical system), we calculate an observer gain vector L and construct an observer to estimate the states for feedback.

The observer error dynamics matrix is $A - LC$. Since the observer error dynamics should be faster than the closed-loop dynamics achieved by state feedback, we choose three desired observer error eigenvalues to be the three control law eigenvalues scaled by 10, namely,

$$\lambda_{1,2,3} = \{-40, -120, -130\}$$

The characteristic polynomial associated with these desired eigenvalues is

$$\alpha_{\text{OBS}}(s) = s^3 + 290s^2 + 25600s + 624000$$

Note that this polynomial is similar to the desired characteristic polynomial $\alpha(s)$ achieved by state feedback in the Chapter 7 Continuing Example 2, but the coefficients of the descending s powers are multiplied by $10^0, 10^1, 10^2$, and 10^3, respectively, because the eigenvalues were each multiplied by 10.

Taking advantage of the duality between state feedback and observer gain design and performing the observer gain vector computation by hand, or using MATLAB functions **place** or **acker**, we find the 3×1 observer gain matrix L to be

$$L = \begin{bmatrix} 287 \\ 24737 \\ 549215 \end{bmatrix}$$

The three plant state variable responses for the observer-based closed-loop system are shown in Figure 8.18 for Continuing Example 2. We initialized the observer with an error of 0.001 rad in the shaft angle $\theta(t)$ estimate [and zero error in $\dot{\theta}(t)$ and $\ddot{\theta}(t)$ estimates]. In Figure 8.18, the output response of the closed-loop system with observer (solid) slightly lags that of the closed-loop system with state feedback (dashed); also, there is significant overshoot in the $\dot{\theta}(t)$ and $\ddot{\theta}(t)$ responses for observer-based closed-loop system. Observer error convergence is obtained in approximately 1.5 s. Since the observer error eigenvalues were obtained by scaling the state feedback eigenvalues by a factor of 10, the observer error transient response goes to zero faster than that of the closed-loop system with state feedback. The dashed and dotted responses in Figure 8.18 are identical to those of Figure 7.22.

8.8   HOMEWORK EXERCISES

Numerical Exercises

NE8.1 For your **NE 7.1** results (all cases) determine acceptable observer error eigenvalues by simply multiplying your **NE 7.1** eigenvalues

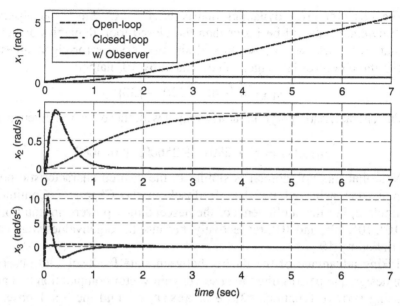

FIGURE 8.18 Open-loop, closed-loop with state feedback, and closed-loop with observer plant state variable responses for Continuing Example 2.

by 10. For each case, co-plot the unit step responses of the generic all-pole transfer functions associated with each set of eigenvalues.

NE8.2 For each (A, C) pair below, use the Bass-Gura formula to calculate the observer gain vector L to place the given eigenvalues for the observer error dynamics matrix $A - LC$. Check your results.

a. $A = \begin{bmatrix} -1 & 0 \\ 0 & -4 \end{bmatrix}$ $C = [1 \quad 1]$ $s_{1,2} = -20 \pm 30i$

b. $A = \begin{bmatrix} 0 & 1 \\ -6 & -8 \end{bmatrix}$ $C = [1 \quad 0]$ $s_{1,2} = -40, -50$

c. $A = \begin{bmatrix} 0 & 1 \\ -6 & 0 \end{bmatrix}$ $C = [1 \quad 0]$ $s_{1,2} = -40, -50$

d. $A = \begin{bmatrix} 0 & 8 \\ 1 & 10 \end{bmatrix}$ $C = [0 \quad 1]$ $s_{1,2} = -10 \pm 10i$

NE8.3 Repeat **NE 8.2** using Ackermann's Formula.

Analytical Exercises

AE8.1 Show that observability is *not* invariant with respect to state feedback.

AE8.2 Show that the closed-loop state equation (8.5) is *not* controllable.

AE8.3 Suppose that the linear state equation

$$\begin{bmatrix} \dot{x}_1(t) \\ \dot{x}_2(t) \end{bmatrix} = \begin{bmatrix} A_{11} & A_{12} \\ A_{21} & A_{22} \end{bmatrix} \begin{bmatrix} x_1(t) \\ x_2(t) \end{bmatrix} + \begin{bmatrix} I \\ 0 \end{bmatrix} u(t)$$

$$y(t) = [\, C_1 \quad C_2 \,] \begin{bmatrix} x_1(t) \\ x_2(t) \end{bmatrix}$$

is controllable and observable. Dualize the reduced-order observer-based compensator construction to derive a stabilizing compensator of dimension $n - m$.

AE8.4 Suppose the linear state equation (8.1) is controllable and observable. Given an $(n - m) \times (n - m)$ matrix F and an $n \times p$ matrix L, consider the feedback compensator defined by

$$\dot{z}(t) = Fz(t) + Gv(t)$$

$$v(t) = y(t) + CHz(t)$$

$$u(t) = Mz(t) + Nv(t)$$

in which the matrices $G, H, M,$ and N satisfy

$$AH - BM = HF$$

$$HG + BN = -L$$

Use the closed-loop state coordinate transformation

$$\begin{bmatrix} w(t) \\ z(t) \end{bmatrix} = \begin{bmatrix} I & H \\ 0 & I \end{bmatrix} \begin{bmatrix} x(t) \\ z(t) \end{bmatrix}$$

to show that the $2n - m$ closed-loop eigenvalues are those of F and those of $A - LC$.

AE 8.5 Suppose that $H_1(s)$ and $H_2(s)$ are two strictly proper transfer functions with observable state-space realizations (A_1, B_1, C_1) and (A_2, B_2, C_2), respectively. Show that

$$\begin{bmatrix} \dot{x}_1(t) \\ \dot{x}_2(t) \end{bmatrix} = \begin{bmatrix} A_1 & 0 \\ 0 & A_2 \end{bmatrix} \begin{bmatrix} x_1(t) \\ x_2(t) \end{bmatrix} + \begin{bmatrix} B_1 \\ B_2 \end{bmatrix} u(t)$$

$$y(t) = [\, C_1 \quad C_2 \,] \begin{bmatrix} x_1(t) \\ x_2(t) \end{bmatrix}$$

is a state-space realization of the parallel interconnection $H_1(s) + H_2(s)$ and that this realization is observable if and only if A_1 and A_2 have no common eigenvalues.

Continuing MATLAB Exercises

CME8.1 For the system given in CME1.1:
   a. Design an observer-based compensator using the control law of CME7.1b. Use observer error eigenvalues that are the desired state feedback eigenvalues of CME7.1a scaled by 10. Compare open-loop, closed-loop with state feedback, and closed-loop with observer responses to a unit step input. Introduce an initial observer error, otherwise, the closed-loop with state feedback and closed-loop with observer responses will be identical.
   b. Design and evaluate in simulation an observer for the servomechanism of CME7.1c.

CME8.2 For the system given in CME1.2:
   a. Design an observer-based compensator using the control law of CME7.2b. Use observer error eigenvalues that are the desired state feedback eigenvalues of CME7.2a scaled by 10. Compare open-loop, closed-loop with state feedback, and closed-loop with observer responses to a unit step input. Introduce an initial observer error, otherwise, the closed-loop with state feedback and closed-loop with observer responses will be identical.
   b. Design and evaluate in simulation an observer for the servomechanism of CME7.2c.

CME8.3 For the system given in CME1.3:

a. Design an observer-based compensator using the control law of CME7.3b. Use observer error eigenvalues that are the desired control law eigenvalues of CME7.3a scaled by 10. Compare open-loop, closed-loop with state feedback, and closed-loop with observer responses to a unit step input. Introduce an initial observer error, otherwise, the closed-loop with state feedback and closed-loop with observer responses will be identical.

b. Design and evaluate in simulation an observer for the servomechanism of CME7.3c.

CME8.4 For the system given in CME1.4:

a. Design an observer-based compensator using the control law of CME7.4b. Use observer error eigenvalues that are the desired state feedback eigenvalues of CME7.4a scaled by 10. Compare open-loop, closed-loop with state feedback, and closed-loop with observer responses to a unit step input. Introduce an initial observer error, otherwise, the closed-loop with state feedback and closed-loop with observer responses will be identical.

b. Design and evaluate in simulation an observer for the servomechanism of CME7.4c.

Continuing Exercises

CE8.1 For the control laws designed in CE7.1b, design observer-based compensators for all three cases. Use observer error eigenvalues that are the desired state feedback eigenvalues scaled by 10 (for case iii, a factor of 2 works better because of numerical conditioning). In each case, evaluate your results: Plot and compare the simulated open-loop, closed-loop with state feedback, and closed-loop with observer output responses for the same cases as in CE2.1; use the same correction factors from CE7.1b. Introduce an initial observer error, otherwise, the closed-loop with state feedback and closed-loop with observer responses will be identical.

CE8.2 For the control laws designed in CE7.2b, design observer-based compensators for all three cases (this is possible only for the observable cases). Use observer error eigenvalues that are the desired state feedback eigenvalues scaled by 10. In each case, evaluate your results: Plot and compare the simulated open-loop, closed-loop with state feedback, and closed-loop with observer output responses for the same cases as in CE2.2; use the same correction factors from CE7.2b. Introduce an initial observer error, otherwise, the closed-loop with state feedback and closed-loop with observer responses will be identical.

CE8.3 For the control laws designed in CE7.3b, design observer-based compensators for both cases. Use observer error eigenvalues that are the desired state feedback eigenvalues scaled by 10. In each case, evaluate your results: Plot and compare the simulated open-loop, closed-loop with state feedback, and closed-loop with observer output responses for the same cases as in CE2.3 (for case ii, use the same correction factor from CE7.3b). Introduce an initial observer error, otherwise, the closed-loop with state feedback and closed-loop with observer responses will be identical.

CE8.4 Design an observer-based compensator for the control law designed in CE7.4b. Use observer error eigenvalues that are your desired state feedback eigenvalues scaled by 10. Evaluate your results: Plot and compare the simulated open-loop, closed-loop with state feedback, and closed-loop with observer output responses for the same case (impulse input, zero initial conditions) as in CE2.4. Introduce an initial observer error, otherwise, the closed-loop with state feedback and closed-loop with observer responses will be identical.

CE8.5 CE4.5 results should indicate that the original system is not observable. Therefore, add a second output: Make $\theta(t)$ an output in addition to $q(t)$; check observability again and proceed. For the control law designed in CE7.5b, design an observer-based compensator for this modified system with two outputs. Use observer error eigenvalues that are your desired state feedback eigenvalues scaled by 10. Evaluate your results: Plot and compare the simulated open-loop, closed-loop with state feedback, and closed-loop with observer output responses for the same case (given initial conditions) as in CE2.5. Introduce an initial observer error; otherwise, the closed-loop with state feedback and closed-loop with observer responses will be identical.

9

INTRODUCTION TO OPTIMAL CONTROL

As we noted in Chapter 1, the ability to formulate and solve optimal control problems was the primary catalyst in the emergence of state-space methods in the late 1950s and early 1960s. It is therefore fitting that we conclude this text with an introduction to the fundamental discoveries of this era that have shaped the theory and practice of control engineering over the past five decades.

We begin by formulating the linear quadratic regulator problem, which is the optimal control problem that is the central focus of this chapter. This is followed by a tutorial on the key mathematical tool that we will employ to solve this problem: the calculus of variations, or variational calculus. Variational calculus can be thought of as a generalization of certain results from multivariable calculus, so we include a brief review of the latter for motivation. We first apply variational calculus to solve the so-called minimum energy control problem before proceeding to the solution of the linear quadratic regulator problem. Since all but the most trivial examples are best solved with the aid of a computer, we next turn our attention to MATLAB functionality that supports the solution to the linear quadratic regulator problem. These tools are applied to the Continuing MATLAB Example and Continuing Example 1.

We must emphasize that this chapter merely scratches the surface of a vast body of basic and applied research on optimal control. For more

Linear State-Space Control Systems, by Robert L. Williams II and Douglas A. Lawrence
Copyright © 2007 John Wiley & Sons, Inc.

in depth study, we refer the interested reader to the many excellent texts on the subject including but by no means limited to Anderson and Moore (1971), Bryson and Ho (1975), Dorato, et al. (1995), Kwakernaak and Sivan (1972), Lewis (1992), and Zhou (1995).

9.1 OPTIMAL CONTROL PROBLEMS

We begin with a somewhat general optimal control problem formulation in order to provide a broader context for the special cases we consider in this chapter. We consider the nonlinear state equation

$$\dot{x}(t) = f[x(t), u(t), t] \quad x(t_0) = x_0$$

and introduce the *performance index, cost function,* or *objective function* to be minimized:

$$J = \int_{t_0}^{t_f} L[x(t), u(t), t]dt + \varphi[x(t_f), t_f]$$

Here $L(x, u, t)$ is called the *loss function* that assigns a penalty to the state and input in a possibly t-dependent way, and the function $\varphi(x, t)$ characterizes the *terminal cost* on the state. The optimization problem at hand is to determine an optimal state trajectory and input signal denoted by $[x^*(t), u^*(t)]$ that minimizes the performance index subject to the constraints imposed by the state equation and the specified initial state. This is, by nature, a regulation problem because the goal is to maintain the state trajectory "close" to the equilibrium at the origin while expending "moderate" control effort. For a particular problem, the performance index quantifies these qualitative notions and further serves to capture the fundamental design tradeoff between the conflicting objectives of regulation performance and control effort.

Additional constraints might be imposed on the state trajectory and input signal of the form

$$C[x(t), u(t), t] = 0 \quad \text{or} \quad C[x(t), u(t), t] \leq 0 \quad \text{for all} \quad t \in (t_0, t_f)$$

In addition, the problem formulation might involve explicit constraints on the terminal state instead of or in addition to the terminal cost term in the performance index.

Necessary conditions in the form of optimality equations that must be satisfied by the optimal solution $[x^*(t), u^*(t)]$ have been derived using the tools of *variational calculus* and *dynamic programming*. However, these equations in general are difficult to solve. Furthermore, by only satisfying a necessary condition for optimality, there is no guarantee that a solution to these equations is in fact optimal. Verifying this may require further nontrivial analysis.

In this chapter we consider the linear time-invariant state equation

$$\dot{x}(t) = Ax(t) + Bu(t) \quad x(t_0) = x_0 \tag{9.1}$$

and seek to minimize the quadratic performance index

$$J = \frac{1}{2} \int_{t_0}^{t_f} \left[x^T(t)Qx(t) + u^T(t)Ru(t) \right] dt + \frac{1}{2} x^T(t_f)Sx(t_f) \tag{9.2}$$

in which the loss function

$$L(x, u) = \frac{1}{2}(x^T Qx + u^T Ru)$$

$$= \frac{1}{2} \begin{bmatrix} x^T & u^T \end{bmatrix} \begin{bmatrix} Q & 0 \\ 0 & R \end{bmatrix} \begin{bmatrix} x \\ u \end{bmatrix}$$

and terminal cost function $\varphi(x) = \frac{1}{2} x^T Sx$ are quadratic forms. We assume that the weighting matrices Q, R, and S are each symmetric, with Q and S positive semidefinite and R positive definite. Regulator design for a linear time-invariant state equation with the goal of minimizing a quadratic performance index naturally is referred to as the linear quadratic regulator (LQR) problem. In contrast to the challenges that arise in the general nonlinear optimal control problem, the linear quadratic regulator problem admits an analytical solution that we will derive in the sequel.

Here we explicitly see the manner in which the quadratic performance index captures a tradeoff between regulation performance and control effort. In particular, if the quadratic terms involving the state are "large" compared with the quadratic term involving the input, then the optimal state trajectory will exhibit "good" regulation performance in that the response to a nonzero initial state will return rapidly to the equilibrium state at the origin, but this may come at the expense of "large" control energy. Conversely, if the input is penalized more heavily than the state, then the optimal control signal may not require a great deal of energy, but the regulation performance may not be acceptable. In essence, the underlying design problem is to translate given performance specifications into choices for the weighting matrices in such a way that the optimal solution meets these specifications. This issue has received a great deal of attention since the linear quadratic regulator problem was introduced in the late 1950s and yet remains something of an art.

9.2 AN OVERVIEW OF VARIATIONAL CALCULUS

We first consider optimization problems on the d-dimensional Euclidean space $\mathbb{R}^d$. As noted earlier, this material should come as a review to the reader and is included here to motivate the elements of variational calculus that we require to solve the optimal control problems posed in this chapter. We consider real-valued functions $f : \mathbb{R}^d \rightarrow \mathbb{R}$ and seek to minimize $f(z)$ over all $z \in \mathbb{R}^d$. We avoid many mathematical technicalities by assuming throughout this chapter that all functions defined on Euclidean spaces have continuous partial derivatives in all arguments over the entire domain. Furthermore, we will only consider optimization problems that have global solutions.

We first derive a necessary condition for $z^* \in \mathbb{R}^d$ to minimize $f(z)$. If z^* minimizes $f(z)$, then $f(z^* + v) > f(z^*)$ for any nonzero vector $v \in \mathbb{R}^d$. This implies that the directional derivative of $f(z)$ at z^* in the direction v, written in terms of the gradient $\nabla f(z)$ as $\nabla f(z^*) \cdot v$, must be zero. To see this, for any $\varepsilon > 0$ we also must have $f(z^* \pm \varepsilon v) > f(z^*)$ from which

$$\nabla f(z^*) \cdot v = \lim_{\varepsilon \to 0^+} \frac{f(z^* + \varepsilon v) - f(z^*)}{\varepsilon} \geq 0$$

$$\nabla f(z^*) \cdot (-v) = \lim_{\varepsilon \to 0^+} \frac{f(z^* - \varepsilon v) - f(z^*)}{\varepsilon} \geq 0$$

Since $\nabla f(z^*) \cdot (-v) = -\nabla f(z^*) \cdot v$, the only way both these inequalities can hold is if $\nabla f(z^*) \cdot v = 0$ as claimed. Since the directional derivative must vanish at z^* for any direction $v \in \mathbb{R}^d$, we conclude that the gradient of $f(z)$ at z^* must be the zero vector. In this case we call z^* a *critical point* of $f(z)$, and we have argued that if z^* minimizes $f(z)$, then necessarily, z^* is a critical point of $f(z)$.

The converse statement does not hold in general. Namely, a critical point of $f(z)$ is not necessarily a minimum. For example, on $\mathbb{R}^2$, $f(z) = -\frac{1}{2}(z_1^2 + z_2^2)$ has $\nabla f(z) = -[z_1, z_2]$, for which the critical point $z^* = (0, 0)$ corresponds to the maximum of $f(z)$. Also, $f(z) = \frac{1}{2}(z_1^2 - z_2^2)$ has $\nabla f(z) = [z_1, -z_2]$, for which the critical point $z^* = (0, 0)$ is a saddle point. Often, second-derivative information is used to classify critical points as being a minimum, maximum, or saddle point. Here we consider an approach that is better suited to the objectives of this chapter.

A real-valued function $f : \mathbb{R}^d \to \mathbb{R}$ is *convex* if

$$f(z + v) - f(z) \geq \nabla f(z) \cdot v \text{ for all } z, v \in \mathbb{R}^d$$

and *strictly convex* if equality holds above when and only when $v = 0 \in \mathbb{R}^d$. It is easy to see that a critical point of a strictly convex function is a minimum point. For if z^* satisfies $\nabla f(z^*) = 0$, then

$$f(z^* + v) - f(z^*) > \nabla f(z^*) \cdot v = 0 \text{ for all } v \neq 0 \in \mathbb{R}^d$$

from which $f(z^* + v) > f(z^*)$ for all nonzero $v \in \mathbb{R}^d$.

We next turn our attention to the problem of minimizing a real-valued function $f(z)$ over all $z \in \mathbb{R}^d$ that satisfy equality constraints of the form

$$g_i(z) = 0 \quad i = 1, \ldots, n$$

or in vector form

$$G(z) = \begin{bmatrix} g_1(z) \\ g_2(z) \\ \vdots \\ g_n(z) \end{bmatrix} = \begin{bmatrix} 0 \\ 0 \\ \vdots \\ 0 \end{bmatrix}$$

We assume that there are fewer constraints n than the Euclidean space dimension d so as not to over constrain the problem. We attack this problem using the method of Lagrange multipliers. The first step is to adjoin the constraints to the original function $f(z)$ using the $n \times 1$ vector of Lagrange multipliers

$$\lambda = \begin{bmatrix} \lambda_1 \\ \lambda_2 \\ \vdots \\ \lambda_n \end{bmatrix}$$

to form the real-valued function

$$\tilde{f}(z) = f(z) + \lambda_1 g_1(z) + \lambda_2 g_2(z) + \cdots + \lambda_n g_n(z)$$
$$= f(z) + \lambda^T G(z)$$

We observe that if z^* minimizes the augmented function $\tilde{f}(z)$ over $\mathbb{R}^d$ so that

$$\tilde{f}(z) > \tilde{f}(z^*) \text{ for all } z \neq z^*$$

then, in particular, for any $z \neq z^*$ that satisfies $G(z) = G(z^*)$, we have for any Lagrange multiplier vector $\lambda \in \mathbb{R}^n$

$$
\begin{aligned}
f(z) - f(z^*) &= f(z) - f(z^*) + \lambda^T [G(z) - G(z^*)] \\
&= [f(z) + \lambda^T G(z)] - [f(z^*) + \lambda^T G(z^*)] \\
&= \tilde{f}(z) - \tilde{f}(z^*) \\
&> 0
\end{aligned}
$$

We conclude that z^* minimizes the original function $f(z)$ over all $z \in \mathbb{R}^d$ that satisfy $G(z) = G(z^*)$. Now, if $\tilde{f}(z)$ is strictly convex (and this in general depends on $\lambda \in \mathbb{R}^n$), then the minimum of $\tilde{f}(z)$ is characterized by $\nabla \tilde{f}(z^*) = 0$. This, along with the constraint equation $G(z^*) = 0$, yields $d + n$ equations that, in principle, can be solved for the $d + n$ unknowns given by the components of z^* and λ.

Note that once z^* and λ have been determined, we must verify that $\tilde{f}(z)$ is strictly convex for this λ in order to conclude that the critical point z^* minimizes $\tilde{f}(z)$ and, in turn, minimizes $f(z)$ over all $z \in \mathbb{R}^d$ that satisfy $G(z) = G(z^*) = 0$ as desired. One situation in which the strict convexity of $\tilde{f}(z)$ is assured for any Lagrange multiplier vector is when the original function $f(z)$ is strictly convex and the equality constraint is given by the linear equation $Cz = d$ so that

$$
G(z) = Cz - d
$$

To see this, we use

$$
\begin{aligned}
\nabla \tilde{f}(z) &= \nabla f(z) + \lambda^T \frac{\partial G}{\partial z}(z) \\
&= \nabla f(z) + \lambda^T C
\end{aligned}
$$

together with the strict convexity of $f(z)$ to write

$$
\begin{aligned}
\tilde{f}(z + v) - \tilde{f}(z) &= \{ f(z + v) + \lambda^T [C(z + v) - d] \} - [f(z) + \lambda^T (Cz - d)] \\
&= [f(z + v) - f(z)] + \lambda^T Cv \\
&\geq \nabla f(z) \cdot v + \lambda^T Cv \\
&= [\nabla f(z) + \lambda^T C] \cdot v \\
&= \nabla \tilde{f}(z) \cdot v
\end{aligned}
$$

for all $z, v \in \mathbb{R}^d$ with equality when and only when $v = 0$, thereby characterizing strict convexity of $\tilde{f}(z)$.

To illustrate these ideas, we consider the problem of minimizing the quadratic form

$$f(z) = \tfrac{1}{2} z^T Q z$$

subject to the equality constraint $Cz = d$. We assume that the $d \times d$ matrix Q is symmetric and positive definite and that the $n \times d$ matrix C has full-row rank n. We can easily check that $f(z)$ is strictly convex. Using the gradient $\nabla f(z) = z^T Q$, we have for all $z, v \in \mathbb{R}^d$

$$
\begin{aligned}
f(z + v) - f(z) &= \tfrac{1}{2}(z + v)^T Q(z + v) - \tfrac{1}{2} z^T Q z \\
&= z^T Q v + \tfrac{1}{2} v^T Q v \\
&= \nabla f(z) \cdot v + \tfrac{1}{2} v^T Q v \\
&\geq \nabla f(z) \cdot v
\end{aligned}
$$

with equality when and only when $v = 0$ because of the assumed positive definiteness of Q. From the preceding discussion, we conclude that

$$\tilde{f}(z) = \tfrac{1}{2} z^T Q z + \lambda^T (Cz - d)$$

is strictly convex as well. We therefore minimize $\tilde{f}(z)$ and, in turn, minimize $f(z)$ over all $z \in \mathbb{R}^d$ that satisfy $Cz = d$ by solving

$$\nabla \tilde{f}(z) = z^T Q + \lambda^T C = 0 \quad \text{and} \quad Cz = d$$

for $z = z^* \in \mathbb{R}^d$ and $\lambda \in \mathbb{R}^n$. This leads to the system of equations

$$
\begin{bmatrix} C & 0 \\ Q & C^T \end{bmatrix} \begin{bmatrix} z \\ \lambda \end{bmatrix} = \begin{bmatrix} d \\ 0 \end{bmatrix}
$$

Since Q is positive definite, it is nonsingular, which gives $z^* = -Q^{-1} C^T \lambda$. On substituting into the constraint equation, we find $-C Q^{-1} C^T \lambda = d$. Since, in addition, C is assumed to have full-row rank n, the symmetric $n \times n$ matrix $C Q^{-1} C^T$ can be shown to be positive definite and hence nonsingular. This ultimately yields the optimal solution

$$\lambda = -(C Q^{-1} C^T)^{-1} d \quad \text{and} \quad z^* = Q^{-1} C^T (C Q^{-1} C^T)^{-1} d$$

We now generalize the preceding discussion for optimization on the Euclidean space $\mathbb{R}^d$ to optimization on the space of d-dimensional vector-valued functions $z(t)$ defined on the interval $[t_0, t_f]$ having a continuous derivative $\dot{z}(t)$. We denote this space of functions by $C^1[t_0, t_f]$, in which the dimension of $z(t)$, i.e. the number of component functions of $z(t)$, is left as understood from the context. We remark that $C^1[t_0, t_f]$ satisfies the definition of a linear vector space given in Appendix B for vector addition and scalar multiplication of (vector-valued) functions defined pointwise in t.

Our discussion will focus on the minimization of real-valued integral functions of the form

$$F(z) = \int_{t_0}^{t_f} f[t, z(t), \dot{z}(t)] \, dt$$

over elements of $C^1[t_0, t_f]$ that, in addition, are assumed to satisfy one of the following possible boundary conditions:

- $z(t_0) = z_0$ is specified.
- $z(t_f) = z_f$ is specified.
- both $z(t_0) = z_0$ and $z(t_f) = z_f$ are specified.

We note that other boundary conditions are possible involving specific components of $z(t)$ at $t = t_0$ and $t = t_f$.

To generalize the previous analysis on $\mathbb{R}^d$, we first need to generalize the notion of a directional derivative that will enable us to establish a necessary condition for minimizing the integral function $F(z)$ and further characterize the requisite convexity property under which the necessary condition becomes sufficient as well. This leads us to consider the *Gâteaux variation* of the integral function $F(z)$ at $z(t)$ in the direction $v(t) \in C^1[t_0, t_f]$, defined as

$$\delta F(z; v) = \lim_{\varepsilon \to 0} \frac{F(z + \varepsilon v) - F(z)}{\varepsilon}$$

If we assume that the integrand function $f(t, z, \dot{z})$ has continuous partial derivatives with respect to the second and third d-dimensional arguments on the entire domain $\mathbb{R} \times \mathbb{R}^d \times \mathbb{R}^d$, we can then compute the Gâteaux variation using the chain rule as follows:

$$\delta F(z; v) = \frac{\partial}{\partial \varepsilon} F(z + \varepsilon v)|_{\varepsilon = 0}$$

$$= \int_{t_0}^{t_f} \frac{\partial}{\partial \varepsilon} f[t, z(t) + \varepsilon v(t), \dot{z}(t) + \varepsilon \dot{v}(t)] \, dt \bigg|_{\varepsilon = 0}$$

$$= \int_{t_0}^{t_f} \left[\frac{\partial f}{\partial z}[t, z + \varepsilon v(t), \dot{z}(t) + \varepsilon \dot{v}(t)]v(t) \right.$$

$$+ \left. \frac{\partial f}{\partial \dot{z}}[t, z(t) + \varepsilon v(t), \dot{z}(t) + \varepsilon \dot{v}(t)]\dot{v}(t) \right] dt \Bigg|_{\varepsilon=0}$$

$$= \int_{t_0}^{t_f} \left\{ \frac{\partial f}{\partial z}[t, z(t), \dot{z}(t)]v(t) + \frac{\partial f}{\partial \dot{z}}[t, z(t), \dot{z}(t)]\dot{v}(t) \right\} dt \quad (9.3)$$

Example 9.1 We consider the integral function $F(z)$ for scalar $z(t)$ with integrand function $f(t, z, \dot{z}) = tz + \frac{1}{2}\dot{z}^2$. The partial derivatives

$$\frac{\partial f}{\partial z}(t, z, \dot{z}) = t \qquad \frac{\partial f}{\partial \dot{z}}(t, z, \dot{z}) = \dot{z}$$

yield the Gâteaux variation

$$\delta F(z; v) = \int_{t_0}^{t_f} [tv(t) + \dot{z}(t)\dot{v}(t)] \, dt \qquad \qquad \square$$

Next, we derive a necessary condition for $z^*(t) \in C^1[t_0, t_f]$ to minimize $F(z)$ over all $z(t) \in C^1[t_0, t_f]$ that satisfy one of the three boundary conditions. If $z^*(t)$ is such a minimizer, then $F(z^* + v) \geq F(z^*)$ for all $v(t) \in C^1[t_0, t_f]$ such that $z^*(t) + v(t)$ also satisfies the specified boundary condition. This requirement characterizes the so-called *admissible directions*:

- If $z(t_0) = z_0$ is specified, then $v(t_0) = 0 \in \mathbb{R}^d$.
- If $z(t_f) = z_f$ is specified, then $v(t_f) = 0 \in \mathbb{R}^d$.
- If both $z(t_0) = z_0$ and $z(t_f) = z_f$ are specified, then $v(t_0) = v(t_f) = 0 \in \mathbb{R}^d$.

Now, if $F(z^* + v) \geq F(z^*)$ for all admissible directions, we see that for any $\varepsilon > 0$, $F(z^* \pm \varepsilon v) \geq F(z^*)$ for all admissible directions, from which we obtain

$$\delta F(z^*; v) = \lim_{\varepsilon \to 0^+} \frac{F(z^* + \varepsilon v) - F(z^*)}{\varepsilon} \geq 0,$$

$$\delta F(z^*; -v) = \lim_{\varepsilon \to 0^+} \frac{F(z^* - \varepsilon v) - F(z^*)}{\varepsilon} \geq 0$$

Moreover since, as with the directional derivative on $\mathbb{R}^d$, the Gâteaux variation satisfies $\delta F(z, -v) = -\delta F(z; v)$, we see that the only way both inequalities can be satisfied is if $\delta F(z^*; v) = 0$ for all admissible directions. Thus, having $\delta F(z^*; v) = 0$ for all admissible directions is a necessary condition for $z^*(t)$ to minimize $F(z)$.

For the integral function $F(z)$, one way to guarantee that the Gâteaux variation (9.3) satisfies $\delta F(z; v) = 0$ for all admissible directions is to first require

$$\frac{d}{dt}\frac{\partial f}{\partial \dot{z}}[t, z(t), \dot{z}(t)] = \frac{\partial f}{\partial z}[t, z(t), \dot{z}(t)] \text{ for all } t \in (t_0, t_f) \qquad (9.4)$$

This is the well-known *Euler-Lagrange equation* whose solutions are called *stationary functions*. Evaluating the Gâteaux variation at a stationary function yields

$$\delta F(z; v) = \int_{t_0}^{t_f} \left\{ \frac{\partial f}{\partial z}[t, z(t), \dot{z}(t)]v(t) + \frac{\partial f}{\partial \dot{z}}[t, z(t), \dot{z}(t)]\dot{v}(t) \right\} dt$$

$$= \int_{t_0}^{t_f} \left(\left\{ \frac{d}{dt}\frac{\partial f}{\partial \dot{z}}[t, z(t), \dot{z}(t)] \right\} v(t) + \frac{\partial f}{\partial \dot{z}}[t, z(t), \dot{z}(t)]\dot{v}(t) \right) dt$$

$$= \int_{t_0}^{t_f} \frac{d}{dt} \left\{ \frac{\partial f}{\partial \dot{z}}[t, z(t), \dot{z}(t)]v(t) \right\} dt$$

$$= \frac{\partial f}{\partial \dot{z}}[t_f, z(t_f), \dot{z}(t_f)]v(t_f) - \frac{\partial f}{\partial \dot{z}}[t_0, z(t_0), \dot{z}(t_0)]v(t_0)$$

Now, we see that the manner in which to achieve $\delta F(z; v) = 0$ for all admissible directions depends on the originally specified boundary conditions. If $z(t_0) = z_0$ is specified but $z(t_f)$ is free, then $v(t)$ is only required to satisfy $v(t_0) = 0$, and in order to achieve $\delta F(z; v) = 0$ for all such directions, we must further impose the *natural boundary condition*

$$\frac{\partial f}{\partial \dot{z}}[t_f, z(t_f), \dot{z}(t_f)] = 0$$

Similarly, if $z(t_f) = z_f$ is specified but $z(t_0)$ is free, then $v(t)$ is only required to satisfy $v(t_f) = 0$, which leads to the natural boundary condition

$$\frac{\partial f}{\partial \dot{z}}[t_0, z(t_0), \dot{z}(t_0)] = 0$$

Finally, if both $z(t_0) = z_0$ and $z(t_f) = z_f$ are specified, then admissible directions must satisfy $v(t_0) = v(t_f) = 0$, from which we obtain $\delta F(z; v) = 0$ without imposing any further conditions. Now we have argued that satisfying the Euler-Lagrange equation together with the appropriate combination of original and natural boundary conditions is one way to achieve $\delta F(z; v) = 0$ for all admissible directions. It can be shown using standard calculus results that this is the only way.   □

Example 9.2 For the integral function of Example 9.1, the Euler-Lagrange equation on the interval $[0, 1]$ is

$$\frac{d}{dt}[\dot{z}(t)] = \ddot{z}(t) = t \quad \text{for all } t \in (0, 1)$$

Integrating twice yields $z(t) = \frac{1}{6}t^3 + c_1 t + c_0$, where the coefficients c_0 and c_1 are determined by the boundary conditions. If we specify $z(0) = 0$ but leave $z(1)$ free, the associated natural boundary condition is $\dot{z}(1) = 0$, which yields $c_0 = 0$ and $c_1 = -\frac{1}{2}$. If we specify $z(1) = 0$ but leave $z(0)$ free, the associated natural boundary condition is $\dot{z}(0) = 0$, which yields $c_0 = -\frac{1}{6}$ and $c_1 = 0$. Finally, if we specify $z(0) = z(1) = 0$, we obtain $c_0 = 0$ and $c_1 = -\frac{1}{6}$.   □

We next extend the notion of convexity to integral functions defined on the space $C^1[t_0, t_f]$ using the Gâteaux variation as a generalization of the directional derivative on Euclidean space. The integral function $F(z)$ is *convex* if

$$F(z + v) - F(z) \geq \delta F(z; v) \text{ for all } z \in C^1[t_0, t_f] \text{ and admissible}$$

$$v \in C^1[t_0, t_f]$$

and *strictly convex* if equality holds above when and only when the only admissible direction is $v(t) \equiv 0$ on $[t_0, t_f]$. It is easy to see that if $\delta F(z^*; v) = 0$ for all admissible directions, i.e., $z^*(t)$ is a stationary function that satisfies the appropriate combination of original and natural boundary conditions, then $z^*(t)$ minimizes the strictly convex integral function $F(z)$.

Example 9.3 For the integral function in the preceding examples, a direct calculation gives

$$F(z + v) - F(z) = \int_0^1 \left\{ t[z(t) + v(t)] + \tfrac{1}{2}[\dot{z}(t) + \dot{v}(t)]^2 \right\} dt$$

$$- \int_0^1 [tz(t) + \tfrac{1}{2}\dot{z}^2(t)] \, dt$$

$$= \int_0^1 [tv(t) + \dot{z}(t)\dot{v}(t)] \, dt + \int_0^1 \tfrac{1}{2}\dot{v}^2(t) \, dt$$

$$= \delta F(z; v) + \int_0^1 \tfrac{1}{2} \dot{v}^2(t) \, dt$$

$$\geq \delta F(z; v)$$

for all $z \in C^1[0, 1]$ and admissible $v \in C^1[0, 1]$ with respect to each of the possible boundary conditions we have considered. We conclude that $F(z)$ is convex. We also see that equality holds when and only when $\dot{v}(t) \equiv 0$, so $v(t)$ is constant on the interval $[0, 1]$. For any one of the three boundary conditions, we have either $v(1) = 0$, $v(0) = 0$, or both. We conclude, because a constant function on $[0, 1]$ that vanishes at a single point is identically zero on $[0, 1]$, that $F(z)$ is strictly convex and that $z^*(t) = \tfrac{1}{6}t^3 + c_1 t + c_0$ minimizes $F(z)$ for the coefficients determined in Example 9.2 for the particular boundary condition. $\qquad \square$

We conclude our overview of variational calculus by incorporating into the problem constraints of the form

$$g_i[t, z(t), \dot{z}(t)] = 0 \quad \text{for all } t \in [t_0, t_f] \quad i = 1, 2, \dots, n$$

or in vector form

$$G[t, z(t), \dot{z}(t)] = \begin{bmatrix} g_1[t, z(t), \dot{z}(t)] \\ g_2[t, z(t), \dot{z}(t)] \\ \vdots \\ g_n[t, z(t), \dot{z}(t)] \end{bmatrix} = \begin{bmatrix} 0 \\ 0 \\ \vdots \\ 0 \end{bmatrix}$$

Following our discussion on constrained minimization on Euclidean space, we introduce a Lagrange multiplier vector that, because the constraints here are imposed pointwise in t, we allow to be a function of t of the form

$$\lambda(t) = \begin{bmatrix} \lambda_1(t) \\ \lambda_2(t) \\ \vdots \\ \lambda_n(t) \end{bmatrix}$$

where, for simplicity, we assume that the Lagrange multiplier functions are also continuously differentiable on the interval $[t_0, t_f]$. From this, we define

$$\tilde{f}(t, z, \dot{z}) = f(t, z, \dot{z}) + \lambda_1(t) g_1(t, z, \dot{z}) + \lambda_2(t) g_2(t, z, \dot{z}) + \cdots$$

$$+ \lambda_n(t) g_n(t, z, \dot{z})$$

$$= f(t, z, \dot{z}) + \lambda^T(t) G(t, z, \dot{z})$$

Now suppose that $z^*(t)$ minimizes the augmented integral function

$$\tilde{F}(z) = \int_{t_0}^{t_f} \tilde{f}[t, z(t), \dot{z}(t)] \, dt$$

over all $z(t) \in C^1[t_0, t_f]$ that satisfy the specified boundary conditions so that

$$\tilde{F}(z) > \tilde{F}(z^*)$$

for all such $z(t) \neq z^*(t)$. In particular, if $z(t)$ in addition satisfies $G[t, z(t), \dot{z}(t)] = G[t, z^*(t), \dot{z}^*(t)]$ for all $t \in [t_0, t_f]$, then for any Lagrange multiplier vector $\lambda(t)$,

$$F(z) - F(z^*) = F(z) - F(z^*)$$
$$+ \int_{t_0}^{t_f} \lambda^T(t) \left\{ G[t, z(t), \dot{z}(t)] - G[t, z^*(t), \dot{z}^*(t)] \right\} dt$$
$$= \left\{ F(z) + \int_{t_0}^{t_f} \lambda^T(t) G[t, z(t), \dot{z}(t)] \, dt \right\}$$
$$- \left\{ F(z^*) + \int_{t_0}^{t_f} \lambda^T(t) G[t, z^*(t), \dot{z}^*(t)] \, dt \right\}$$
$$= \tilde{F}(z) - \tilde{F}(z^*)$$
$$> 0$$

We conclude that $z^*(t)$ minimizes $F(z)$ over all $z(t) \in C^1[t_0, t_f]$ that satisfy the specified boundary conditions and $G[t, z(t), \dot{z}(t)] = G[t, z^*(t), \dot{z}^*(t)]$ for all $t \in [t_0, t_f]$. If $\tilde{F}(z)$ is strictly convex [which in general depends on $\lambda(t)$], then the minimum of $\tilde{F}(z)$ is characterized by $\delta \tilde{F}(z^*; v) = 0$ for all admissible directions as determined by the boundary conditions. We can then characterize $z^*(t)$ as the solution to the Euler-Lagrange equation

$$\frac{d}{dt} \frac{\partial \tilde{f}}{\partial \dot{z}}[t, z(t), \dot{z}(t)] = \frac{\partial \tilde{f}}{\partial z}[t, z(t), \dot{z}(t)] \quad \text{for all } t \in (t_0, t_f) \qquad (9.5)$$

with the accompanying combination of originally specified boundary conditions and natural boundary conditions that now involve $(\partial \tilde{f}/\partial \dot{z})[t, z(t), \dot{z}(t)]$ at $t = t_0$ and/or $t = t_f$.

Our hope is that this, along with the constraint equation $G[t, z(t), \dot{z}(t)] = 0$, enables us to determine $z^*(t)$ and $\lambda(t)$ and, furthermore, for $\lambda(t)$ so obtained that $\tilde{F}(z)$ is strictly convex. We then can conclude that $z^*(t)$ minimizes $F(z)$ over all $z(t) \in C^1[t_0, t_f]$ that satisfy the specified boundary conditions and the constraint $G[t, z(t), \dot{z}(t)] = 0$ for all $t \in [t_0, t_f]$.

One situation in which strict convexity of $\tilde{F}(z)$ is assured for any Lagrange multiplier vector is when $F(z)$ is strictly convex and $z(t)$ is constrained to satisfy the linear ordinary differential equation

$$E(t)\dot{z}(t) = C(t)z(t) + d(t)$$

This leads to a constraint equation of the form

$$G(t, z(t), \dot{z}(t)) = C(t)z(t) + d(t) - E(t)\dot{z}(t) = 0$$

In this case, we have

$$\frac{\partial \tilde{f}}{\partial z}(t, z, \dot{z}) = \frac{\partial f}{\partial z}(t, z, \dot{z}) + \lambda^T(t)\frac{\partial G}{\partial z}(t, z, \dot{z}) = \frac{\partial f}{\partial z}(t, z, \dot{z}) + \lambda^T(t)C(t)$$

$$\frac{\partial \tilde{f}}{\partial \dot{z}}(t, z, \dot{z}) = \frac{\partial f}{\partial \dot{z}}(t, z, \dot{z}) + \lambda^T(t)\frac{\partial G}{\partial \dot{z}}(t, z, \dot{z}) = \frac{\partial f}{\partial \dot{z}}(t, z, \dot{z}) - \lambda^T(t)E(t)$$

which gives

$$\delta \tilde{F}(z; v) = \int_{t_0}^{t_f} \left\{ \frac{\partial \tilde{f}}{\partial z}[t, z(t), \dot{z}(t)]v(t) + \frac{\partial \tilde{f}}{\partial \dot{z}}[t, z(t), \dot{z}(t)]\dot{v}(t) \right\} dt$$

$$= \int_{t_0}^{t_f} \left(\left\{ \frac{\partial f}{\partial z}[t, z(t), \dot{z}(t)] + \lambda^T(t)C(t) \right\} v(t) \right.$$

$$\left. + \left\{ \frac{\partial f}{\partial \dot{z}}[t, z(t), \dot{z}(t)] - \lambda^T(t)E(t) \right\} \dot{v}(t) \right) dt$$

$$= \delta F(z; v) + \int_{t_0}^{t_f} \lambda^T(t)[C(t)v(t) - E(t)\dot{v}(t)] dt$$

This allows us to conclude that for any $\lambda(t)$,

$$\tilde{F}(z + v) - \tilde{F}(z) = \left(F(z + v) + \int_{t_0}^{t_f} \lambda^T(t)\{C(t)[z(t) + v(t)] + d(t) \right.$$

$$\left. - E(t)[\dot{z}(t) + \dot{v}(t)]\} dt \right)$$

$$- \left\{ F(z) + \int_{t_0}^{t_f} \lambda^T(t)[C(t)z(t) + d(t) - E(t)\dot{z}(t)] dt \right\}$$

$$= F(z+v) - F(z) + \int_{t_0}^{t_f} \lambda^T(t)[C(t)v(t) - E(t)\dot{v}(t)]\,dt$$

$$\geq \delta F(z;v) + \int_{t_0}^{t_f} \lambda^T(t)[C(t)v(t) - E(t)\dot{v}(t)]\,dt$$

$$= \delta \tilde{F}(z;v)$$

for all admissible directions with equality when and only when the only admissible direction is $v(t) \equiv 0$ on $[t_0, t_f]$. Thus, for this class of constraints, (strict) convexity of $F(z)$ implies (strict) convexity of $\tilde{F}(z)$ for any Lagrange multiplier vector $\lambda(t)$.

It is important to note that strict convexity of $\tilde{F}(z)$ is sufficient but not necessary for this process to yield the minimizer of $F(z)$ subject to the specified boundary conditions and equality constraints. We will see that for the optimal control problems of this chapter, strict convexity fails to hold, and yet direct problem-specific arguments allow us to reach the desired conclusions nonetheless.

9.3 MINIMUM ENERGY CONTROL

Before attacking the linear quadratic regulator problem, we first investigate a related optimal control problem that involves computing the input signal of minimum energy among those that transfer the state trajectory from a specified initial state to a specified final state. This has practical significance in vehicle trajectory planning problems, such as launch vehicle ascent or satellite orbit transfer, in which minimizing control energy can be related to minimizing fuel consumption.

We begin under the assumption that the state equation (9.1) is controllable. The objective is to minimize the control "energy"

$$J = \tfrac{1}{2} \int_{t_0}^{t_f} \|u(t)\|^2 dt \tag{9.6}$$

over all control signals $u(t)$ on the time interval $[t_0, t_f]$ that meet the following boundary conditions:

$$x(t_0) = x_0 \qquad x(t_f) = x_f \tag{9.7}$$

for specified initial state x_0 and final state x_f. We observe that even though the objective function (9.6) does not directly involve the state, a solution to this problem consists of the minimum energy control signal and the associated state trajectory because the two are linked by the state equation (9.1). In other words, we seek to minimize Equation (9.6) over all $z(t) = [x^T(t), u^T(t)]^T$ that satisfy the boundary conditions (9.7) and the equality constraint

$$G[t, z(t), \dot{z}(t)] = Ax(t) + Bu(t) - \dot{x}(t)$$

$$= \begin{bmatrix} A & B \end{bmatrix} \begin{bmatrix} x(t) \\ u(t) \end{bmatrix} - \begin{bmatrix} I & 0 \end{bmatrix} \begin{bmatrix} \dot{x}(t) \\ \dot{u}(t) \end{bmatrix}$$

$$= 0 \tag{9.8}$$

Adapting the methodology of Section 9.2 to the problem at hand, we first characterize admissible directions associated with the boundary condition (9.7) as $v(t) = [\xi^T(t), \mu^T(t)]^T$ with $\xi(t_0) = \xi(t_f) = 0$, and because $u(t_0)$ and $u(t_f)$ are unconstrained, $\mu(t_0)$ and $\mu(t_f)$ are free. We adjoin the constraint (9.8) to the original objective function using the vector of Lagrange multipliers

$$\lambda(t) = \begin{bmatrix} \lambda_1(t) \\ \lambda_2(t) \\ \vdots \\ \lambda_n(t) \end{bmatrix}$$

to form the augmented objective function

$$\tilde{J}(z) = \int_{t_0}^{t_f} \left\{ \tfrac{1}{2}\|u(t)\|^2 + \lambda^T(t)[Ax(t) + Bu(t) - \dot{x}(t)] \right\} dt$$

We directly check the convexity of the augmented objective function. On identifying

$$\tilde{f}(t, z, \dot{z}) = \tfrac{1}{2}\|u\|^2 + \lambda^T(t)(Ax + Bu - \dot{x})$$

we have

$$\frac{\partial \tilde{f}}{\partial z}(t, z, \dot{z}) = \begin{bmatrix} \dfrac{\partial \tilde{f}}{\partial x}(t, z, \dot{z}) & \dfrac{\partial \tilde{f}}{\partial u}(t, z, \dot{z}) \end{bmatrix}$$

$$= \begin{bmatrix} \lambda^T(t)A & u^T + \lambda^T(t)B \end{bmatrix}$$

and

$$\frac{\partial \tilde{f}}{\partial \dot{z}}(t, z, \dot{z}) = \begin{bmatrix} \dfrac{\partial \tilde{f}}{\partial \dot{x}}(t, z, \dot{z}) & \dfrac{\partial \tilde{f}}{\partial \dot{u}}(t, z, \dot{z}) \end{bmatrix}$$

$$= \begin{bmatrix} -\lambda^T(t) & 0 \end{bmatrix}$$

which yields the Gâteaux variation

$$\delta\tilde{J}(z;v) = \int_{t_0}^{t_f} \left\{ \begin{bmatrix} \lambda^T(t)A & u^T(t)+\lambda^T(t)B \end{bmatrix} \begin{bmatrix} \xi(t) \\ \mu(t) \end{bmatrix} \right.$$

$$+ \left. \begin{bmatrix} -\lambda^T(t) & 0 \end{bmatrix} \begin{bmatrix} \dot{\xi}(t) \\ \dot{\mu}(t) \end{bmatrix} \right\} dt$$

$$= \int_{t_0}^{t_f} \left\{ [u^T(t)+\lambda^T(t)B]\mu(t) + \lambda^T(t)[A\xi(t) - \dot{\xi}(t)] \right\} dt$$

In terms of this, we find

$$\tilde{J}(z+v) - \tilde{J}(z)$$

$$= \left(\int_{t_0}^{t_f} \tfrac{1}{2}\|u(t) + \mu(t)\|^2 + \lambda^T(t)\{A[x(t) + \xi(t)]\right.$$

$$\left. + B[u(t) + \mu(t)] - [\dot{x}(t) + \dot{\xi}(t)]\} dt \right)$$

$$- \left\{ \int_{t_0}^{t_f} \tfrac{1}{2}\|u(t)\|^2 + \lambda^T(t)[A(t)x(t) + Bu(t) - \dot{x}(t)]dt \right\}$$

$$= \delta\tilde{J}(z;v) + \int_{t_0}^{t_f} \tfrac{1}{2}\|\mu(t)\|^2 dt$$

$$\geq \delta\tilde{J}(z;v)$$

from which we conclude that $\tilde{J}(z)$ is convex. Unfortunately, we cannot conclude that $\tilde{J}(z)$ is strictly convex because equality holds for any admissible direction of the form $v(t) = [\xi^T(t), 0]^T$. However, we can argue directly that if equality holds above for $z(t)$ and $v(t) = [\xi^T(t), 0]^T$ in which $\xi(t_0) = \xi(t_f) = 0$, and if both $z(t)$ and $z(t) + v(t)$ satisfy the constraint (9.8), equivalently, the linear state equation (9.1), we must then have $\xi(t) \equiv 0$. For if $z(t) = [x^T(t), u^T(t)]^T$ satisfies

$$\dot{x}(t) = Ax(t) + Bu(t)$$

and $z(t) + v(t) = \left[(x(t) + \xi(t))^T, u^T(t) \right]^T$ satisfies

$$\dot{x}(t) + \dot{\xi}(t) = A[x(t) + \xi(t)] + Bu(t)$$

then subtracting these identities yields the homogeneous linear state equation

$$\dot{\xi}(t) = A\xi(t)$$

from which either the initial condition $\xi(t_0) = 0$ or the final condition $\xi(t_f) = 0$ is enough to force $\xi(t) \equiv 0$ on $[t_0, t_f]$ as claimed. We therefore conclude that if $z^*(t)$ satisfies $\delta \tilde{J}(z; v) = 0$ for all admissible directions along with the constraint (9.8), then $z^*(t)$ minimizes $J(z)$ over all $z(t)$ that satisfy the boundary conditions (9.7) and the constraint (9.8), thereby determining the minimum energy control signal and corresponding state trajectory.

We now turn our attention to the Euler-Lagrange equation (9.5), which takes the form

$$\frac{d}{dt} \begin{bmatrix} -\lambda^T(t) & 0 \end{bmatrix} = \begin{bmatrix} \lambda^T(t)A & u^T(t) + \lambda^T(t)B \end{bmatrix}$$

and can be reformulated as

$$\dot{\lambda}(t) = -A^T \lambda(t)$$

$$u(t) = -B^T \lambda(t) \tag{9.9}$$

We note that the natural boundary conditions associated with Equation (9.7) are

$$\frac{\partial \tilde{f}}{\partial \dot{u}}(t, z(t_0), \dot{z}(t_0)) = 0 \quad \text{and} \quad \frac{\partial \tilde{f}}{\partial \dot{u}}(t, z(t_f), \dot{z}(t_f)) = 0$$

which are automatically satisfied because $\tilde{f}(t, z, \dot{z})$ is independent of $\dot{u}$.

The homogeneous linear state equation in Equation (9.9) implies that for any $t_a \in [t_0, t_f]$, on specifying $\lambda(t_a) = \lambda_a$, $\lambda(t) = e^{-A^T(t-t_a)}\lambda_a$ is uniquely determined. Here we choose $t_a = t_0$, which, in terms of $\lambda(t_0) = \lambda_0$, yields

$$\lambda(t) = e^{-A^T(t-t_0)}\lambda_0 = e^{A^T(t_0-t)}\lambda_0 \quad \text{and} \quad u(t) = -B^T e^{A^T(t_0-t)}\lambda_0$$

This input signal uniquely determines the state trajectory $x(t)$ given the specified initial state $x(t_0) = x_0$, and it remains to choose λ_0 so that

$$x(t_f) = e^{A(t_f-t_0)}x_0 + \int_{t_0}^{t_f} e^{A(t_f-\tau)} Bu(\tau)d\tau$$

$$= x_0 - \int_{t_0}^{t_f} e^{A(t_f-\tau)} BB^T e^{A^T(t_0-\tau)}d\tau \lambda_0$$

$$= e^{A(t_f-t_0)}\left[x_0 - \int_{t_0}^{t_f} e^{A(t_0-\tau)} BB^T e^{A^T(t_0-\tau)}d\tau \lambda_0 \right]$$

$$= x_f$$

We recall from Chapter 3 that the controllability Gramian

$$W(t_0, t_f) = \int_{t_0}^{t_f} e^{A(t_0-\tau)} B B^T e^{A^T(t_0-\tau)} d\tau$$

is nonsingular for any $t_f > t_0$ because the state equation (9.1) is assumed to be controllable. We then can solve for λ_0 to obtain

$$\lambda_0 = W^{-1}(t_0, t_f) \left(x_0 - e^{A(t_0-t_f)} x_f\right)$$

which, in turn, now completely specifies the minimum energy control signal

$$u^*(t) = -B^T e^{A^T(t_0-t)} W^{-1}(t_0, t_f) \left(x_0 - e^{A(t_0-t_f)} x_f\right) \tag{9.10}$$

We observe that this is precisely the input signal used in Chapter 3 to demonstrate that the trajectory of a controllable state equation can be steered between any initial and final states over a finite time interval. Also, the reader is invited to verify that the minimum energy control signal also can be expressed in terms of the reachability Gramian introduced in AE3.4, that is,

$$W_R(t_0, t_f) = \int_{t_0}^{t_f} e^{A(t_f-\tau)} B B^T e^{A^T(t_f-\tau)} d\tau$$

as

$$u^*(t) = B^T e^{A^T(t_f-t)} W_R^{-1}(t_0, t_f) \left(x_f - e^{A(t_f-t_0)} x_0\right)$$

We can compute the minimum energy value achieved by this control signal via

$$J^* = \tfrac{1}{2} \int_{t_0}^{t_f} \|u^*(t)\|^2 dt$$

$$= \tfrac{1}{2} \int_{t_0}^{t_f} \left[-B^T e^{A^T(t_0-t)} W^{-1}(t_0, t_f) \left(x_0 - e^{A(t_0-t_f)} x_f\right) \right]^T$$

$$\times \left[-B^T e^{A^T(t_0-t)} W^{-1}(t_0, t_f) \left(x_0 - e^{A(t_0-t_f)} x_f\right) \right] dt$$

$$= \tfrac{1}{2} \left(x_0 - e^{A(t_0-t_f)} x_f\right)^T W^{-1}(t_0, t_f) \left(\int_{t_0}^{t_f} e^{A(t_0-t)} B B^T e^{A^T(t_0-t)} dt \right)$$

$$\times W^{-1}(t_0, t_f) \left(x_0 - e^{A(t_0-t_f)} x_f\right)$$

$$= \frac{1}{2} \left(x_0 - e^{A(t_0 - t_f)} x_f \right)^T W^{-1}(t_0, t_f) W(t_0, t_f)$$
$$\times W^{-1}(t_0, t_f) \left(x_0 - e^{A(t_0 - t_f)} x_f \right)$$
$$= \frac{1}{2} \left(x_0 - e^{A(t_0 - t_f)} x_f \right)^T W^{-1}(t_0, t_f) \left(x_0 - e^{A(t_0 - t_f)} x_f \right)$$

Alternatively, the minimum energy can be represented in terms of the reachability Gramian as

$$J^* = \frac{1}{2} \left(x_f - e^{A(t_f - t_0)} x_0 \right)^T W_R^{-1}(t_0, t_f) \left(x_f - e^{A(t_f - t_0)} x_0 \right)$$

Example 9.4 In this example we solve the minimum energy control problem for the controllable state equation

$$\begin{bmatrix} \dot{x}_1(t) \\ \dot{x}_2(t) \end{bmatrix} = \begin{bmatrix} 0 & 1 \\ 0 & 0 \end{bmatrix} \begin{bmatrix} x_1(t) \\ x_2(t) \end{bmatrix} + \begin{bmatrix} 0 \\ 1 \end{bmatrix} u(t)$$

that describes a double integrator. The associated matrix exponential is

$$e^{At} = \begin{bmatrix} 1 & t \\ 0 & 1 \end{bmatrix}$$

in terms of which we compute the controllability Gramian

$$W(t_0, t_f) = \int_{t_0}^{t_f} \begin{bmatrix} 1 & t_0 - \tau \\ 0 & 1 \end{bmatrix} \begin{bmatrix} 0 \\ 1 \end{bmatrix} \begin{bmatrix} 0 & 1 \end{bmatrix} \begin{bmatrix} 1 & 0 \\ t_0 - \tau & 1 \end{bmatrix} d\tau$$
$$= \begin{bmatrix} \frac{1}{3}(t_f - t_0)^3 & -\frac{1}{2}(t_f - t_0)^2 \\ -\frac{1}{2}(t_f - t_0)^2 & (t_f - t_0) \end{bmatrix}$$

as well as the reachability gramian

$$W_R(t_0, t_f) = \int_{t_0}^{t_f} \begin{bmatrix} 1 & t_f - \tau \\ 0 & 1 \end{bmatrix} \begin{bmatrix} 0 \\ 1 \end{bmatrix} \begin{bmatrix} 0 & 1 \end{bmatrix} \begin{bmatrix} 1 & 0 \\ t_f - \tau & 1 \end{bmatrix} d\tau$$
$$= \begin{bmatrix} \frac{1}{3}(t_f - t_0)^3 & \frac{1}{2}(t_f - t_0)^2 \\ \frac{1}{2}(t_f - t_0)^2 & (t_f - t_0) \end{bmatrix}$$

For $t_0 = 0$, $t_f = 1$, and

$$x_0 = \begin{bmatrix} 0 \\ 0 \end{bmatrix} \qquad x_f = \begin{bmatrix} 1 \\ 1 \end{bmatrix}$$

the minimum energy control signal given by Equation (9.10) is

$$u^*(t) = -\begin{bmatrix} 0 & 1 \end{bmatrix} \begin{bmatrix} 1 & 0 \\ -t & 1 \end{bmatrix} \begin{bmatrix} \frac{1}{3} & -\frac{1}{2} \\ -\frac{1}{2} & 1 \end{bmatrix}^{-1} \left(\begin{bmatrix} 0 \\ 0 \end{bmatrix} - \begin{bmatrix} 1 & -1 \\ 0 & 1 \end{bmatrix} \begin{bmatrix} 1 \\ 1 \end{bmatrix} \right)$$

$$= \begin{bmatrix} t & -1 \end{bmatrix} \begin{bmatrix} 12 & 6 \\ 6 & 4 \end{bmatrix} \begin{bmatrix} 0 \\ -1 \end{bmatrix}$$

$$= 4 - 6t$$

The reader is invited to check that the alternate formula in terms of the reachability Gramian yields the same result. The associated state trajectory is given by

$$x^*(t) = \int_0^t e^{A(t-\tau)} B u^*(\tau) d\tau$$

$$= \int_0^t \begin{bmatrix} 1 & t - \tau \\ 0 & 1 \end{bmatrix} \begin{bmatrix} 0 \\ 1 \end{bmatrix} (4 - 6\tau) d\tau$$

$$= \begin{bmatrix} -t^3 + 2t^2 \\ -3t^2 + 4t \end{bmatrix}$$

Note that $x_2^*(t) = \int_0^t u^*(\tau) d\tau$ and $x_1^*(t) = \int_0^t x_2^*(\tau) d\tau$ as required for this double integrator system. The minimum achievable control energy, expressed in terms of the reachability Gramian, is

$$J^* = \frac{1}{2} \left(\begin{bmatrix} 1 \\ 1 \end{bmatrix} - \begin{bmatrix} 1 & 1 \\ 0 & 1 \end{bmatrix} \begin{bmatrix} 0 \\ 0 \end{bmatrix} \right)^T \begin{bmatrix} \frac{1}{3} & \frac{1}{2} \\ \frac{1}{2} & 1 \end{bmatrix}^{-1} \left(\begin{bmatrix} 1 \\ 1 \end{bmatrix} - \begin{bmatrix} 1 & 1 \\ 0 & 1 \end{bmatrix} \begin{bmatrix} 0 \\ 0 \end{bmatrix} \right)$$

$$= \frac{1}{2} \begin{bmatrix} 1 & 1 \end{bmatrix} \begin{bmatrix} 12 & -6 \\ -6 & 4 \end{bmatrix} \begin{bmatrix} 1 \\ 1 \end{bmatrix}$$

$$= 2$$

The minimum control input $u^*(t)$ is plotted versus time in Figure 9.1, and the state trajectory is plotted versus time in Figure 9.2. In addition, the phase portrait $[x_2(t)$ versus $x_1(t)]$ is shown in Figure 9.3. □

9.4 THE LINEAR QUADRATIC REGULATOR

We now return to the linear quadratic regulator problem formulated at the beginning of the chapter. As with the minimum energy control problem

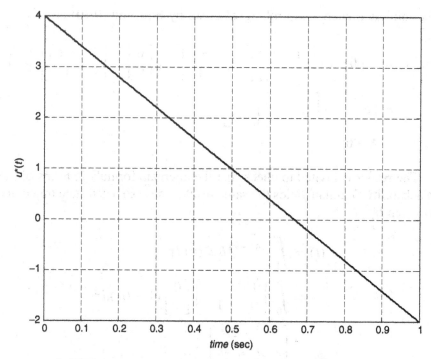

FIGURE 9.1 Minimum energy control signal for Example 9.4.

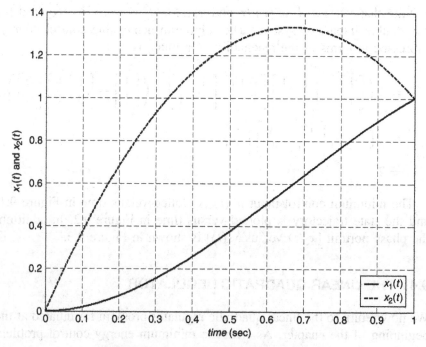

FIGURE 9.2 State trajectory for Example 9.4.

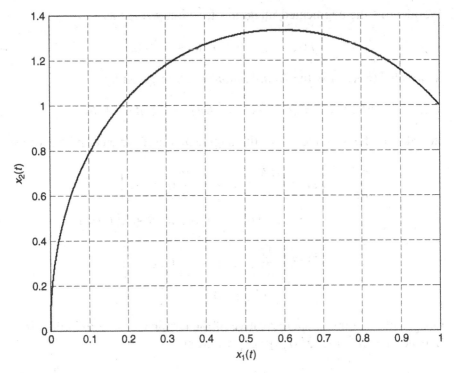

FIGURE 9.3 Phase portrait for Example 9.4.

of the preceding section, the state equation and specified initial state in Equation (9.1) serve as an equality constraint and boundary condition, respectively, as we seek to minimize the performance index (9.2). Note that in contrast to treating the final state as a boundary condition, as we did in the minimum energy control problem, here the final state is free but is penalized via the quadratic form representing the terminal cost term in Equation (9.2). We will see the impact of this as we apply the techniques of Section 9.2 to this problem. First, in terms of $z(t) = [x^T(t), u^T(t)]^T$, the admissible directions $v(t) = [\xi^T(t), \mu^T(t)]^T$ now must just satisfy $\xi(t_0) = 0$, with $\xi(t_f)$, $\mu(t_0)$, and $\mu(t_f)$ free.

Next, we adjoin the same equality constraint (9.8) to the original performance index via the Lagrange multiplier vector $\lambda(t)$ to yield the augmented performance index

$$\tilde{J}(z) = \int_{t_0}^{t_f} \left\{ \tfrac{1}{2} \left[x^T(t) Q x(t) + u^T(t) R u(t) \right] \right.$$
$$\left. + \lambda^T(t) [A x(t) + B u(t) - \dot{x}(t)] \right\} dt + \tfrac{1}{2} x^T(t_f) S\, x(t_f)$$

We identify the augmented integrand function written as

$$\tilde{f}(t, z, \dot{z}) = h(t, x, u) - \lambda^T(t)\dot{x}$$

in which $h(t, x, u)$ is the *Hamiltonian function* given by

$$h(t, x, u) = \tfrac{1}{2}(x^T Q x + u^T R u) + \lambda^T(t)(Ax + Bu)$$

In order to check convexity of the augmented performance index, we use

$$\frac{\partial \tilde{f}}{\partial z}(t, z, \dot{z}) = \left[\frac{\partial h}{\partial x}(t, x, u) \quad \frac{\partial h}{\partial u}(t, x, u) \right]$$

$$= \left[x^T Q + \lambda^T(t)A \quad u^T R + \lambda^T(t)B \right]$$

$$\frac{\partial \tilde{f}}{\partial \dot{z}}(t, z, \dot{z}) = \left[\frac{\partial \tilde{f}}{\partial \dot{x}}(t, z, \dot{z}) \quad \frac{\partial \tilde{f}}{\partial \dot{u}}(t, z, \dot{z}) \right]$$

$$= \left[-\lambda^T(t) \quad 0 \right]$$

along with the directional derivative of the terminal cost term in the direction $v(t_f) = [\xi^T(t_f), \mu^T(t_f)]^T$ expressed as

$$\frac{\partial}{\partial z(t_f)} \left[\tfrac{1}{2} x^T(t_f) S \, x(t_f) \right] \cdot v(t_f) = \left[x^T(t_f)S \quad 0 \right] \begin{bmatrix} \xi(t_f) \\ \mu(t_f) \end{bmatrix}$$

$$= x^T(t_f)S\xi(t_f)$$

to obtain the Gâteaux variation

$$\delta \tilde{J}(z; v) = \int_{t_0}^{t_f} \left[\left[x^T(t)Q + \lambda^T(t)A \quad u^T(t)R + \lambda^T(t)B \right] \begin{bmatrix} \xi(t) \\ \mu(t) \end{bmatrix} \right.$$

$$+ \left[-\lambda^T(t) \quad 0 \right] \begin{bmatrix} \dot{\xi}(t) \\ \dot{\mu}(t) \end{bmatrix} \bigg] dt + x^T(t_f)S\xi(t_f)$$

$$= \int_{t_0}^{t_f} \left[(u^T(t)R + \lambda^T(t)B) \, \mu(t) \right.$$

$$+ (x^T(t)Q + \lambda^T(t)A) \, \xi(t) - \lambda^T(t)\dot{\xi}(t) \big] dt + x^T(t_f)S\xi(t_f)$$

In terms of this, we see that

$$\tilde{J}(z + v) - \tilde{J}(z)$$

$$= \left(\int_{t_0}^{t_f} \tfrac{1}{2} \{ [x(t) + \xi(t)]^T Q[x(t) + \xi(t)] \right.$$

$$+ [u(t) + \mu(t)]^T R[u(t) + \mu(t)]\}$$

$$+ \lambda^T(t) \{A[x(t) + \xi(t)] + B[u(t) + \mu(t)] - [\dot{x}(t) + \dot{\xi}(t)]\} dt$$

$$+ \tfrac{1}{2} [x(t_f) + \xi(t_f)]^T S [x(t_f) + \xi(t_f)] \Big)$$

$$- \Big(\int_{t_0}^{t_f} \tfrac{1}{2} [x^T(t)Qx(t) + u^T(t)Ru(t)]$$

$$+ \lambda^T(t)[A(t)x(t) + Bu(t) - \dot{x}(t)]dt \Big) + \tfrac{1}{2}x^T(t_f)S\, x(t_f) \Big)$$

$$= \delta \tilde{J}(z; v) + \int_{t_0}^{t_f} \tfrac{1}{2} [\xi^T(t)Q\xi(t) + \mu^T(t)R\mu(t)] dt + \tfrac{1}{2}\xi^T(t_f)S\xi(t_f)$$

$$\geq \delta \tilde{J}(z; v)$$

from which we conclude that $\tilde{J}(z)$ is convex, and equality holds when and only when the quadratic forms each satisfy

$$\xi^T(t)Q\xi(t) \equiv 0, \quad \mu^T(t)R\mu(t) \equiv 0 \text{ for all } t \in [t_0, t_f] \text{ and}$$

$$\xi^T(t_f)S\xi(t_f) = 0$$

Now, under the assumption that R is positive definite, the second equality holds if and only if $\mu(t) \equiv 0$ for all $t \in [t_0, t_f]$. However, having only assumed that Q and S are positive semidefinite, the first and third identities may hold for nonzero $\xi(t)$. We therefore cannot conclude that $\tilde{J}(z)$ is strictly convex because equality holds above for any admissible direction of the form $v(t) = [\xi^T(t), 0]^T$ with $\xi^T(t)Q\xi(t) \equiv 0$ and $\xi^T(t_f)S\xi(t_f) = 0$. Undaunted, we employ the same argument used in the preceding section to conclude that if equality holds above for $z(t)$ and $v(t) = [\xi^T(t), 0]^T$ in which $\xi(t_0) = 0$, and if both $z(t)$ and $z(t) + v(t)$ satisfy the equality constraint derived from the linear state equation (9.1), then we must have $\xi(t) \equiv 0$. We then conclude that if $z^*(t)$ satisfies $\delta \tilde{J}(z; v) = 0$ for all admissible directions along with the constraint (9.8), then $z^*(t)$ minimizes the performance index (9.2) over all $z(t)$ that satisfy the initial condition $x(t_0) = x_0$ and the constraint (9.8), thereby determining the optimal control signal and corresponding state trajectory.

We now turn our attention to the Euler-Lagrange equation (9.5), which here takes the form

$$\frac{d}{dt} \begin{bmatrix} -\lambda^T(t) & 0 \end{bmatrix} = \begin{bmatrix} x^T(t)Q + \lambda^T(t)A & u^T(t)R + \lambda^T(t)B \end{bmatrix}$$

and can be repackaged as

$$\dot{\lambda}(t) = -A^T \lambda(t) - Qx(t)$$
$$u(t) = -R^{-1}B^T \lambda(t) \tag{9.11}$$

in which we have used the fact that R is nonsingular because it is assumed to be positive definite. In order to achieve $\delta \tilde{J}(z; v) = 0$ for all admissible directions, we use the Euler-Lagrange equation to write

$$\delta \tilde{J}(z; v) = \int_{t_0}^{t_f} \left\{ \begin{bmatrix} -\dot{\lambda}^T(t) & 0 \end{bmatrix} \begin{bmatrix} \xi(t) \\ \mu(t) \end{bmatrix} + \begin{bmatrix} -\lambda^T(t) & 0 \end{bmatrix} \begin{bmatrix} \dot{\xi}(t) \\ \dot{\mu}(t) \end{bmatrix} \right\} dt$$

$$+ x^T(t_f)S\xi(t_f)$$

$$= -\int_{t_0}^{t_f} \frac{d}{dt}\left[\lambda^T(t)\xi(t)\right] dt + x^T(t_f)S\xi(t_f)$$

$$= -\lambda^T(t_f)\xi(t_f) + \lambda^T(t_0)\xi(t_0) + x^T(t_f)S\xi(t_f)$$

As noted earlier, admissible directions $v(t) = [\xi^T(t), \mu^T(t)]^T$ must satisfy $\xi(t_0) = 0$, but $\xi(t_f)$ is free. This leads to

$$\delta \tilde{J}(z; v) = \left[-\lambda^T(t_f) + x^T(t_f)S\right]\xi(t_f)$$

which holds for any $\xi(t_f)$ if and only if $\lambda(t_f) = Sx(t_f)$.

We now have everything in place to characterize the solution to the linear quadratic regulator problem. The optimal state trajectory and control signal, along with the Lagrange multiplier vector, are governed by the homogeneous $2n$-dimensional state equation

$$\begin{bmatrix} \dot{x}(t) \\ \dot{\lambda}(t) \end{bmatrix} = \begin{bmatrix} A & -BR^{-1}B^T \\ -Q & -A^T \end{bmatrix} \begin{bmatrix} x(t) \\ \lambda(t) \end{bmatrix} \tag{9.12}$$

$$u(t) = -R^{-1}B^T \lambda(t)$$

with *mixed* boundary condition

$$x(t_0) = x_0 \qquad \lambda(t_f) = S\,x(t_f) \tag{9.13}$$

so named because $x(t)$ is specified at the initial time, and $\lambda(t)$ is specified at the final time. Consequently, this is referred to as a *two-point boundary value problem* that, at first glance, appears to pose a severe computational dilemma; for if we pick a value for $\lambda(t_0)$, then the state equation in Equation (9.12) can be solved forward in time to yield $x(t)$ and $\lambda(t)$ for all $t \in [t_0, t_f]$, but there is no guarantee that $x(t_f)$ and $\lambda(t_f)$ will satisfy the second relationship in Equation (9.13). Alternatively, if we pick $x(t_f)$ and set $\lambda(t_f) = S\,x(t_f)$, then the state equation in Equation (9.12) can be solved backward in time to yield $x(t)$ and $\lambda(t)$ for all $t \in [t_0, t_f]$, but there is no guarantee that $x(t)$ will achieve the specifed initial value.

Fortunately, this apparent predicament can be resolved using the so-called sweep method. For this we assume that $x(t)$ and $\lambda(t)$ are linearly related according to

$$\lambda(t) = P(t)\, x(t) \tag{9.14}$$

The second identity in Equation (9.13) is satisfied by setting $P(t_f) = S$, and our task is specify the t-dependent $n \times n$ matrix $P(t)$ on the interval $[t_0, t_f]$ that, in effect, *sweeps* this terminal condition backward in time to yield the correct initial condition $\lambda(t_0)$ that together with $x(t_0) = x_0$ would allow the state equation (9.12) to be solved forward in time. Even better, once $P(t)$ is determined, the linear relationship (9.14) allows the optimal control signal to be realized as the state feedback law

$$u(t) = -K(t)x(t) \tag{9.15}$$

in which the time-varying feedback gain matrix is given by

$$K(t) = -R^{-1}B^T P(t) \tag{9.16}$$

On differentiating Equation (9.14) and substituting previous identities, we find

$$\dot{\lambda}(t) = \dot{P}(t)x(t) + P(t)\dot{x}(t)$$
$$= \dot{P}(t)x(t) + P(t)[Ax(t) - BR^{-1}B^T\lambda(t)]$$
$$= \left[\dot{P}(t) + P(t)A - P(t)BR^{-1}B^T P(t)\right]x(t)$$

to which we equate

$$\dot{\lambda}(t) = -A^T\lambda(t) - Qx(t)$$
$$= \left[-A^T P(t) - Q\right]x(t)$$

Since equality must hold for any state trajectory $x(t)$, we therefore require that $P(t)$ satisfy the matrix differential equation

$$-\dot{P}(t) = A^T P(t) + P(t)A - P(t)BR^{-1}B^T P(t) + Q \tag{9.17}$$

with boundary condition $P(t_f) = S$. This is the celebrated *differential Riccati equation* named in honor of the Italian mathematician Count J. F. Riccati (1676–1754). We note that a solution to Equation (9.17) with boundary condition specified by a symmetric matrix is necessarily a symmetric matrix at each $t \in [t_0, t_f]$.

At this point we have completely specified the solution to the linear quadratic regulator problem. The differential Riccati equation solution yields the optimal control signal, as determined by the state feedback law (9.15) with time-varying gain matrix (9.16). The optimal state trajectory then is generated by the closed-loop state equation

$$\dot{x}(t) = \left[A - BR^{-1}B^T P(t)\right] x(t) \qquad x(t_0) = x_0$$

We then can compute the optimal value of the performance index by substituting Equations (9.15) and (9.16) into Equation (9.2) to yield

$$J^* = \tfrac{1}{2} \int_{t_0}^{t_f} x^T(t)\left[Q + P(t)BR^{-1}B^T P(t)\right] x(t)dt + \tfrac{1}{2}x^T(t_f)S\, x(t_f)$$

$$= -\tfrac{1}{2} \int_{t_0}^{t_f} x^T(t)\left[\dot{P}(t) + A^T P(t) + P(t)A - 2P(t)BR^{-1}B^T P(t)\right]$$

$$\times x(t)dt + \tfrac{1}{2}x^T(t_f)S\, x(t_f)$$

$$= -\tfrac{1}{2} \int_{t_0}^{t_f} x^T(t)\left\{\dot{P}(t) + \left[A - BR^{-1}B^T P(t)\right]^T P(t)\right.$$

$$\left. + P(t)\left[A - BR^{-1}B^T P(t)\right]\right\} x(t)dt + \tfrac{1}{2}x^T(t_f)S\, x(t_f)$$

$$= -\tfrac{1}{2} \int_{t_0}^{t_f} \frac{d}{dt}\left[x^T(t)P(t)x(t)\right]dt + \tfrac{1}{2}x^T(t_f)S\, x(t_f)$$

$$= -\tfrac{1}{2}x^T(t_f)P(t_f)x(t_f) + \tfrac{1}{2}x^T(t_0)P(t_0)x(t_0) + \tfrac{1}{2}x^T(t_f)S\, x(t_f)$$

$$= \tfrac{1}{2}x^T(t_0)P(t_0)x(t_0)$$

which, remarkably enough, is given by a quadratic form involving the initial state and the differential Riccati equation solution evaluated at the initial time.

Riccati Equation Solution

The differential Riccati equation is a *nonlinear* matrix differential equation as a result of the quadratic term $P(t)BR^{-1}B^T P(t)$. Solving this equation to yield the optimal control law therefore may pose a sizable challenge. It turns out, however, that the differential Riccati equation solution can be obtained from the solution to the $2n$-dimensional homogeneous linear matrix differential equation

$$\begin{bmatrix} \dot{X}(t) \\ \dot{\Lambda}(t) \end{bmatrix} = \begin{bmatrix} A & -BR^{-1}B^T \\ -Q & -A^T \end{bmatrix} \begin{bmatrix} X(t) \\ \Lambda(t) \end{bmatrix} \qquad (9.18)$$

in which $X(t)$ and $\Lambda(t)$ are each $n \times n$ matrices that satisfy the boundary condition

$$\begin{bmatrix} X(t_f) \\ \Lambda(t_f) \end{bmatrix} = \begin{bmatrix} I \\ S \end{bmatrix} \tag{9.19}$$

The $2n \times 2n$ matrix appearing above and in Equation (9.12), namely,

$$H = \begin{bmatrix} A & -BR^{-1}B^T \\ -Q & -A^T \end{bmatrix} \tag{9.20}$$

is called the *Hamiltonian matrix*. We claim that if $X(t)$ is nonsingular at each $t \in [t_0, t_f]$, the differential Riccati equation solution can be expressed as

$$P(t) = \Lambda(t)X^{-1}(t)$$

It is clear that the boundary condition (9.19) implies $P(t_f) = S$, and we show next that $\Lambda(t)X^{-1}(t)$ satisfies Equation (9.17). First, the product rule for differentiating matrix-valued functions yields

$$\frac{d}{dt}\left[\Lambda(t)X^{-1}(t)\right] = \frac{d\Lambda(t)}{dt}X^{-1}(t) + \Lambda(t)\frac{dX^{-1}(t)}{dt}$$

Also, differentiating the identity $X(t)X^{-1}(t) = I$ gives

$$\frac{dX^{-1}(t)}{dt} = -X^{-1}(t)\frac{dX(t)}{dt}X^{-1}(t)$$

Using these relationships, along with expressions for the derivatives of $X(t)$ and $\Lambda(t)$ extracted from Equation (9.18), we see that

$$-\frac{d}{dt}\left[\Lambda(t)X^{-1}(t)\right] = -\left[-QX(t) - A^T\Lambda(t)\right]X^{-1}(t) + \Lambda(t)X^{-1}(t)$$
$$\times \left[AX(t) - BR^{-1}B^T\Lambda(t)\right]X^{-1}(t)$$
$$= Q + A^T\left[\Lambda(t)X^{-1}(t)\right] + \left[\Lambda(t)X^{-1}(t)\right]A$$
$$- \left[\Lambda(t)X^{-1}(t)\right]BR^{-1}B^T\left[\Lambda(t)X^{-1}(t)\right]$$

so that $\Lambda(t)X^{-1}(t)$ satisfies Equation (9.17) as claimed.

Example 9.5 We solve the linear quadratic regulator problem for the one-dimensional state equation

$$\dot{x}(t) = u(t) \quad x(0) = x_0$$

describing a single integrator and the performance index

$$J = \frac{1}{2}\int_0^1 \left[x^2(t) + u^2(t)\right]dt + \frac{1}{2}\sigma x^2(1)$$

We see by inspection that the parameters for this example are $A = 0$, $B = 1$, $Q = 1$, $R = 1$, $S = \sigma$, $t_0 = 0$, and $t_f = 1$. The Hamiltonian matrix (9.20) is given by

$$H = \begin{bmatrix} 0 & -1 \\ -1 & 0 \end{bmatrix}$$

from which we compute the matrix exponential

$$e^{Ht} = \begin{bmatrix} \frac{1}{2}(e^{-t} + e^{t}) & \frac{1}{2}(e^{-t} - e^{t}) \\ \frac{1}{2}(e^{-t} - e^{t}) & \frac{1}{2}(e^{-t} + e^{t}) \end{bmatrix}$$

This yields the solution to Equation (9.18) for the boundary condition (9.19)

$$\begin{bmatrix} X(t) \\ \Lambda(t) \end{bmatrix} = \begin{bmatrix} \frac{1}{2}(e^{-(t-1)} + e^{(t-1)}) & \frac{1}{2}(e^{-(t-1)} - e^{(t-1)}) \\ \frac{1}{2}(e^{-(t-1)} - e^{(t-1)}) & \frac{1}{2}(e^{-(t-1)} + e^{(t-1)}) \end{bmatrix} \begin{bmatrix} 1 \\ \sigma \end{bmatrix}$$

$$= \begin{bmatrix} \frac{1}{2}\big[(1+\sigma)e^{-(t-1)} + (1-\sigma)e^{(t-1)}\big] \\ \frac{1}{2}\big[(1+\sigma)e^{-(t-1)} - (1-\sigma)e^{(t-1)}\big] \end{bmatrix}$$

from which we construct

$$P(t) = \frac{\Lambda(t)}{X(t)} = \frac{(1+\sigma)e^{-(t-1)} - (1-\sigma)e^{(t-1)}}{(1+\sigma)e^{-(t-1)} + (1-\sigma)e^{(t-1)}}$$

$$= \frac{(1+\sigma) - (1-\sigma)e^{2(t-1)}}{(1+\sigma) + (1-\sigma)e^{2(t-1)}}$$

The reader is invited to check that the same solution is obtained from the scalar differential Riccatti equation

$$-\dot{P}(t) = 1 - P^2(t) \qquad P(1) = \sigma$$

The associated feedback gain is $K(t) = -P(t)$, yielding the time-varying closed-loop state equation

$$\dot{x}(t) = -P(t)x(t) \qquad x(0) = x_0$$

The closed-loop state equation was simulated with the initial state $x(0) = 1$ to yield the optimal state trajectory for the following values of the terminal cost weighting parameter $\sigma = 0, 1$, and *10*. We see from Figure 9.4 that as the penalty on the terminal state increases, the regulation

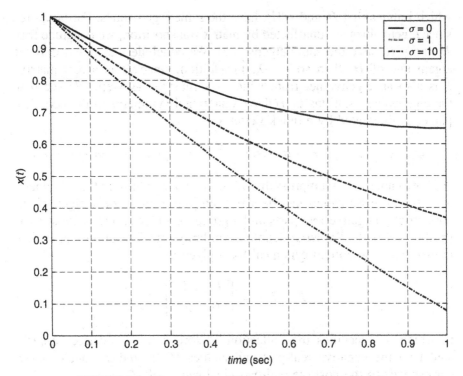

FIGURE 9.4 Closed-loop state response for Example 9.5.

performance improves in the sense that the state trajectory is closer to zero at the final time.   □

The Hamiltonian matrix (9.20) has special eigenstructure properties that allow us to be even more explicit in representing the solution to the differential Riccati equation (9.17). We observe that the $2n \times 2n$ matrix

$$J = \begin{bmatrix} 0 & I \\ -I & 0 \end{bmatrix}$$

is nonsingular with inverse $J^{-1} = -J$. A direct computation reveals that

$$J^{-1}HJ = -JHJ$$

$$= -\begin{bmatrix} 0 & I \\ -I & 0 \end{bmatrix} \begin{bmatrix} A & -BR^{-1}B^T \\ -Q & -A^T \end{bmatrix} \begin{bmatrix} 0 & I \\ -I & 0 \end{bmatrix}$$

$$= -\begin{bmatrix} A^T & -Q \\ -BR^{-1}B^T & -A \end{bmatrix}$$

$$= -H^T$$

which implies that H and $-H^T$ have the same eigenvalues. Furthermore, since eigenvalues are unaffected by matrix transposition, we conclude that H and $-H$ have the same eigenvalues. We conclude that if $\lambda \in \mathbb{C}$ is an eigenvalue of H, then so is $-\lambda$. In addition, since H is a real matrix, $\bar{\lambda}$ is also an eigenvalue, from which $-\bar{\lambda}$ must be as well. We see that the eigenvalue locations of H have quadrantal symmetry in the complex plane, as illustrated by a quick sketch of

$$\lambda = \sigma + j\omega, \quad -\lambda = -\sigma - j\omega, \quad \bar{\lambda} = \sigma - j\omega, \quad -\bar{\lambda} = -\sigma + j\omega$$

Now, under the assumption that H has no eigenvalues on the imaginary axis, the $2n$ eigenvalues can be separated into a group of n eigenvalues with strictly negative real parts and a group of n eigenvalues with strictly positve real parts. We let T denote a similarity transformation matrix that transforms H to Jordan canonical form given by

$$T^{-1}H\,T = \begin{bmatrix} J^- & 0 \\ 0 & J^+ \end{bmatrix}$$

in which J^- specifies the collection of Jordan block matrices associated with the negative real-part eigenvalues of H, and J^+ analogously corresponds to the positive real-part eigenvalues of H.

We use the similarity transformation T partitioned conformably with H into four $n \times n$ blocks as

$$T = \begin{bmatrix} T_{11} & T_{12} \\ T_{21} & T_{22} \end{bmatrix}$$

to transform Equation (9.18) via

$$\begin{bmatrix} X(t) \\ \Lambda(t) \end{bmatrix} = \begin{bmatrix} T_{11} & T_{12} \\ T_{21} & T_{22} \end{bmatrix} \begin{bmatrix} \hat{X}(t) \\ \hat{\Lambda}(t) \end{bmatrix}$$

into

$$\begin{bmatrix} \dot{\hat{X}}(t) \\ \dot{\hat{\Lambda}}(t) \end{bmatrix} = \begin{bmatrix} J^- & 0 \\ 0 & J^+ \end{bmatrix} \begin{bmatrix} \hat{X}(t) \\ \hat{\Lambda}(t) \end{bmatrix}$$

The boundary condition (9.19) transforms according to

$$\begin{bmatrix} I \\ S \end{bmatrix} = \begin{bmatrix} T_{11} & T_{12} \\ T_{21} & T_{22} \end{bmatrix} \begin{bmatrix} \hat{X}(t_f) \\ \hat{\Lambda}(t_f) \end{bmatrix}$$

Although this uniquely determines both $\hat{X}(t_f)$ and $\hat{\Lambda}(t_f)$, for our purposes, it will suffice to express $\hat{\Lambda}(t_f) = M \hat{X}(t_f)$, in which

$$M = -[T_{22} - S \, T_{12}]^{-1}[T_{21} - S \, T_{11}]$$

For then we have

$$\begin{bmatrix} \hat{X}(t) \\ \hat{\Lambda}(t) \end{bmatrix} = \begin{bmatrix} e^{J^-(t-t_f)} \\ e^{J^+(t-t_f)} M \end{bmatrix} \hat{X}(t_f)$$

$$= \begin{bmatrix} I \\ e^{J^+(t-t_f)} M e^{-J^-(t-t_f)} \end{bmatrix} e^{J^-(t-t_f)} \hat{X}(t_f)$$

which leads to

$$\begin{bmatrix} X(t) \\ \Lambda(t) \end{bmatrix} = \begin{bmatrix} T_{11} + T_{12} e^{J^+(t-t_f)} M e^{-J^-(t-t_f)} \\ T_{21} + T_{22} e^{J^+(t-t_f)} M e^{-J^-(t-t_f)} \end{bmatrix} e^{J^-(t-t_f)} \hat{X}(t_f)$$

Finally, we can combine these identities into the desired end result

$$P(t) = \Lambda(t)X^{-1}(t) = \left(T_{21} + T_{22} e^{J^+(t-t_f)} M e^{-J^-(t-t_f)} \right)$$

$$\times \left(T_{11} + T_{12} e^{J^+(t-t_f)} M e^{-J^-(t-t_f)} \right)^{-1} \quad (9.21)$$

Example 9.6 For the setup in Example 9.5, the similarity transformation matrix

$$T = \begin{bmatrix} 1 & 1 \\ 1 & -1 \end{bmatrix}$$

yields

$$T^{-1}HT = \frac{1}{2} \begin{bmatrix} 1 & 1 \\ 1 & -1 \end{bmatrix} \begin{bmatrix} 0 & -1 \\ -1 & 0 \end{bmatrix} \begin{bmatrix} 1 & 1 \\ 1 & -1 \end{bmatrix}$$

$$= \begin{bmatrix} -1 & 0 \\ 0 & 1 \end{bmatrix}$$

from which we identify $J^- = -1$ and $J^+ = 1$. Recalling that $S = \sigma$, we have

$$M = -\frac{1-\sigma}{-1-\sigma} = \frac{1-\sigma}{1+\sigma}$$

from which we compute

$$P(t) = \frac{1 - \dfrac{1-\sigma}{1+\sigma}e^{2(t-1)}}{1 + \dfrac{1-\sigma}{1+\sigma}e^{2(t-1)}} = \frac{(1+\sigma) - (1-\sigma)e^{2(t-1)}}{(1+\sigma) + (1-\sigma)e^{2(t-1)}}$$

which agrees with the result obtained in Example 9.5.   □

Steady-State Linear Quadratic Regulator Problem

We now turn our attention to the steady-state linear quadratic regulator problem, in which the performance index becomes

$$J = \tfrac{1}{2} \int_0^\infty \left[x^T(t)Qx(t) + u^T(t)Ru(t) \right] dt \qquad (9.22)$$

We interpret this as arising from the original performance index (9.2) by setting $t_0 = 0$ and $S = 0$ and letting t_f tend to infinity. To treat this as a limiting case of the preceding analysis, we use the fact that both matrix exponentials

$$e^{-J^-(t-t_f)} = e^{J^-(t_f-t)} \quad \text{and} \quad e^{J^+(t-t_f)} = e^{-J^+(t_f-t)}$$

tend to $n \times n$ zero matrices as t_f tends to infinity because J^- and $-J^+$ have negative real-part eigenvalues. This allows us to conclude from Equation (9.21) that

$$\lim_{t_f \to \infty} P(t) = T_{21}T_{11}^{-1} \triangleq \overline{P}$$

That is, as the final time t_f tends to infinity, the differential Riccati equation solution tends to a constant steady-state value $\overline{P}$. We therefore expect that $\overline{P}$ characterizes an equilibrium solution to Equation (9.17), that is,

$$A^T\overline{P} + \overline{P}A - \overline{P}BR^{-1}\overline{P} + Q = 0 \qquad (9.23)$$

This is naturally referred to as the *algebraic Riccati equation*. We can directly verify that $\overline{P} = T_{21}T_{11}^{-1}$ satisfies Equation (9.23) as follows. First, the similarity transformation matrix T yielding the Jordan canonical form of the Hamiltonian matrix H satisfies

$$\begin{bmatrix} A & -BR^{-1}B^T \\ -Q & -A^T \end{bmatrix} \begin{bmatrix} T_{11} \\ T_{21} \end{bmatrix} = \begin{bmatrix} T_{11} \\ T_{21} \end{bmatrix} J^- \qquad (9.24)$$

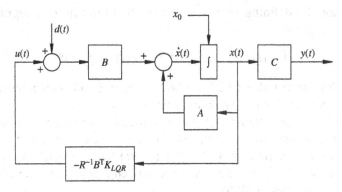

FIGURE 9.5   Linear quadratic regulator closed-loop block diagram.

This can be further manipulated to yield

$$
\begin{bmatrix} \overline{P} & -I \end{bmatrix}
\begin{bmatrix} A & -BR^{-1}B^T \\ -Q & -A^T \end{bmatrix}
\begin{bmatrix} I \\ P \end{bmatrix}
= \begin{bmatrix} \overline{P} & -I \end{bmatrix}
\begin{bmatrix} I \\ P \end{bmatrix} T_{11} J^- T_{11}^{-1}
$$

$$
= 0
$$

On expanding the left hand side, we recover Equation (9.23).

We conclude that the steady-state linear quadratic regulator problem can be solved if the differential Riccati equation has a well-defined constant steady-state solution $\overline{P}$ that satisfies the algebraic Riccati equation. In this case, the optimal control feedback law is time-invariant, given by

$$
u(t) = -K_{\text{LQR}} x(t) \quad \text{with} \quad K_{\text{LQR}} = R^{-1} B^T \overline{P} \tag{9.25}
$$

This yields the optimal time-invariant closed-loop state equation

$$
\dot{x}(t) = (A - BR^{-1}B^T \overline{P}) x(t) \quad x(0) = x_0 \tag{9.26}
$$

depicted by the block diagram of Figure 9.5.

The existence of $\overline{P}$, in turn, relies on the previous assumption that H has no eigenvalues on the imaginary axis and that the similarity transformation matrix yielding the Jordan canonical form of H can be chosen so that T_{11} is nonsingular. We therefore desire explicit conditions expressed in terms of the linear state equation (9.1) and the performance index (9.22) that ensure that the steady-state linear quadratic regulator problem has a solution. As a preliminary step toward deriving such conditions, we note that when Q is only positive semidefinite or, in other words, $q \triangleq \text{rank } Q < n$, a basic fact from linear algebra gives that Q can be factored as $Q = C^T C$ in which the $q \times n$ matrix C has full-row rank. We note that with this factorization, by defining $y(t) = Cx(t)$ the first term in the integral (9.22) can be written as

$$x^T(t)Qx(t) = x^T(t)C^TC\,x(t) = \|y(t)\|^2$$

so the performance index (9.22) captures a tradeoff between control energy and regulation of this newly defined output $y(t)$.

We are now prepared to strike a somewhat remarkable connection between the ability to solve the steady-state linear quadratic regulator problem and controllability of the pair (A, B) together with observability of the pair (A, C). The following theorem and proof largely follow the development in Zhou (1995).

Theorem 9.1 If the pair (A, B) is controllable and the pair (A, C) is observable, then the algebraic Riccati equation (9.23) has a unique symmetric positive-definite solution, and the closed-loop state-equation (9.26) is asymptotically stable.

Proof. We divide the proof into several steps:

1. Show that joint controllability of (A, B) and observability of (A, C) imply that the Hamiltonian matrix H has no eigenvalues on the imaginary axis.

2. Show that T_{11} and T_{21} in Equation (9.24) are such that $T_{21}^*T_{11}$ is symmetric and T_{11} is nonsingular.

3. Show that the closed-loop state equation (9.26) is asymptotically stable and that $\overline{P} = T_{21}T_{11}^{-1}$ is the unique symmetric positive definite solution to Equation (9.23).

For the first step, we assume that H has an imaginary eigenvalue $j\omega$ and let

$$\begin{bmatrix} v \\ w \end{bmatrix} \in \mathbb{C}^{2n}$$

denote a corresponding eigenvector. We note that $v, w \in \mathbb{C}^n$ cannot both be zero vectors. From

$$\begin{bmatrix} A & -BR^{-1}B^T \\ -C^TC & -A^T \end{bmatrix} \begin{bmatrix} v \\ w \end{bmatrix} = j\omega \begin{bmatrix} v \\ w \end{bmatrix}$$

we obtain, after some algebra,

$$(j\omega I - A)v = -BR^{-1}B^T w$$

$$(j\omega I - A)^*w = C^TC\,v$$

Multiplying the first identity on the left by w^* and the second identity on the left by v^* gives

$$w^*(j\omega I - A)v = -w^* B R^{-1} B^T w$$

$$v^*(j\omega I - A)^* w = v^* C^T C\, v$$

Now each right-hand side is a quadratic form involving a real symmetric matrix and therefore is a real quantity. The left-hand sides are conjugates of each other and therefore are equal. This allows us to write

$$-w^* B R^{-1} B^T w = v^* C^T C\, v$$

Since the quadratic forms $w^* B R^{-1} B^T w$ and $v^* C^T C\, v$ are each nonnegative, they both must be zero. This, in turn, implies $B^T w = 0$ and $Cv = 0$ so

$$(j\omega I - A)v = 0$$

$$(j\omega I - A)^* w = 0$$

These results can be reorganized to yield

$$w^* \begin{bmatrix} j\omega I - A & B \end{bmatrix} = \begin{bmatrix} 0 & 0 \end{bmatrix} \quad \text{and} \quad \begin{bmatrix} C \\ j\omega I - A \end{bmatrix} v = \begin{bmatrix} 0 \\ 0 \end{bmatrix}$$

We conclude that since v and w cannot both be zero vectors, the Popov-Belevitch-Hautus rank tests for controllability and observability imply that we cannot simultaneously have (A, B) controllable and (A, C) observable. This completes the first step of the proof.

Proceeding to the second step, we first show that $T_{21}^* T_{11}$ is symmetric. Multiplying Equation (9.24) on the left by $\begin{bmatrix} T_{11}^* & T_{21}^* \end{bmatrix} J$ gives

$$\begin{bmatrix} T_{11}^* & T_{21}^* \end{bmatrix} J H \begin{bmatrix} T_{11} \\ T_{21} \end{bmatrix} = \begin{bmatrix} T_{11}^* & T_{21}^* \end{bmatrix} J \begin{bmatrix} T_{11} \\ T_{21} \end{bmatrix} J^-$$

Now, since the product JH is real and symmetric, the left-hand side is Hermitian and so must the right-hand side be. Expanding the right-hand side and equating the result to its conjugate transpose, we find

$$(-T_{21}^* T_{11} + T_{11}^* T_{21}) J^- = \left[\left(-T_{21}^* T_{11} + T_{11}^* T_{21} \right) J^- \right]^*$$

$$= \left(J^- \right)^* \left(-T_{21}^* T_{11} + T_{11}^* T_{21} \right)^*$$

$$= \left(J^- \right)^* \left(-T_{11}^* T_{21} + T_{21}^* T_{11} \right)$$

$$= - \left(J^- \right)^* \left(-T_{21}^* T_{11} + T_{11}^* T_{21} \right)$$

This can be rearranged into the Lyapunov matrix equation

$$(-T_{21}^* T_{11} + T_{11}^* T_{21}) J^- + (J^-)^* (-T_{21}^* T_{11} + T_{11}^* T_{21}) = 0$$

which, since J^- has negative real-part eigenvalues, has the unique solution

$$-T_{21}^* T_{11} + T_{11}^* T_{21} = 0$$

It follows that $T_{21}^* T_{11} = T_{11}^* T_{21}$, so $T_{21}^* T_{11}$ is symmetric.

We next show that T_{11} in Equation (9.24) is nonsingular. Suppose that T_{11} is singular, and let $x \in \mathbb{C}^n$ be any nonzero vector lying in Ker T_{11}. Premultiplying the first block row Equation in (9.24) by $x^* T_{21}^*$ and postmultiplying by x gives

$$x^* T_{21}^* A(T_{11}x) - x^* T_{21}^* B R^{-1} B^T T_{21}x = x^* T_{21}^* T_{11} J^- x$$
$$= x^* T_{11}^* T_{21} J^- x$$
$$= (T_{11}x)^* T_{21} J^- x$$
$$= 0$$

This implies that $x^* T_{21}^* B R^{-1} B^T T_{21}x = 0$, from which $B^T T_{21}x = 0$. Now, by simply postmultiplying the first-block row in Equation (9.24) by x, we find

$$0 = A(T_{11}x) - B R^{-1}(B^T T_{21}x) = T_{11} J^- x$$

Thus $J^- x \in$ Ker T_{11} for any $x \in$ Ker T_{11}. We let d denote the dimension of the subspace Ker T_{11} and let $X = \begin{bmatrix} x_1 & x_2 & \cdots & x_d \end{bmatrix}$ denote an $n \times d$ matrix whose columns form a basis for Ker T_{11}. It follows that there exists a $d \times d$ matrix J_{11} satisfying

$$J^- X = X J_{11}$$

We let μ denote an eigenvalue of J_{11} and $\alpha \in \mathbb{C}^d$ an associated right eigenvector. We see by construction that

$$X\alpha = \begin{bmatrix} x_1 & x_2 & \cdots & x_d \end{bmatrix} \begin{bmatrix} \alpha_1 \\ \alpha_2 \\ \vdots \\ \alpha_d \end{bmatrix}$$
$$= \alpha_1 x_1 + \alpha_2 x_2 + \cdots + \alpha_d x_d$$
$$\in \text{Ker } T_{11}$$

Also, $x \triangleq X\alpha \neq 0$ because $\alpha \neq 0$ (it is an eigenvector) and X has full column rank. Thus postmultiplying the second block identity in

Equation (9.24) by x gives

$$-Q(T_{11}x) - A^T(T_{21}x) = T_{21}J^-(X\alpha)$$
$$= T_{21}X\ J_{11}\alpha$$
$$= \mu T_{21}X\alpha$$
$$= \mu T_{21}x$$

Using $T_{11}x = 0$, we see that if $T_{21}x \neq 0$ then $T_{21}x$ is a right eigenvector of A^T with associated eigenvalue $-\mu$. Furthermore, $x \in \text{Ker } T_{11}$ implies that $B^T T_{21}x = 0$, which by the Popov-Belevitch-Hautus eigenvector test implies that the pair (A^T, B^T) is not observable; equivalently, the pair (A, B) is not controllable. This contradiction implies that $T_{21}x = 0$, which together with $T_{11}x = 0$ gives

$$\begin{bmatrix} T_{11} \\ T_{21} \end{bmatrix} x = \begin{bmatrix} 0 \\ 0 \end{bmatrix}$$

This, in turn, implies that the first n columns of T are linearly dependent, which contradicts the nonsingularity of T. We therefore conclude that $\text{Ker } T_{11} = 0$; equivalently, T_{11} is nonsingular, which completes the second step of the proof.

For the last step, we first use the fact that $T_{21}^* T_{11} = T_{11}^* T_{21}$, along with the identity

$$\overline{P} = T_{21}T_{11}^{-1} = (T_{11}^{-1})^*(T_{11}^* T_{21})T_{11}^{-1}$$

to conclude that $\overline{P}$ is symmetric. Next, Equation (9.24) and the nonsingularity of T_{11} allow us to write

$$\begin{bmatrix} A & -BR^{-1}B^T \\ -Q & -A^T \end{bmatrix} \begin{bmatrix} I \\ P \end{bmatrix} = \begin{bmatrix} I \\ P \end{bmatrix} T_{11}J^- T_{11}^{-1}$$

from which the first-block row yields

$$A - BR^{-1}B^T\overline{P} = T_{11}J^- T_{11}^{-1}$$

and we therefore conclude that the eigenvalues of $A - BR^{-1}B^T\overline{P}$ coincide with the eigenvalues of J^-, which have negative real parts. This implies that the closed-loop state equation (9.26) is asymptotically stable.

We next rewrite the algebraic Riccati equation (9.23) as the following Lyapunov matrix equation:

$$(A - BR^{-1}B^T\overline{P})^T\overline{P} + \overline{P}(A - BR^{-1}B^T\overline{P}) = -(C^T C + \overline{P}BR^{-1}B^T\overline{P})$$

from which $\overline{P}$ can be represented as

$$\overline{P} = \int_0^\infty e^{(A-BR^{-1}B^T\overline{P})^T t}(C^T C + \overline{P}BR^{-1}B^T\overline{P})e^{(A-BR^{-1}B^T\overline{P})t}\,dt$$

This indicates that $\overline{P}$ is positive semidefinite because $C^T C + \overline{P}BR^{-1}\overline{P}$ is positive semidefinite. To argue that $\overline{P}$ is in fact positive definite, we take any $x_0 \in \text{Ker } \overline{P}$ and write

$$x_0^T \overline{P} x_0 = x_0^T \int_0^\infty e^{(A-BR^{-1}B^T\overline{P})^T t}(C^T C + \overline{P}BR^{-1}B^T\overline{P})e^{(A-BR^{-1}B^T\overline{P})t}\,dt\, x_0$$

$$= x_0^T \int_0^\infty e^{(A-BR^{-1}B^T\overline{P})^T t}C^T C e^{(A-BR^{-1}B^T\overline{P})t}\,dt\, x_0$$

$$+ x_0^T \int_0^\infty e^{(A-BR^{-1}B^T\overline{P})^T t}\overline{P}BR^{-1}B^T\overline{P}e^{(A-BR^{-1}B^T\overline{P})t}\,dt\, x_0$$

$$= 0$$

Since each term is nonnegative, both must vanish. From the second term, we conclude that

$$R^{-1}B^T\overline{P}e^{(A-BR^{-1}B^T\overline{P})t} \equiv 0 \quad \text{for all } t \geq 0$$

which, using a matrix exponential identity, allows us to write

$$e^{(A-BR^{-1}B^T\overline{P})t}x_0 = e^{At}x_0 - \int_0^t e^{A(t-\tau)}BR^{-1}B^T\overline{P}e^{(A-BR^{-1}B^T\overline{P})\tau}x_0\,d\tau$$

$$= e^{At}x_0$$

From the first term, we see that

$$Ce^{At}x_0 = Ce^{(A-BR^{-1}B^T\overline{P})t}x_0 \equiv 0 \quad \text{for all } t \geq 0$$

Now, by definition, if $x_0 \neq 0$, then (A, C) is an unobservable pair. Thus the only $x_0 \in \text{Ker } \overline{P}$ is $x_0 = 0$, from which $\overline{P}$ is nonsingular and hence positive definite. Finally, uniqueness can be argued by showing that $\overline{P} = T_{21}T_{11}^{-1}$ is the only solution to the algebraic Riccati equation from which a stabilizing state feedback gain matrix (9.25) can be constructed. This is pursued in AE9.1. □

Example 9.7 We again return to the scalar state equation of Example 9.5, now with the performance index

$$J = \tfrac{1}{2}\int_0^\infty \left[x^2(t) + u^2(t)\right]\,dt$$

The solution of the differential Riccati equation for arbitrary final time t_f and terminal cost weight σ is easily adapted from previous computations:

$$P(t) = \frac{1 - \dfrac{1-\sigma}{1+\sigma}e^{2(t-t_f)}}{1 + \dfrac{1-\sigma}{1+\sigma}e^{2(t-t_f)}} = \frac{(1+\sigma) - (1-\sigma)e^{2(t-t_f)}}{(1+\sigma) + (1-\sigma)e^{2(t-t_f)}}$$

Since $e^{2(t-t_f)} = e^{-2(t_f-t)}$ tends to zero as t_f tends to infinity, $P(t)$ approaches the steady-state value

$$\overline{P} = \frac{1+\sigma}{1+\sigma} = 1$$

which is independent of σ.

With $A = 0$ and $B = 1$, the pair (A, B) is controllable, and for $C = \sqrt{Q} = 1$, the pair (A, C) is observable. The algebraic Riccati equation

$$1 - P^2 = 0$$

has two solutions, $\overline{P} = \pm 1$. Of these, only $\overline{P} = 1$ is positive (definite) for which the state feedback gain $K_{\text{LQR}} = 1$ yields the asymptotically stable closed-loop state equation

$$\dot{x}(t) = -x(t) \qquad\qquad \square$$

9.5 MATLAB FOR OPTIMAL CONTROL

MATLAB Functions for Linear Quadratic Regulator Design

The following MATLAB functions are useful for design of state feedback control laws that solve the linear quadratic regulator problem:

Pbar = are(A,BB,Q)   Returns the positive definite (stabilizing) solution to the algebraic Riccati equation, given the system dynamics matrix **A**, the coefficient matrix of the quadratic term **BB = B*inv(R)*B'**, and weighting matrix **Q**.

Klqr = lqr(A,B,Q,R)   Directly calculates the optimal linear quadratic regulator gain matrix **Klqr**, given the system dynamics matrix **A**, the input matrix **B**, and the weighting matrices **Q** and **R**.

Continuing MATLAB Example

For the Continuing MATLAB Example (rotational mechanical system), we now design an optimal linear quadratic regulator state feedback control law by determining the gain matrix K_{LQR}. Here we choose the weights

$$Q = \begin{bmatrix} 20 & 0 \\ 0 & 20 \end{bmatrix} \text{ and } R = 1$$

The following MATLAB code, in combination with the m-files from previous chapters, performs the required computations:

```
%------------------------------------------------------------
%  Chapter 9. Linear Quadratic Regulator Design
%------------------------------------------------------------

Q = 20*eye(2);            % Weighting matrix for
                          % state error
R = [1];                  % Weighting matrix for
                          % input effort
BB = B*inv(R)*B';

Pbar = are(A,BB,Q);       % Solve algebraic Riccati
                          % equation
KLQR = inv(R)*B'*Pbar;    % Computer state feedback
    gain
ALQR = A-B*KLQR;          % Compute closed-loop
                          % system dynamics matrix
JbkRLQR = ss(ALQR,B,C,D); % Create LQR closed-loop
                          % state equation

% Compare open and closed-loop zero-input responses
[YLQR,t,XLQR] = initial(JbkRLQR,XO,t);

figure;
subplot(211)
plot(t,Xo(:,1),'--',t,Xc(:,1),'-.',t,XLQR(:,1));
grid; axis([0 4 -0.2 0.5]);
set(gca,'FontSize',18);
legend('Open-loop','Closed-loop','LQR');
ylabel('\itx_1 (rad)')
subplot(212)
plot(t,Xo(:,2),'--',t,Xc(:,2),'-.',t,XLQR(:,2));
```

```
grid; axis([0 4 -2 1]);
set(gca,'FontSize',18);
xlabel('\ittime (sec)'); ylabel('\itx_2 (rad/sec)');

% Calculate and plot to compare closed-loop and LQR
% input efforts required
Uc = -K*Xc';  % Chapter 7 input effort
ULQR = -inv(R)*B'*Pbar*XLQR'; % LQR input effort

figure;
plot(t,Uc,'--',t,ULQR); grid; axis([0 4 -10 6]);
set(gca,'FontSize',18);
legend('Closed-loop','LQR');
xlabel('\ittime (sec)'); ylabel('\itU (Nm)');
```

The solution to the algebraic Riccati equation is computed to be:

```
Pbar =
83.16     0.25
0.25      2.04
```

The associated state feedback gain is

```
KLQR =

0.2492  2.0414
```

The closed-loop eigenvalues are located at $-3.0207 \pm j5.5789$.

We see in Figure 9.6 that the linear quadratic regulator design exhibits improved regulation performance as compared with the open-loop response and the closed-loop response resulting from the eigenvalue placement design presented in Chapter 7. The zero-input response for the linear quadratic regulator design decays to zero faster with less oscillation.

Figure 9.7 compares the control signals produced by the linear quadratic regulator design and eigenvalue placement design. Interestingly enough, the control signal for the linear quadratic regulator design has a smaller amplitude than that for the eigenvalue placement design. These comparisons indicate that the linear quadratic regulator optimal controller is superior in terms of both regulation performance and control effort.

9.6 CONTINUING EXAMPLE 1: LINEAR QUADRATIC REGULATOR

For Continuing Example 1 (two-mass translational mechanical system), we now design and evaluate a linear quadratic regulator state feedback

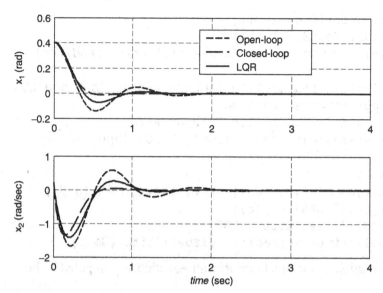

FIGURE 9.6 State responses for the Continuing MATLAB Example: Open-loop, eigen-value placement design and linear quadratic regulator design.

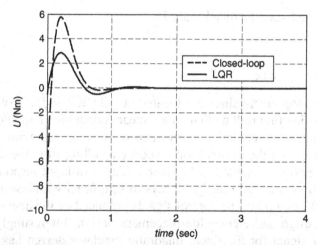

FIGURE 9.7 Control signal responses for the Continuing MATLAB Example: eigen-value placement Design versus linear quadratic regulator design.

control law for case b [for zero input $u_2(t)$, initial state $x(0) = [0.1, 0, 0.2, 0]^T$, and output $y_1(t)$].

Case b. To account for the relative scaling of the state variables and input signals, the weighting matrices for the linear quadratic regulator performance index were chosen to be:

$$Q = 300 I_4 \quad R = 1$$

We use the MATLAB function **are** to solve the algebraic Riccati equation, yielding

$$\overline{P} = \begin{bmatrix} 3618 & 51 & -345 & 62 \\ 51 & 326 & -21 & 130 \\ -345 & -21 & 1712 & 4 \\ 62 & 130 & 4 & 235 \end{bmatrix}$$

The corresponding state feedback control law (9.26) yields the closed-loop system dynamics matrix

$$A - BR^{-1}B^{\mathrm{T}}\overline{P} = \begin{bmatrix} 0 & 1 & 0 & 0 \\ -15 & -0.75 & 5 & 0.25 \\ 0 & 0 & 0 & 1 \\ 9.85 & 0.17 & -10.01 & -1.09 \end{bmatrix}$$

We see in Figure 9.8 that the linear quadratic regulator design exhibits improved regulation performance as compared with the underdamped open-loop response. However, the linear quadratic regulator-based response is more oscillatory as compared with the response achieved by the eigenvalue placement design described in Chapter 7, which decays faster with less oscillation. Thus, based on regulation performance alone, it appears that the *optimal* response does not outperform the Chapter 7 design.

However, the regulation performance of the Chapter 7 design does not come without a price. Large control signal amplitude is required to achieve the fast, well-damped transient response for this design. In contrast, the control signal amplitude for the linear quadratic regulator is significantly smaller, as shown in Figure 9.9. We observe that the linear quadratic regulator formulation provides a convenient mechanism for assessing the tradeoff between regulation performance and control signal amplitude. In this example, regulation performance can be improved by increasing the scale factor appearing in the weighting matrix Q (currently set to 300) while keeping R fixed. This will result in an accompanying increase in control signal amplitude as the interested reader is invited to explore.

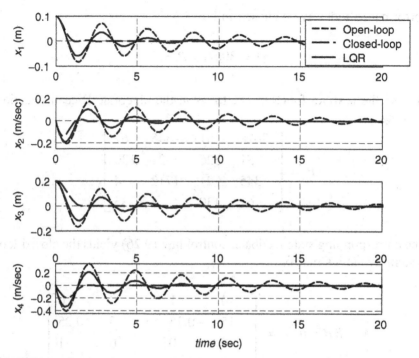

FIGURE 9.8   State responses for Continuing Example 1 (case *b*): Open-loop, eigen-value placement design, and linear quadratic regulator design.

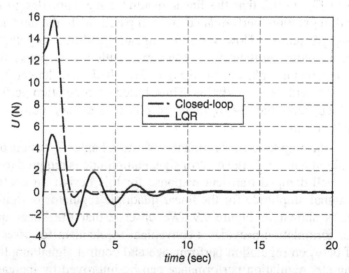

FIGURE 9.9   Control signal responses for Continuing Example 1 (case *b*): Eigen-value placement design versus linear quadratic regulator design.

9.7 HOMEWORK EXERCISES

Numerical Exercises

NE9.1 For each matrix pair (A, B) below, calculate the algebraic Riccati equation solution $\overline{P}$ and the linear quadratic regulator gain matrix K_{LQR} given $Q = I$ and $R = 1$.

a. $A = \begin{bmatrix} -1 & 0 \\ 0 & -4 \end{bmatrix} \quad B = \begin{bmatrix} 1 \\ 1 \end{bmatrix}$

b. $A = \begin{bmatrix} 0 & 1 \\ -6 & -8 \end{bmatrix} \quad B = \begin{bmatrix} 0 \\ 1 \end{bmatrix}$

c. $A = \begin{bmatrix} 0 & 1 \\ -6 & 0 \end{bmatrix} \quad B = \begin{bmatrix} 0 \\ 1 \end{bmatrix}$

d. $A = \begin{bmatrix} 0 & 8 \\ 1 & 10 \end{bmatrix} \quad B = \begin{bmatrix} 1 \\ 0 \end{bmatrix}$

NE9.2 Repeat **NE 9.1** using: (i) $Q = I$ and $R = 0.5$, and (ii) $Q = I$ and $R = 2$. Compare and discuss your results.

NE9.3 Repeat the analysis conducted in Examples 9.5 though 9.7 for fixed $\sigma = 1$ and variable $R = \rho^2$.

Analytical Exercises

AE9.1 Suppose that $\overline{P}_1$ and $\overline{P}_2$ are two symmetric positive definite solutions to the algebraic Riccati equation. Show that $\overline{P}_1$ and $\overline{P}_2$ satisfy

$$
\begin{aligned}
(A - BR^{-1}B^T\overline{P}_2)^T(\overline{P}_1 - \overline{P}_2) \\
+ (\overline{P}_1 - \overline{P}_2)(A - BR^{-1}B^T\overline{P}_1) = 0
\end{aligned}
$$

and argue that $\overline{P}_1 = \overline{P}_2$, thereby verifying the uniqueness claim of Theorem 9.1.

AE9.2 Suppose that the pair (A, B) is controllable. Show that the controllability Gramian satisfies

$$
-\frac{d}{dt}W^{-1}(t, t_f) = A^T W^{-1}(t, t_f) + W^{-1}(t, t_f)A \\
- W^{-1}(t, t_f)BB^T W^{-1}(t, t_f)
$$

Use this to relate the solutions of the minimum energy control problem and the linear quadratic regulator problem with $Q = 0$ and $R = I$ and with the terminal cost term replaced by the constraint $x(t_f) = x_f$.

AE9.3 Suppose that the algebraic Riccati equation has a unique symmetric positive-definite solution $\overline{P}$. Determine the $n \times n$ matrix X for which

$$T = \begin{bmatrix} I & 0 \\ \overline{P} & I \end{bmatrix}$$

satisfies

$$T^{-1} H T = \begin{bmatrix} A - BR^{-1}B^T\overline{P} & X \\ 0 & -(A - BR^{-1}B^T\overline{P})^T \end{bmatrix}$$

Use this to relate the eigenvalues of H to the eigenvalues of the closed-loop state equation (9.26).

AE9.4 Derive a solution to the optimal control problem involving the weighted performance index

$$J_\alpha = \tfrac{1}{2} \int_0^\infty [x^T(t)Qx(t) + u^T(t)Ru(t)]e^{2\alpha t}\,dt$$

and show that the associated closed-loop eigenvalues have real parts less than $-\alpha$.

AE9.5 For the single-input, single-output case, show that

$$|1 + K_{\text{LQR}}(j\omega - A)^{-1}B| \geq 1 \qquad \text{for all } -\infty < \omega < \infty$$

As a result, show that the Nyquist plot of $K_{\text{LQR}}(sI - A)^{-1}B$ never enters the circle of radius one centered at $-1 + j0$ in the complex plane. Argue that the linear quadratic regulator achieves infinite gain margin and a phase margin of at least $60°$.

Continuing MATLAB Exercises

CME9.1 For the CME1.1 system, design and evaluate an optimal linear quadratic regulator with equal weighting for the state and input. Plot the output response to a unit step input, and compare on the same graph the open-loop and closed-loop responses obtained in CME7.1b and the linear quadratic regulator responses (assuming zero initial conditions). Also plot the associated control signals for the CME7.1b design and the linear quadratic regulator design.

CME9.2 For the CME1.2 system, design and evaluate an optimal linear quadratic regulator with the state weighting ten times greater than the input weighting. Plot the output response to a unit step input, and compare on the same graph the open-loop and closed-loop responses obtained in CME7.2b and the linear quadratic regulator responses (assuming zero initial conditions). Also plot the associated control signals for the CME7.2b design and the linear quadratic regulator design.

CME9.3 For the CME1.3 system, design and evaluate an optimal linear quadratic regulator with the state weighting 300 times greater than the input weighting. Plot the output response to a unit step input and compare on the same graph the open-loop and the closed-loop response obtained in CME7.3b and the linear quadratic regulator responses (assuming zero initial conditions). Also plot the associated control signals for the CME7.3b design and the linear quadratic regulator design.

CME9.4 For the CME1.4 system, design and evaluate an optimal linear quadratic regulator with the state weighting 100 times greater than the input weighting. Plot the output response to a unit step input and compare on the same graph the open-loop and the closed-loop response obtained in CME7.4b and the linear quadratic regulator responses (assuming zero initial conditions). Also plot the associated control signals for the CME7.4b design and the linear quadratic regulator design.

Continuing Exercises

CE9.1 Design and evaluate a linear quadratic regulator for the CE2.1.i.b system. Use equal weighting on the state and input. In addition to plotting the open-loop and closed-loop linear quadratic regulator state responses, separately plot the input signals. Compare with the CE7.1 results.

CE9.2 Design and evaluate a linear quadratic regulator for the CE2.2.i.b system. Use equal weighting on the state and input. In addition to plotting the open-loop and closed-loop linear quadratic regulator state responses, separately plot the input signals. Compare with the CE7.2 results.

CE9.3 Design and evaluate a linear quadratic regulator for the CE2.3.i.b system. Use equal weighting on the state and input. In addition to plotting the open-loop and closed-loop linear quadratic regulator state responses, separately plot the input signals. Compare with the CE7.3 results.

CE9.4 Design and evaluate a linear quadratic regulator for the CE2.4 system. Weight the state twice as much as the input. In addition to plotting the open-loop and closed-loop linear quadratic regulator state responses, separately plot the input signals. Compare with the CE7.4 results.

APPENDIX A

MATRIX INTRODUCTION

This appendix presents a matrix introduction to support the material of the textbook. Topics covered are matrix basics, matrix arithmetic, matrix determinants, and matrix inversion.

A.1 BASICS

A matrix is a two-dimensional array of real or complex numbers. Here a few examples:

$$
\begin{bmatrix} 2 & 0 & -1 \\ 1 & 5 & 3 \end{bmatrix}
\qquad
\begin{bmatrix} -1+j2 & 3 \\ 0 & -j \\ 2-j & 1 \\ 1+j3 & 0 \end{bmatrix}
\qquad
\begin{bmatrix} -3 & 1 \\ 2 & -4 \end{bmatrix}
\qquad
\begin{bmatrix} 0 & 2 & 4 \end{bmatrix}
$$

$$
\begin{bmatrix} -1 \\ 0 \\ -2 \\ 7 \end{bmatrix}
$$

The number of rows and columns specify the dimension of the matrix. That is, a matrix with m rows and n columns is said to have dimension $m \times n$ or is referred to as an $m \times n$ matrix. The preceding matrices have dimension 2×3, 4×2, 2×2, 1×3, and 4×1, respectively. Note that a row vector can be regarded as a matrix with a single row, and a column vector can be regarded a matrix with a single column. Matrices with the same number of rows and columns are called *square*.

Linear State-Space Control Systems, by Robert L. Williams II and Douglas A. Lawrence
Copyright © 2007 John Wiley & Sons, Inc.

An $m \times n$ matrix will be represented generically as

$$A = \begin{bmatrix} a_{11} & a_{12} & \cdots & a_{1n} \\ a_{21} & a_{22} & \cdots & a_{2n} \\ \vdots & \vdots & \ddots & \vdots \\ a_{m1} & a_{m2} & \cdots & a_{mn} \end{bmatrix}$$

or, using the shorthand notation, $A = [a_{ij}]$, where a_{ij} denotes the element in the ith row and jth column. Hence the first subscript indexes the row in which the element lies, and the second subscript indexes the column in which the element lies.

The $n \times n$ *identity matrix*, denoted by I_n (or simply I when the dimension is clear from the context), is the square matrix with ones along the main diagonal and zeros elsewhere. For example,

$$I_3 = \begin{bmatrix} 1 & 0 & 0 \\ 0 & 1 & 0 \\ 0 & 0 & 1 \end{bmatrix}$$

The $m \times n$ matrix with a zero in every element is called a *zero matrix* and will be denoted either by $0_{m \times n}$ or just 0, again when the dimension is clear from the context.

The *transpose* of an $m \times n$ matrix $A = [a_{ij}]$ is the $n \times m$ matrix given by $A^T = [a_{ji}]$. That is, the transpose is obtained by interchanging rows and columns. The *conjugate transpose* of an $m \times n$ matrix $A = [a_{ij}]$ is the $n \times m$ matrix given by $A^* = [\bar{a}_{ji}]$. Hence the conjugate transpose is obtained by combining matrix transposition with element-wise conjugation. The conjugate transpose is also referred to as the *Hermitian transpose*.

Example A.1   For

$$A = \begin{bmatrix} -1 + j2 & 3 \\ 0 & -j \\ 2 - j & 1 \\ 1 + j3 & 0 \end{bmatrix}$$

we obtain

$$A^T = \begin{bmatrix} -1 + j2 & 0 & 2 - j & 1 + j3 \\ 3 & -j & 1 & 0 \end{bmatrix} \quad \text{and}$$

$$A^* = \begin{bmatrix} -1 - j2 & 0 & 2 + j & 1 - j3 \\ 3 & j & 1 & 0 \end{bmatrix}$$

If A is a real matrix, meaning that all elements are real numbers, then $A^* = A^T$. □

A.2 MATRIX ARITHMETIC

Matrix addition and subtraction operations are performed on an element-by-element basis and therefore are defined only for matrices of the same dimension. For $m \times n$ matrices $A = [a_{ij}]$ and $B = [b_{ij}]$, their sum $C = A + B$ is specified by

$$[c_{ij}] = [a_{ij} + b_{ij}]$$

and their difference $D = A - B$ is specified by

$$[d_{ij}] = [a_{ij} - b_{ij}]$$

Multiplication of a matrix A of any dimension by a real or complex scalar α is specified by

$$\alpha A = [\alpha a_{ij}]$$

that is, every element of A is multiplied by α to produce αA. Note that matrix subtraction can be represented as $A - B = A + (-1)B$.

Example A.2 For

$$A = \begin{bmatrix} -1+j2 & 3 \\ 0 & -j \\ 2-j & 1 \\ 1+j3 & 0 \end{bmatrix} \quad \text{and} \quad B = \begin{bmatrix} 0 & 1+j \\ 1 & 2 \\ j & -1 \\ 5 & -1-j4 \end{bmatrix}$$

we obtain

$$A + B = \begin{bmatrix} -1+j2 & 4+j \\ 1 & 2-j \\ 2 & 0 \\ 6+j3 & -1-j4 \end{bmatrix} \quad \text{and}$$

$$A - B = \begin{bmatrix} -1+j2 & 2-j \\ -1 & -2-j \\ 2-j2 & 2 \\ -4+j3 & 1+j4 \end{bmatrix}$$

Also,

$$
2A = \begin{bmatrix} -2+j4 & 6 \\ 0 & -j2 \\ 4-j2 & 2 \\ 2+j6 & 0 \end{bmatrix} \quad \text{and} \quad -jB = \begin{bmatrix} 0 & 1-j \\ -j & -j2 \\ 1 & j \\ -j5 & -4+j \end{bmatrix}
$$

$\square$

Based on the definitions, it is clear that matrix addition and scalar multiplication have general properties very similar to their counterparts for real and complex numbers: For any $m \times n$ matrices A, B, and C and any pair of scalars α and β,

1. Commutativity
   $A + B = B + A$
2. Associativity
   a. $(A + B) + C = A + (B + C)$
   b. $(\alpha\beta)A = \alpha(\beta A)$
3. Distributivity
   a. $\alpha(A + B) = \alpha A + \alpha B$
   b. $(\alpha + \beta)A = \alpha A + \beta A$
4. Identity elements
   a. The zero matrix 0 satisfies $A + 0 = A$
   b. $0A = 0_{m \times n}$ and $1A = A$

The matrix product $C = AB$ of an $m \times n$ matrix A and a $p \times q$ matrix B can be defined only when $n = p$; i.e., the number of columns of the left factor A matches the number of rows of the right factor B. When this holds, the matrices A and B are said to be *conformable*, and the matrix product is the $m \times q$ matrix given by

$$
C = [c_{ij}] \quad c_{ij} = \sum_{k=1}^{n} a_{ik}b_{kj}
$$

It is often helpful to visualize the computation of c_{ij} as the inner or dot product of the ith row of A and the jth column of B, which are vectors of equal length because of conformability.

Example A.3 Given

$$A = \begin{bmatrix} -1 + j2 & 3 \\ 0 & -j \\ 2 - j & 1 \\ 1 + j3 & 0 \end{bmatrix} \quad \text{and} \quad B = \begin{bmatrix} -3 & 1 \\ 2 & -4 \end{bmatrix}$$

we obtain

$$AB = \begin{bmatrix} (-1 + j2)(-3) + (3)(2) & (-1 + j2)(1) + (3)(-4) \\ 0(-3) + (-j)(2) & 0(1) + (-j)(-4) \\ (2 - j)(-3) + (1)(2) & (2 - j)(1) + (1)(-4) \\ (1 + j3)(-3) + (0)(2) & (1 + j3)(1) + (0)(-4) \end{bmatrix}$$

$$= \begin{bmatrix} 9 - j6 & -13 + j2 \\ -j2 & j4 \\ -4 + j3 & -2 - j \\ -3 - j9 & 1 + j3 \end{bmatrix}$$

In this example, it is not possible to define the product BA because the number of columns of B does not match the number of rows of A. □

It is important to note that matrix multiplication is not commutative. Even when both products AB and BA can be formed and have the same dimension (which happens when A and B are square matrices of the same size), in general, $AB \neq BA$. Consequently, the order of the factors is very important. Matrix multiplication does satisfy other properties reminiscent of the scalar case. For matrices A, B, and C with dimensions such that every matrix product below is defined:

1. Associativity
   $(AB)C = A(BC)$
2. Distributivity
   a. $A(B + C) = AB + AC$
   b. $(B + C)A = BA + CA$

Finally, matrix multiplication and matrix transposition are related as follows:

$$(AB)^{\mathrm{T}} = B^{\mathrm{T}} A^{\mathrm{T}}$$

and this can be extended to any number of factors. In words, the transpose of a product is the product of the transposes with the factors arranged in reverse order. An analogous result holds for conjugate transposition.

A.3 DETERMINANTS

Determinants are only defined for square matrices. The determinant of an $n \times n$ matrix A is denoted by $|A|$. For dimensions $n = 1, 2, 3$, the determinant is computed according to

$$n = 1: \quad |A| = a_{11}$$

$$n = 2: \quad |A| = a_{11}a_{22} - a_{12}a_{21}$$

$$n = 3: \quad |A| = a_{11}a_{22}a_{33} + a_{12}a_{23}a_{31} + a_{13}a_{32}a_{21} - a_{31}a_{22}a_{13}$$
$$- a_{32}a_{23}a_{11} - a_{33}a_{12}a_{21}$$

For $n > 3$, the determinant is computed via the Laplace expansion specified as follows: First, the cofactor C_{ij} of the matrix element a_{ij} is defined as

$$C_{ij} = (-1)^{i+j} M_{ij}$$

in which M_{ij} is the minor of the matrix element a_{ij}, which, in turn, is defined as the determinant of the $(n-1) \times (n-1)$ submatrix of A obtained by deleting the ith row and jth column. In terms of these definitions, the determinant is given by the formula(s):

$$|A| = \sum_{j=1}^{n} a_{ij} C_{ij} \quad \text{for fixed } 1 \leq i \leq n$$

$$|A| = \sum_{i=1}^{n} a_{ij} C_{ij} \quad \text{for fixed } 1 \leq j \leq n$$

A few remarks are in order. First, there are multiple ways to compute the determinant. In the first summation, the row index i is fixed, and the sum ranges over the column index j. This is referred to as *expanding the determinant along the ith row*, and there is freedom to expand the determinant along any row in the matrix. Alternatively, the second summation has the column index j fixed, and the sum ranges over the row index i. This is referred to as *expanding the determinant along the jth column*, and there is freedom to expand the determinant along any column in the matrix. Second, the Laplace expansion specifies a recursive computation in that the determinant of an $n \times n$ matrix A involves determinants of $(n-1) \times (n-1)$ matrices, namely, the minors of A, which, in turn, involve determinants of $(n-2) \times (n-2)$ matrices, and so on. The recursion terminates when one of the cases above is reached ($n = 1, 2, 3$) for which a closed-form expression is available. Third, observe that whenever $a_{ij} = 0$, computation of the cofactor C_{ij} is not necessary. It therefore makes sense to expand the determinant along the row or column containing the greatest number of zero elements to simplify the calculation as much as possible.

Example A.4   For the 4×4 matrix

$$A = \begin{bmatrix} 1 & 0 & 0 & -3 \\ 0 & 5 & 2 & 1 \\ -1 & 0 & 0 & -1 \\ 3 & 1 & 0 & 0 \end{bmatrix}$$

there are several choices for a row or column with two zero elements but only one choice, namely, the third column, for which there are three zero elements. Expanding the determinant about the third column ($j = 3$) gives

$$|A| = \sum_{i=1}^{4} a_{i3} C_{i3}$$
$$= (0)C_{13} + (2)C_{23} + (0)C_{33} + (0)C_{43}$$
$$= (2)C_{23}$$

so the only cofactor required is $C_{23} = (-1)^{2+3} M_{23}$. The associated minor is the determinant of the 3×3 submatrix of A obtained by deleting row 2 and column 3:

$$M_{23} = \begin{vmatrix} 1 & 0 & -3 \\ -1 & 0 & -1 \\ 3 & 1 & 0 \end{vmatrix}$$
$$= (1)(0)(0) + (0)(-1)(3) + (-3)(1)(-1) - (3)(0)(-3)$$
$$\quad - (1)(-1)(1) - (0)(0)(-1)$$
$$= 4$$

so $C_{23} = (-1)^{2+3}(4) = -4$. Finally,

$$|A| = 2C_{23}$$
$$= -8 \qquad \qquad \Box$$

Several useful properties of determinants are collected below:

1. $|A| = |A^T|$
2. $|AB| = |A||B| = |BA|$
3. $|\alpha A| = \alpha^n |A|$
4. If any row or column of A is multiplied by a scalar α to form B, then $|B| = \alpha |A|$.
5. If any two rows or columns of A are interchanged to form B, then $|B| = -|A|$.
6. If a scalar multiple of any row (respectively, column) of A is added to another row (respectively, column) of A to form B, then $|B| = |A|$.

A.4 MATRIX INVERSION

An $n \times n$ matrix A is called *invertible* or *nonsingular* if there is another $n \times n$ matrix B that satisfies the relationship

$$AB = BA = I_n$$

In such cases, B is called the *inverse* of A and is instead written as A^{-1}. If A has an inverse, the inverse is unique. If A has no inverse, then A is called *singular*. The following basic fact provides a test for invertibility (nonsingularity).

Proposition A.1 *A is invertible (nonsingular) if and only if $|A| \neq 0$.* $\square$

The inverse of an invertible matrix is specified by the formula

$$A^{-1} = \frac{\text{adj}(A)}{|A|}$$

in which adj(A) is the *adjugate* or *adjoint* of A and is given by

$$\text{adj}(A) = [C_{ij}]^T$$

where C_{ij} is the cofactor of the (i, j)th element of A. That is, the adjugate of A is the transpose of the matrix of cofactors. The fraction appearing in the preceding formula for the inverse should be interpreted as multiplication of the matrix adj(A) by the scalar $1/|A|$.

Example A.5 The 4×4 matrix A from Example A.4 is invertible (nonsingular) because $|A| = -8 \neq 0$. Construction of the 4×4 adjugate matrix requires the calculation of a total of 16 3×3 determinants. From a previous calculation, $C_{23} = -4$, and as another sample calculation

$$M_{11} = \begin{vmatrix} 5 & 2 & 1 \\ 0 & 0 & -1 \\ 1 & 0 & 0 \end{vmatrix}$$

$$= (5)(0)(0) + (2)(-1)(1) + (1)(0)(0) - (1)(0)(1)$$
$$- (0)(-1)(5) - (0)(2)(0)$$
$$= -2$$

from which $C_{23} = (-1)^{1+1} M_{11} = -2$. After 14 more such calculations,

$$\text{adj}(A) = \begin{bmatrix} -2 & 6 & -16 & 2 \\ 0 & 0 & -4 & 0 \\ 6 & -18 & 44 & 2 \\ 0 & -8 & 20 & 0 \end{bmatrix}^{T}$$

$$= \begin{bmatrix} -2 & 0 & 6 & 0 \\ 6 & 0 & -18 & -8 \\ -16 & -4 & 44 & 20 \\ 2 & 0 & 2 & 0 \end{bmatrix}$$

which leads to

$$A^{-1} = \frac{1}{-8} \begin{bmatrix} -2 & 0 & 6 & 0 \\ 6 & 0 & -18 & -8 \\ -16 & -4 & 44 & 20 \\ 2 & 0 & 2 & 0 \end{bmatrix}$$

$$= \begin{bmatrix} \frac{1}{4} & 0 & \frac{-3}{4} & 0 \\ \frac{-3}{4} & 0 & \frac{9}{4} & 1 \\ 2 & \frac{1}{2} & \frac{-11}{2} & \frac{-5}{2} \\ \frac{-1}{4} & 0 & \frac{-1}{4} & 0 \end{bmatrix}$$

The correctness of an inverse calculation can always be verified by checking that $AA^{-1} = A^{-1}A = I$. This is left as an exercise for the reader.

□

The transpose of a matrix is invertible if and only if the matrix itself is invertible. This follows from Proposition A.1 and the first determinant property listed in Section A.3. When A is invertible,

$$(A^T)^{-1} = (A^{-1})^T$$

In general, an asserted expression for the inverse of a matrix can be verified by checking that the product of the matrix and the asserted inverse (in either order) yields the identity matrix. Here, this check goes as follows:

$$A^T(A^T)^{-1} = A^T(A^{-1})^T$$
$$= (A^{-1}A)^T$$
$$= (I)^T$$
$$= I$$

This result shows that matrix transposition and matrix inversion can be interchanged, and this permits the unambiguous notation A^{-T}.

Also, the product of two $n \times n$ matrices is invertible if and only if each factor is invertible. This follows from Proposition A.1 and the second determinant property listed in Section A.3. When the product is invertible, the inverse is easily verified to be

$$(AB)^{-1} = B^{-1}A^{-1}$$

That is, the inverse of a product is the product of the inverses with the factors arranged in reverse order.

LINEAR ALGEBRA

This appendix presents an overview of selected topics from linear algebra that support state-space methods for linear control systems.

B.1 VECTOR SPACES

Definition B.1 *A **linear vector space** $\mathbb{X}$ over a field of scalars $\mathbb{F}$ is a set of elements (called vectors) that is closed under two operations: vector addition and scalar multiplication. That is,*

$$x_1 + x_2 \in \mathbb{X} \quad \text{for all } x_1, x_2 \in \mathbb{X}$$

and

$$\alpha x \in \mathbb{X} \quad \text{for all } x \in \mathbb{X} \quad \text{and for all } \alpha \in \mathbb{F}$$

In addition, the following axioms are satisfied for all $x, x_1, x_2, x_3 \in \mathbb{X}$ and for all $\alpha, \alpha_1, \alpha_2 \in \mathbb{F}$:

1. *Commutativity*
   $x_1 + x_2 = x_2 + x_1$
2. *Associativity*
   a. $(x_1 + x_2) + x_3 = x_1 + (x_2 + x_3)$
   b. $(\alpha_1 \alpha_2)x = \alpha_1(\alpha_2 x)$
3. *Distributivity*
   a. $\alpha(x_1 + x_2) = \alpha x_1 + \alpha x_2$
   b. $(\alpha_1 + \alpha_2)x = \alpha_1 x + \alpha_2 x$

Linear State-Space Control Systems, by Robert L. Williams II and Douglas A. Lawrence
Copyright © 2007 John Wiley & Sons, Inc.

4. *Identity elements*

   a. *There is a zero vector* $0 \in \mathbb{X}$ *such that* $x + 0 = x$

   b. *For additive and multiplicative identity elements in* $\mathbb{F}$, *denoted* 0 *and* 1, *respectively,*

$$0x = 0 \text{ and } 1x = x \qquad \square$$

Example B.1

$$\mathbb{F}^n = \{x = (x_1, x_2, \ldots, x_n), \quad x_i \in \mathbb{F}, \quad i = 1, \ldots, n\}$$

i.e., the set of all n-tuples with elements in $\mathbb{F}$. When $\mathbb{F}$ is either the real field $\mathbb{R}$ or the complex field $\mathbb{C}$, $\mathbb{R}^n$ and $\mathbb{C}^n$ denote real and complex Euclidean space, respectively. $\qquad \square$

Example B.2

$$\mathbb{F}^{m \times n} = \{A = [a_{ij}], \quad a_{ij} \in \mathbb{F}, \quad i = 1, \ldots, m, \quad j = 1, \ldots, n\}$$

i.e., the set of all $m \times n$ matrices with elements in $\mathbb{F}$. Again, typically $\mathbb{F}$ is either $\mathbb{R}$ or $\mathbb{C}$, yielding the set of real or complex $m \times n$ matrices, respectively. That $\mathbb{C}^{m \times n}$ is a vector space over $\mathbb{C}$ is a direct consequence of the discussion in Section A.2. Note that by stacking columns of an $m \times n$ matrix on top of one another to form a vector, we can identify $\mathbb{F}^{m \times n}$ with $\mathbb{F}^{mn}$. $\qquad \square$

Example B.3 $C[a, b]$, the set of continuous functions $f : [a, b] \to \mathbb{F}$, with vector addition and scalar multiplication defined in a pointwise sense as follows:

$$(f + g)(x) := f(x) + g(x) \quad (\alpha f)(x) := \alpha f(x) \quad \text{for all} \quad x \in [a, b]$$

for all $f, g \in C[a, b]$ and $\alpha \in \mathbb{F}$. $\qquad \square$

Definition B.2   Let $x_1, x_2, \ldots, x_k$ be vectors in $\mathbb{X}$. Their **span** is defined as

$$\text{span}\{x_1, x_2, \ldots, x_k\} \quad := \quad \{x = \alpha_1 x_1 + \alpha_2 x_2 + \cdots + \alpha_k x_k, \quad \alpha_i \in \mathbb{F}\}$$

i.e., *the set of all linear combinations of* $x_1, x_2, \ldots, x_k$. $\qquad \square$

Definition B.3 *A set of vectors $\{x_1, x_2, \ldots, x_k\}$ is **linearly independent** if the relation*

$$\alpha_1 x_1 + \alpha_2 x_2 + \cdots + \alpha_k x_k = 0$$

implies that $\alpha_1 = \alpha_2 = \cdots = \alpha_k = 0$. $\quad\square$

Lemma B.4 *If $\{x_1, x_2, \ldots, x_k\}$ is a linearly independent set of vectors and $x \in \mathrm{span}\{x_1, x_2, \ldots, x_k\}$, then the relation*

$$x = \alpha_1 x_1 + \alpha_2 x_2 + \cdots + \alpha_k x_k$$

is unique.

Proof. Suppose that $x = \beta_1 x_1 + \beta_2 x_2 + \cdots + \beta_k x_k$ is another representation with $\beta_i \neq \alpha_i$ for at least one $i \in \{1, \ldots, k\}$. Then

$$
\begin{aligned}
0 &= x - x \\
&= (\alpha_1 x_1 + \alpha_2 x_2 + \cdots + \alpha_k x_k) - (\beta_1 x_1 + \beta_2 x_2 + \cdots + \beta_k x_k) \\
&= (\alpha_1 - \beta_1)x_1 + (\alpha_2 - \beta_2)x_2 + \cdots + (\alpha_k - \beta_k)x_k
\end{aligned}
$$

By assumption, $\alpha_i - \beta_i \neq 0$ for some i, which contradicts the linear independence hypothesis. $\quad\square$

B.2 SUBSPACES

Definition B.5 *A **linear subspace** $\mathbb{S}$ of a linear vector space $\mathbb{X}$ is a subset of $\mathbb{X}$ that is itself a linear vector space under the vector addition and scalar multiplication defined on $\mathbb{X}$.* $\quad\square$

Definition B.6 *A **basis** for a linear subspace $\mathbb{S}$ is a linearly independent set of vectors $\{x_1, x_2, \ldots, x_k\}$ such that*

$$\mathbb{S} = \mathrm{span}\{x_1, x_2, \ldots, x_k\}$$
$\quad\square$

A basis for $\mathbb{S}$ is not unique; however, all bases for $\mathbb{S}$ have the same number of elements, which defines the *dimension* of the linear subspace $\mathbb{S}$. Since a linear vector space $\mathbb{X}$ can be viewed as a subspace of itself, the concepts of basis and dimension apply to $\mathbb{X}$ as well.

On $\mathbb{R}^3$:

1. $\{0\}$ is a zero-dimensional subspace called the *zero subspace*.
2. Any line through the origin is a one-dimensional subspace, and any nonzero vector on the line is a valid basis.
3. Any plane through the origin is a two-dimensional subspace, and any two noncollinear vectors in the plane form a valid basis.
4. $\mathbb{R}^3$ is a three-dimensional subspace of itself, and any three noncoplanar vectors form a valid basis.

Example B.4 The set of all 2×2 real matrices $\mathbb{R}^{2 \times 2}$ is a four-dimensional vector space, and a valid basis is

$$\left\{ \begin{bmatrix} 1 & 0 \\ 0 & 0 \end{bmatrix}, \begin{bmatrix} 0 & 1 \\ 0 & 0 \end{bmatrix}, \begin{bmatrix} 0 & 0 \\ 1 & 0 \end{bmatrix}, \begin{bmatrix} 0 & 0 \\ 0 & 1 \end{bmatrix} \right\}$$

because

$$A = \begin{bmatrix} a_{11} & a_{12} \\ a_{21} & a_{22} \end{bmatrix}$$

$$= a_{11} \begin{bmatrix} 1 & 0 \\ 0 & 0 \end{bmatrix} + a_{12} \begin{bmatrix} 0 & 1 \\ 0 & 0 \end{bmatrix} + a_{21} \begin{bmatrix} 0 & 0 \\ 1 & 0 \end{bmatrix} + a_{22} \begin{bmatrix} 0 & 0 \\ 0 & 1 \end{bmatrix}$$

The subset of all upper triangular matrices

$$A = \begin{bmatrix} a_{11} & a_{12} \\ 0 & a_{22} \end{bmatrix}$$

is a three-dimensional subspace, and a valid basis is obtained by omitting

$$\begin{bmatrix} 0 & 0 \\ 1 & 0 \end{bmatrix}$$

in the basis for $\mathbb{R}^{2 \times 2}$ given above.

The subset of symmetric matrices $A = A^T$ (so that $a_{12} = a_{21}$) is a three-dimensional subspace, and a valid basis is

$$\left\{ \begin{bmatrix} 1 & 0 \\ 0 & 0 \end{bmatrix}, \begin{bmatrix} 0 & 1 \\ 1 & 0 \end{bmatrix}, \begin{bmatrix} 0 & 0 \\ 0 & 1 \end{bmatrix} \right\} \qquad \square$$

Example B.5 The set of all degree-k polynomials is a $(k+1)$-dimensional subspace of the infinite dimensional vector space $C[a, b]$, and a valid basis for this subspace is:

$$\{1, x, x^2, \ldots, x^k\} \qquad \square$$

B.3 STANDARD BASIS

As we have already noted, a basis for a linear subspace or vector space is not unique. On $\mathbb{R}^n$ or $\mathbb{C}^n$, the *standard basis* $\{e_1, e_2, \ldots, e_n\}$ is defined by

$$e_i = \begin{bmatrix} 0 \\ \vdots \\ 0 \\ 1 \\ 0 \\ \vdots \\ 0 \end{bmatrix} \leftarrow i^{\text{th}} \text{ position} \qquad i = 1, \ldots, n$$

Equivalently, $[e_1 \quad e_2 \quad \cdots \quad e_n]$ forms the $n \times n$ identity matrix I.
On $\mathbb{R}^3$ or $\mathbb{C}^3$,

$$e_1 = \begin{bmatrix} 1 \\ 0 \\ 0 \end{bmatrix} \qquad e_2 = \begin{bmatrix} 0 \\ 1 \\ 0 \end{bmatrix} \qquad e_3 = \begin{bmatrix} 0 \\ 0 \\ 1 \end{bmatrix}$$

For any $x \in \mathbb{R}^3$, we have the unique representation in terms of $\{e_1, e_2, e_3\}$:

$$x = x_1 e_1 + x_2 e_2 + x_3 e_3$$

$$= x_1 \begin{bmatrix} 1 \\ 0 \\ 0 \end{bmatrix} + x_2 \begin{bmatrix} 0 \\ 1 \\ 0 \end{bmatrix} + x_3 \begin{bmatrix} 0 \\ 0 \\ 1 \end{bmatrix}$$

$$= \begin{bmatrix} x_1 \\ x_2 \\ x_3 \end{bmatrix}$$

B.4 CHANGE OF BASIS

Let $\{x_1, x_2, \ldots, x_n\}$ and $\{y_1, y_2, \ldots, y_n\}$ be bases for an n-dimensional linear vector space $\mathbb{X}$ over $\mathbb{F}$. Each basis vector y_j can be uniquely represented in terms of the basis $\{x_1, x_2, \ldots, x_n\}$ as

$$y_j = t_{1j}x_1 + t_{2j}x_2 + \cdots + t_{nj}x_n$$

$$= \sum_{i=1}^{n} t_{ij}x_i$$

$$t_{ij} \in \mathbb{F} \quad \text{for } i, j = 1, \cdots, n$$

Next, let $x \in \mathbb{X}$ be an arbitrary element with unique representation in each basis

$$x = \alpha_1 x_1 + \alpha_2 x_2 + \cdots + \alpha_n x_n$$

$$= \beta_1 y_1 + \beta_2 y_2 + \cdots + \beta_n y_n$$

The n-tuple $(\alpha_1, \alpha_2, \ldots, \alpha_n)$ defines the *coordinates* of the vector x in the basis $\{x_1, x_2, \ldots, x_n\}$, and analogously, the n-tuple $(\beta_1, \beta_2, \ldots, \beta_n)$ defines the *coordinates* of the vector x in the other basis $\{y_1, y_2, \ldots, y_n\}$. We establish a connection between these two coordinate representations by writing

$$x = \beta_1 \left(\sum_{i=1}^{n} t_{i1}x_i \right) + \beta_2 \left(\sum_{i=1}^{n} t_{i2}x_i \right) + \cdots + \beta_n \left(\sum_{i=1}^{n} t_{in}x_i \right)$$

$$= \sum_{j=1}^{n} \beta_j \left(\sum_{i=1}^{n} t_{ij}x_i \right)$$

$$= \sum_{i=1}^{n} \left(\sum_{j=1}^{n} t_{ij}\beta_j \right) x_i$$

$$= \sum_{i=1}^{n} \alpha_i x_i$$

from which it follows that

$$\alpha_i = \sum_{j=1}^{n} t_{ij}\beta_j \qquad i = 1, \cdots, n$$

In matrix form,

$$
\begin{bmatrix} \alpha_1 \\ \alpha_2 \\ \vdots \\ \alpha_n \end{bmatrix} = \begin{bmatrix} t_{11} & t_{12} & \cdots & t_{1n} \\ t_{21} & t_{22} & \cdots & t_{2n} \\ \vdots & \vdots & \ddots & \vdots \\ t_{n1} & t_{n2} & \cdots & t_{nn} \end{bmatrix} \begin{bmatrix} \beta_1 \\ \beta_2 \\ \vdots \\ \beta_n \end{bmatrix}
$$

or, more compactly,

$$
\alpha = T\beta
$$

The matrix T must be invertible (nonsingular), so this relationship can be reversed to obtain

$$
\beta = T^{-1}\alpha
$$

That is, the matrix T allows us to transform the coordinate representation of any vector $x \in \mathbb{X}$ in the basis $\{y_1, y_2, \ldots, y_n\}$ into an equivalent coordinate representation in the basis $\{x_1, x_2, \ldots, x_n\}$. This transformation was defined originally in terms of the unique representation of each basis vector in $\{y_1, y_2, \ldots, y_n\}$ in the basis $\{x_1, x_2, \ldots, x_n\}$. Conversely, the matrix T^{-1} allows us to go from a coordinate representation in the basis $\{x_1, x_2, \ldots, x_n\}$ into an equivalent coordinate representation in the basis $\{y_1, y_2, \ldots, y_n\}$.

Example B.6 Consider the standard basis for $\mathbb{R}^3$, $\{e_1, e_2, e_3\}$, and a second basis $\{y_1, y_2, y_3\}$ defined via

$$
y_1 = (1)e_1 + (-1)e_2 + (0)e_3
$$
$$
y_2 = (1)e_1 + (0)e_2 + (-1)e_3
$$
$$
y_3 = (0)e_1 + (1)e_2 + (0)e_3.
$$

It is customary to instead write

$$
y_1 = \begin{bmatrix} 1 \\ -1 \\ 0 \end{bmatrix} \qquad y_2 = \begin{bmatrix} 1 \\ 0 \\ -1 \end{bmatrix} \qquad y_3 = \begin{bmatrix} 0 \\ 1 \\ 0 \end{bmatrix}
$$

These relationships allow us to specify the transformation matrix relating these two bases as

$$
T = \begin{bmatrix} 1 & 1 & 0 \\ -1 & 0 & 1 \\ 0 & -1 & 0 \end{bmatrix}
$$

We now seek to find the coordinate representation of the vector

$$x = (2)e_1 + (3)e_2 + (8)e_3$$

$$= \begin{bmatrix} 2 \\ 3 \\ 8 \end{bmatrix}$$

$$= \begin{bmatrix} \alpha_1 \\ \alpha_2 \\ \alpha_3 \end{bmatrix}$$

in the basis $\{y_1, y_2, y_3\}$. In our setup we need to solve $\alpha = T\beta$ for β. That is,

$$\begin{bmatrix} \beta_1 \\ \beta_2 \\ \beta_3 \end{bmatrix} = \begin{bmatrix} 1 & 1 & 0 \\ -1 & 0 & 1 \\ 0 & -1 & 0 \end{bmatrix}^{-1} \begin{bmatrix} 2 \\ 3 \\ 8 \end{bmatrix}$$

$$= \begin{bmatrix} 1 & 0 & 1 \\ 0 & 0 & -1 \\ 1 & 1 & 1 \end{bmatrix} \begin{bmatrix} 2 \\ 3 \\ 8 \end{bmatrix}$$

$$= \begin{bmatrix} 10 \\ -8 \\ 13 \end{bmatrix}$$

so that

$$x = (10)y_1 + (-8)y_2 + (13)y_3 \qquad \square$$

B.5 ORTHOGONALITY AND ORTHOGONAL COMPLEMENTS

We focus on the linear vector space $\mathbb{C}^n$ with obvious specialization to $\mathbb{R}^n$.

Definition B.7 *For vectors* $x = (x_1, x_2, \ldots, x_n)$ *and* $y = (y_1, y_2, \ldots, y_n)$

1. *The **inner product** of x and y is defined as*

$$\langle x, y \rangle := x^* y = \sum_{i=1}^{n} \overline{x}_i y_i$$

where the *-superscript denotes conjugate transpose. Note that

$$\langle y, x \rangle = y^* x = (x^* y)^* = \langle x, y \rangle^*. \text{ For } x, y \in \mathbb{R}^n, \langle x, y \rangle := x^T y$$
$$= y^T x = \langle y, x \rangle.$$

2. *Vectors* $x, y \in \mathbb{C}^n$ *are* **orthogonal** *if* $\langle x, y \rangle = 0$.
3. *The* **Euclidian norm** *of* $x \in \mathbb{C}^n$ *is given by*

$$\|x\| = \langle x, x \rangle^{\frac{1}{2}} = \left(\sum_{i=1}^{n} |x_i|^2 \right)^{\frac{1}{2}}$$

4. *A set of vectors* $\{x_1, x_2, \ldots, x_k\}$ *is* **orthogonal** *if* $\langle x_i, x_j \rangle = 0$ *for* $i \neq j$ *and* **orthonormal** *if, in addition,* $\|x_i\| = 1, i = 1, \ldots, k$.

For a subspace $\mathbb{S} \subset \mathbb{C}^n$

1. *A set of vectors* $\{x_1, x_2, \cdots, x_k\}$ *is called an* **orthonormal basis** *for* $\mathbb{S}$ *if it is an orthonormal set and is a basis for* $\mathbb{S}$.
2. *The* **orthogonal complement** *of S is defined as*

$$\mathbb{S}^\perp = \{ y \in \mathbb{C}^n | \langle y, x \rangle = 0 \quad \forall x \in \mathbb{S} \} \qquad \square$$

It follows from the definition that $\mathbb{S}^\perp$ is also a subspace of $\mathbb{C}^n$. Moreover, if $\{x_1, x_2, \ldots, x_k\}$ is a basis for $\mathbb{S}$, then an equivalent though simpler characterization of $\mathbb{S}^\perp$ is

$$\mathbb{S}^\perp = \{ y \in \mathbb{C}^n | \langle y, x_i \rangle = 0; \quad i = 1, \ldots, k \}$$

It also can be shown that $\dim(\mathbb{S}^\perp) = \dim(\mathbb{C}^n) - \dim(\mathbb{S}) = n - k$ and

$$\mathbb{S}^\perp = \text{span}\{y_1, y_2, \ldots, y_{n-k}\}$$

for any linearly independent set of vectors $\{y_1, y_2, \ldots, y_{n-k}\}$ that satisfy

$$\langle y_j, x_i \rangle = 0 \qquad i = 1, \ldots, k \qquad j = 1, \ldots, n - k$$

Example B.7   On $\mathbb{R}^3$,

1. Suppose that $\mathbb{S} = \text{span}\{x_1\}$ with

$$x_1 = \begin{bmatrix} 1 \\ 0 \\ -1 \end{bmatrix}$$

Then $\mathbb{S}^\perp = \text{span}\{y_1, y_2\}$ with

$$y_1 = \begin{bmatrix} 1 \\ 0 \\ 1 \end{bmatrix} \qquad y_2 = \begin{bmatrix} 0 \\ 1 \\ 0 \end{bmatrix}$$

2. Suppose that $\mathbb{S} = \text{span}\{x_1, x_2\}$ with

$$x_1 = \begin{bmatrix} 1 \\ 1 \\ 1 \end{bmatrix} \qquad x_2 = \begin{bmatrix} 0 \\ 1 \\ 1 \end{bmatrix}$$

Then $\mathbb{S}^\perp = \text{span}\{y_1\}$ with

$$y_1 = \begin{bmatrix} 0 \\ -1 \\ 1 \end{bmatrix}$$ $\qquad\qquad\Box$

B.6   LINEAR TRANSFORMATIONS

Definition B.8   *Let* $\mathbb{X}$ *and* $\mathbb{Y}$ *be linear vector spaces over the same field* $\mathbb{F}$. *A transformation* $A : \mathbb{X} \rightarrow \mathbb{Y}$ *is linear if*

$$A(\alpha_1 x_1 + \alpha_2 x_2) = \alpha_1 A x_1 + \alpha_2 A x_2 \qquad \textit{for all } x_1, x_2 \in \mathbb{X}$$
$$\textit{and for all } \alpha_1, \alpha_2 \in \mathbb{F} \qquad\qquad\Box$$

Suppose that $\mathbb{X} = \mathbb{C}^n$ and $\mathbb{Y} \in \mathbb{C}^m$. A linear transformation $A : \mathbb{C}^n \rightarrow \mathbb{C}^m$ is specified in terms of its action on a basis for $\mathbb{C}^n$, represented in terms of a basis for $\mathbb{C}^m$.

Let $\{x_1, x_2, \ldots, x_n\}$ be a basis for $\mathbb{C}^n$ and $\{y_1, y_2, \ldots, y_m\}$ be a basis for $\mathbb{C}^m$. Then $A x_j \in \mathbb{C}^m$, $j = 1, \ldots, n$, has the unique representation

$$A x_j = a_{1j} y_1 + a_{2j} y_2 + \cdots + a_{mj} y_m$$

As we have seen previously, the m-tuple $(a_{1j}, a_{2j}, \ldots, a_{mj})$ defines the coordinates of Ax_j in the basis $\{y_1, y_2, \ldots, y_m\}$.

Next, for every $x \in \mathbb{C}^n$ and $y := Ax$, we have the unique representation in the appropriate basis

$$x = \alpha_1 x_1 + \alpha_2 x_2 + \cdots + \alpha_n x_n$$
$$y = \beta_1 y_1 + \beta_2 y_2 + \cdots + \beta_m y_m$$

Again, the n-tuple $(\alpha_1, \alpha_2, \ldots, \alpha_n)$ defines the coordinates of the vector x in the basis $\{x_1, x_2, \ldots, x_n\}$ and the m-tuple $(\beta_1, \beta_2, \ldots, \beta_m)$ defines the coordinates of the vector y in the basis $\{y_1, y_2, \ldots, y_m\}$.

Putting everything together and using linearity of the transformation A, leads to

$$y = Ax$$

$$= A\left(\sum_{j=1}^{n} \alpha_j x_j\right)$$

$$= \sum_{j=1}^{n} \alpha_j A x_j$$

$$= \sum_{j=1}^{n} \alpha_j \left(\sum_{i=1}^{m} a_{ij} y_i\right)$$

$$= \sum_{i=1}^{m} \left(\sum_{j=1}^{n} a_{ij}\alpha_j\right) y_i$$

Comparing this with the unique representation for $y = Ax$ given previously, we therefore require

$$\beta_i = \sum_{j=1}^{n} a_{ij}\alpha_j \qquad i = 1, \cdots, m$$

or more compactly, in matrix notation,

$$\begin{bmatrix} \beta_1 \\ \beta_2 \\ \vdots \\ \beta_m \end{bmatrix} = \begin{bmatrix} a_{11} & a_{12} & \cdots & a_{1n} \\ a_{21} & a_{22} & \cdots & a_{2n} \\ \vdots & \vdots & \ddots & \vdots \\ a_{m1} & a_{m2} & \cdots & a_{mn} \end{bmatrix} \begin{bmatrix} \alpha_1 \\ \alpha_2 \\ \vdots \\ \alpha_n \end{bmatrix}$$

Thus, with respect to the basis $\{x_1, x_2, \ldots, x_n\}$ for $\mathbb{C}^n$ and the basis $\{y_1, y_2, \ldots, y_m\}$ for $\mathbb{C}^m$, the linear transformation $A : \mathbb{C}^n \rightarrow \mathbb{C}^m$ has the $m \times n$ matrix representation

$$A = \begin{bmatrix} a_{11} & a_{12} & \cdots & a_{1n} \\ a_{21} & a_{22} & \cdots & a_{2n} \\ \vdots & \vdots & \ddots & \vdots \\ a_{m1} & a_{m2} & \cdots & a_{mn} \end{bmatrix}.$$

If different bases are used for $\mathbb{C}^n$ and/or $\mathbb{C}^m$, the *same* linear transformation will have a *different* matrix representation. Often a matrix is used to represent a linear transformation with the implicit understanding that the standard bases for $\mathbb{C}^n$ and $\mathbb{C}^m$ are being used.

Example B.8   Let $A : \mathbb{R}^3 \rightarrow \mathbb{R}^2$ be the linear transformation having the matrix representation with respect to the standard bases for $\mathbb{R}^3$ and $\mathbb{R}^2$ given by

$$A = \begin{bmatrix} 1 & 2 & 3 \\ 4 & 5 & 6 \end{bmatrix}$$

That is,

$$A \begin{bmatrix} 1 \\ 0 \\ 0 \end{bmatrix} = \begin{bmatrix} 1 \\ 4 \end{bmatrix} = (1) \begin{bmatrix} 1 \\ 0 \end{bmatrix} + (4) \begin{bmatrix} 0 \\ 1 \end{bmatrix}$$

$$A \begin{bmatrix} 0 \\ 1 \\ 0 \end{bmatrix} = \begin{bmatrix} 2 \\ 5 \end{bmatrix} = (2) \begin{bmatrix} 1 \\ 0 \end{bmatrix} + (5) \begin{bmatrix} 0 \\ 1 \end{bmatrix}$$

$$A \begin{bmatrix} 0 \\ 0 \\ 1 \end{bmatrix} = \begin{bmatrix} 3 \\ 6 \end{bmatrix} = (3) \begin{bmatrix} 1 \\ 0 \end{bmatrix} + (6) \begin{bmatrix} 0 \\ 1 \end{bmatrix}.$$

Suppose instead that we consider the following basis for $\mathbb{R}^3$:

$$\left\{ x_1 = \begin{bmatrix} 1 \\ 1 \\ 1 \end{bmatrix}, \quad x_2 = \begin{bmatrix} 0 \\ 2 \\ 2 \end{bmatrix}, \quad x_3 = \begin{bmatrix} 0 \\ 0 \\ 3 \end{bmatrix} \right\}$$

The linear transformation acts on the new basis vectors according to

$$Ax_1 = A\left[(1)\begin{bmatrix}1\\0\\0\end{bmatrix} + (1)\begin{bmatrix}0\\1\\0\end{bmatrix} + (1)\begin{bmatrix}0\\0\\1\end{bmatrix}\right]$$

$$= (1)A\begin{bmatrix}1\\0\\0\end{bmatrix} + (1)A\begin{bmatrix}0\\1\\0\end{bmatrix} + (1)A\begin{bmatrix}0\\0\\1\end{bmatrix}$$

$$= (1)\begin{bmatrix}1\\4\end{bmatrix} + (1)\begin{bmatrix}2\\5\end{bmatrix} + (1)\begin{bmatrix}3\\6\end{bmatrix}$$

$$= \begin{bmatrix}6\\15\end{bmatrix}$$

$$Ax_2 = A\left[(0)\begin{bmatrix}1\\0\\0\end{bmatrix} + (2)\begin{bmatrix}0\\1\\0\end{bmatrix} + (2)\begin{bmatrix}0\\0\\1\end{bmatrix}\right]$$

$$= (0)A\begin{bmatrix}1\\0\\0\end{bmatrix} + (2)A\begin{bmatrix}0\\1\\0\end{bmatrix} + (2)A\begin{bmatrix}0\\0\\1\end{bmatrix}$$

$$= (0)\begin{bmatrix}1\\4\end{bmatrix} + (2)\begin{bmatrix}2\\5\end{bmatrix} + (2)\begin{bmatrix}3\\6\end{bmatrix}$$

$$= \begin{bmatrix}10\\22\end{bmatrix}$$

$$Ax_3 = A\left[(0)\begin{bmatrix}1\\0\\0\end{bmatrix} + (0)\begin{bmatrix}0\\1\\0\end{bmatrix} + (3)\begin{bmatrix}0\\0\\1\end{bmatrix}\right]$$

$$= (0)A\begin{bmatrix}1\\0\\0\end{bmatrix} + (0)A\begin{bmatrix}0\\1\\0\end{bmatrix} + (3)A\begin{bmatrix}0\\0\\1\end{bmatrix}$$

$$= (0)\begin{bmatrix}1\\4\end{bmatrix} + (0)\begin{bmatrix}2\\5\end{bmatrix} + (3)\begin{bmatrix}3\\6\end{bmatrix}$$

$$= \begin{bmatrix}9\\18\end{bmatrix}$$

Thus, with a different basis for $\mathbb{R}^3$, the same linear transformation now has the matrix representation

$$A = \begin{bmatrix} 6 & 10 & 9 \\ 15 & 22 & 18 \end{bmatrix}.$$

The preceding calculations can be cast in matrix terms as

$$\begin{bmatrix} 6 & 10 & 9 \\ 15 & 22 & 18 \end{bmatrix} = \begin{bmatrix} 1 & 2 & 3 \\ 4 & 5 & 6 \end{bmatrix} \begin{bmatrix} 1 & 0 & 0 \\ 1 & 2 & 0 \\ 1 & 2 & 3 \end{bmatrix}$$

which illustrates that the matrix representation for the linear transformation with respect to the new $\{x_1, x_2, x_3\}$ basis for $\mathbb{R}^3$ and the standard basis for $\mathbb{R}^2$ is given by the matrix representation for the linear transformation with respect to the original standard basis for $\mathbb{R}^3$ and the standard basis for $\mathbb{R}^2$ multiplied on the right by a 3×3 matrix characterizing the change of basis on $\mathbb{R}^3$ from the $\{x_1, x_2, x_3\}$ basis to the standard basis $\{e_1, e_2, e_3\}$ as described in Section B.4. If, in addition, we had considered a change of basis on $\mathbb{R}^2$, the original matrix representation for the linear transformation would have been multiplied on the left by a 2×2 matrix relating the two bases for $\mathbb{R}^2$ to yield the new matrix representation with respect to the different bases for both $\mathbb{R}^2$ and $\mathbb{R}^3$. $\qquad\square$

B.7   RANGE AND NULL SPACE

Definition B.9   *For the linear transformation* $A : \mathbb{C}^n \rightarrow \mathbb{C}^m$

1. *The range space or image of A is defined by*

$$\mathbb{R}(A) = \text{Im } A = \{y \in \mathbb{C}^m | \exists x \in \mathbb{C}^n \text{ such that } y = Ax\}.$$

2. *The null space or kernel of A is defined by*

$$\mathbb{N}(A) = \text{Ker } A = \{x \in \mathbb{C}^n | Ax = 0\}. \qquad\square$$

It is a direct consequence of the definitions and linearity of A that Im A is a subspace of $\mathbb{C}^m$ and that Ker A is a subspace of $\mathbb{C}^n$. For instance, it is clear that the zero vector of $\mathbb{C}^m$ is an element of Im A and the zero vector of $\mathbb{C}^n$ is an element of Ker A. Further, letting $\{a_1, a_2, \ldots, a_n\}$

denote the columns of a matrix representation for A, it also follows from the definition that

$$\mathbb{R}(A) = \text{Im } A = \text{span}\{a_1, a_2, \cdots, a_n\}$$

We next let rank(A) denote the dimension of Im A and nullity(A) denote the dimension of Ker A. The rank of a linear transformation leads to the notion of matrix rank, for which the following proposition collects several equivalent characterizations:

Proposition B.10 *The rank of an $m \times n$ matrix A is characterized by any one of the following:*

1. *The maximal number of linearly independent columns in A*
2. *The maximal number of linearly independent rows in A*
3. *The size of the largest nonsingular submatrix that can be extracted from A* □

The last characterization of A needs to be interpreted carefully: rank$(A) = r$ if and only if there is *at least one* nonsingular $r \times r$ submatrix of A and every larger square submatrix of A is singular. Also of interest is the following relationship between rank and nullity:

Proposition B.11 (Sylvester's Law of Nullity)
For the linear transformation $A : \mathbb{C}^n \to \mathbb{C}^m$

$$\text{rank}(A) + \text{nullity}(A) = n \qquad \qquad □$$

The following numerical example illustrates how these important subspaces associated with a linear transformation can be characterized.

Example B.9 Let $A : \mathbb{R}^4 \to \mathbb{R}^3$ be the linear transformation having the matrix representation with respect to the standard bases for $\mathbb{R}^4$ and $\mathbb{R}^3$ given by

$$A = \begin{bmatrix} 0 & -2 & -4 & 6 \\ 0 & 1 & 1 & -1 \\ 0 & -2 & -5 & 8 \end{bmatrix}$$

We seek to find the rank and nullity of A along with bases for Im A and Ker A. A key computational tool is the application of elementary row operations:

1. Multiply any row by a nonzero scalar,
2. Interchange any two rows, and
3. Add a scalar multiple of any row to another row.

to yield the so-called row-reduced echelon form of the matrix A, denoted A_R. For this example, the steps proceed as follows:

Step 1:

$$\begin{bmatrix} 0 & -2 & -4 & 6 \\ 0 & 1 & 1 & -1 \\ 0 & -2 & -5 & 8 \end{bmatrix} \quad \text{row 1} \leftarrow (-\tfrac{1}{2}) \times \text{row 1}$$

$$\Rightarrow \begin{bmatrix} 0 & 1 & 2 & -3 \\ 0 & 1 & 1 & -1 \\ 0 & -2 & -5 & 8 \end{bmatrix}$$

Step 2:

$$\begin{bmatrix} 0 & 1 & 2 & -3 \\ 0 & 1 & 1 & -1 \\ 0 & -2 & -5 & 8 \end{bmatrix} \quad \begin{array}{l} \text{row 2} \leftarrow (-1) \times \text{row 1} + \text{row 2} \\ \text{row 3} \leftarrow (2) \times \text{row 1} + \text{row 3} \end{array}$$

$$\Rightarrow \begin{bmatrix} 0 & 1 & 2 & -3 \\ 0 & 0 & -1 & 2 \\ 0 & 0 & -1 & 2 \end{bmatrix}$$

Step 3:

$$\begin{bmatrix} 0 & 1 & 2 & -3 \\ 0 & 0 & -1 & 2 \\ 0 & 0 & -1 & 2 \end{bmatrix} \quad \text{row 2} \leftarrow (-1) \times \text{row 2}$$

$$\Rightarrow \begin{bmatrix} 0 & 1 & 2 & -3 \\ 0 & 0 & 1 & -2 \\ 0 & 0 & -1 & 2 \end{bmatrix}$$

Step 4:

$$\begin{bmatrix} 0 & 1 & 2 & -3 \\ 0 & 0 & 1 & -2 \\ 0 & 0 & -1 & 2 \end{bmatrix} \quad \begin{array}{l} \text{row 1} \leftarrow (-2) \times \text{row 2} + \text{row 1} \\ \\ \text{row 3} \leftarrow (1) \times \text{row 2} + \text{row 3} \end{array}$$

$$\Rightarrow \begin{bmatrix} 0 & 1 & 0 & 1 \\ 0 & 0 & 1 & -2 \\ 0 & 0 & 0 & 0 \end{bmatrix} = A_R$$

We then have

$$\text{rank}(A) = \text{the number of linearly independent columns in } A \text{ or } A_R$$
$$= \text{the number of linearly independent rows in } A \text{ or } A_R$$
$$= \text{the number of nonzero rows in } A_R$$
$$= 2$$

By Sylvester's law of nullity:

$$\text{nullity}(A) = n - \text{rank}(A)$$
$$= 4 - 2$$
$$= 2$$

A basis for Im A can be formed by any $\text{rank}(A) = 2$ linearly independent columns of A. (A_R cannot be used here because elementary row operations affect the image.) Possible choices are

$$\left\{ \begin{bmatrix} -2 \\ 1 \\ -2 \end{bmatrix}, \begin{bmatrix} -4 \\ 1 \\ -5 \end{bmatrix} \right\} \quad \left\{ \begin{bmatrix} -2 \\ 1 \\ -2 \end{bmatrix}, \begin{bmatrix} 6 \\ -1 \\ 8 \end{bmatrix} \right\}, \quad \left\{ \begin{bmatrix} -4 \\ 1 \\ -5 \end{bmatrix}, \begin{bmatrix} 6 \\ -1 \\ 8 \end{bmatrix} \right\}$$

Linear independence of each vector pair can be verified by checking that the associated 3×2 matrix has a nonsingular 2×2 submatrix.

A basis for Ker A can be formed by any $\text{nullity}(A) = 2$ linearly independent solutions to the homogeneous matrix equation $Ax = 0$. Since the solution space in not affected by elementary row operations, instead we may seek to characterize linearly independent solutions to $A_R x = 0$. Writing

$$\begin{bmatrix} 0 & 1 & 0 & 1 \\ 0 & 0 & 1 & -2 \\ 0 & 0 & 0 & 0 \end{bmatrix} \begin{bmatrix} x_1 \\ x_2 \\ x_3 \\ x_4 \end{bmatrix} = \begin{bmatrix} x_2 + x_4 \\ x_3 - 2x_4 \\ 0 \end{bmatrix} = \begin{bmatrix} 0 \\ 0 \\ 0 \end{bmatrix}$$

we must satisfy $x_2 = -x_4$ and $x_3 = 2x_4$, with x_1 and x_4 treated as free parameters. The combination $x_1 = 1, x_4 = 0$ yields $x_2 = x_3 = 0$, and we readily see that

$$\begin{bmatrix} 1 \\ 0 \\ 0 \\ 0 \end{bmatrix} \in \text{Ker } A$$

The combination $x_1 = 0, x_4 = 1$ yields $x_2 = -1, x_3 = 2$ which gives

$$\begin{bmatrix} 0 \\ -1 \\ 2 \\ 1 \end{bmatrix} \in \text{Ker } A$$

The set

$$\left\{ \begin{bmatrix} 1 \\ 0 \\ 0 \\ 0 \end{bmatrix}, \begin{bmatrix} 0 \\ -1 \\ 2 \\ 1 \end{bmatrix} \right\}$$

is clearly linearly independent by virtue of our choices for x_1 and x_4 because the associated 4×2 matrix has the 2×2 identity matrix as the submatrix obtained by extracting the first and fourth rows. This set therefore qualifies as a basis for Ker A. $\square$

Lemma B.12   *For the linear transformation $A: \quad \mathbb{C}^n \to \mathbb{C}^m$*

1. $[\text{Im } A]^{\perp} = \text{Ker } A^*$
2. $[\text{Ker } A]^{\perp} = \text{Im } A^*$ $\qquad\qquad\qquad\qquad\qquad\qquad\qquad\qquad$ $\square$

It is a worthwhile exercise to prove the first part of Lemma B.12, which asserts the equality of two subspaces. In general, the equality of two subspaces $\mathbb{S} = \mathbb{T}$ can be verified by demonstrating the pair of containments $\mathbb{S} \subset \mathbb{T}$ and $\mathbb{T} \subset \mathbb{S}$. To show a containment relationship, say, $\mathbb{S} \subset \mathbb{T}$, it is enough to show that an arbitrary element of $\mathbb{S}$ is also an element of $\mathbb{T}$.

We first show that $[\text{Im } A]^{\perp} \subset \text{Ker } A^*$. Let y be an arbitrary element of $[\text{Im } A]^{\perp}$ so that, by definition, $\langle y, z \rangle = 0$ for all $z \in \text{Im } A$. Consequently $\langle y, Ax \rangle = 0$ for all $x \in \mathbb{C}^n$, from which we conclude that

$$\langle A^*y, x \rangle = (A^*y)^*x = y^*Ax = \langle y, Ax \rangle = 0$$

Since the only vector orthogonal to every $x \in \mathbb{C}^n$ is the zero vector, we must have $A^*y = 0$, that is, $y \in \text{Ker } A^*$. The desired containment $[\text{Im } A]^{\perp} \subset \text{Ker } A^*$ follows because $y \in [\text{Im } A]^{\perp}$ was arbitrary.

We next show that Ker $A^* \subset [\text{Im } A]^\perp$. Let y be an arbitrary element of Ker A^* so that $A^*y = 0$. Since the zero vector is orthogonal to every vector in $\mathbb{C}^n$, $\langle A^*y, x \rangle = 0$, for all $x \in \mathbb{C}^n$, from which we conclude that

$$\langle y, Ax \rangle = y^*Ax = (A^*y)^*x = \langle A^*y, x \rangle = 0$$

Since for every $z \in \text{Im } A$, there must exist some $x \in \mathbb{C}^n$ such that $z = Ax$, it follows that $\langle y, z \rangle = 0$, that is, $y \in [\text{Im } A]^\perp$. The desired containment Ker $A^* \subset [\text{Im } A]^\perp$ follows because $y \in$ Ker A^* was arbitrary. Having established the required subspace containments, we may conclude that $[\text{Im } A]^\perp = $ Ker A^*. $\qquad\qquad\qquad\qquad\qquad\qquad\qquad \square$

Note that since the orthogonal complements satisfy $\mathbb{S} = \mathbb{T}$ if and only if $\mathbb{S}^\perp = \mathbb{T}^\perp$ and $(\mathbb{S}^\perp)^\perp = \mathbb{S}$, the second subspace identity in Proposition B.4 is equivalent to $[\text{Im} A^*]^\perp = $ Ker A. This, in turn, follows by applying the first subspace identity in Proposition B.4 to the linear transformation $A^* : \mathbb{C}^m \to \mathbb{C}^n$.

B.8 EIGENVALUES, EIGENVECTORS, AND RELATED TOPICS

Eigenvalues and Eigenvectors

Definition B.13 *For a matrix $A \in \mathbb{C}^{n \times n}$*

1. *The **characteristic polynomial** of A is defined as*

$$|\lambda I - A| = \lambda^n + a_{n-1}\lambda^{n-1} + \cdots + a_1\lambda + a_0$$

2. *The **eigenvalues** of A are the n roots[1] of the characteristic equation*

$$|\lambda I - A| = 0$$

3. *The **spectrum** of A is its set of eigenvalues, denoted*

$$\sigma(A) = \{\lambda_1, \lambda_2, \ldots, \lambda_n\}$$

4. *For each $\lambda_i \in \sigma(A)$, any nonzero vector $v \in \mathbb{C}^n$ that satisfies*

$$(\lambda_i I - A)v = 0 \quad \text{equivalently} \quad Av = \lambda_i v$$

*is called a **right eigenvector** of A associated with the eigenvalue λ_i.*

---

[1]As indicated, the characteristic polynomial of an $n \times n$ matrix is guaranteed to be a monic degree-n polynomial that, by the fundamental theorem of algebra, must have n roots. If A is real, the roots still may be complex but must occur in conjugate pairs.

5. *For each* $\lambda_i \in \sigma(A)$, *any nonzero vector* $w \in \mathbb{C}^n$ *that satisfies*

$$w^*(\lambda_i I - A) = 0 \quad \text{equivalently} \quad w^* A = \lambda_i w^*$$

*is called a **left eigenvector** of A associated with the eigenvalue* λ_i.

For each eigenvalue λ_i of A, by definition, $|\lambda_i I - A| = 0$, so that $\lambda_i I - A$ is a singular matrix and therefore has a nontrivial null space. Right eigenvectors, viewed as nontrivial solutions to the homogeneous equation $(\lambda_i I - A)v = 0$ are therefore nonzero vectors lying in $\text{Ker}(\lambda_i I - A)$. As such, if v is an eigenvector of A associated the eigenvalue λ_i, then so is any nonzero scalar multiple of v.

By taking the conjugate transpose of either identity defining a left eigenvector w associated with the eigenvalue λ_i, we see that w can be interpreted as a right eigenvector of A^* associated with the conjugate eigenvalue $\overline{\lambda}_i$.

Example B.10 For

$$A = \begin{bmatrix} 1 & 0 & 0 \\ 2 & 3 & -1 \\ 2 & 2 & 1 \end{bmatrix}$$

1. The characteristic polynomial of A can be obtained by expanding $|\lambda I - A|$ about the first row to yield

$$|\lambda I - A| = \begin{bmatrix} \lambda - 1 & 0 & 0 \\ -2 & \lambda - 3 & 1 \\ -2 & -2 & \lambda - 1 \end{bmatrix}$$

$$= (\lambda - 1)[(\lambda - 3)(\lambda - 1) - (-2)(1)]$$

$$= (\lambda - 1)[\lambda^2 - 4\lambda + 5]$$

$$= \lambda^3 - 5\lambda^2 + 9\lambda - 5$$

2. Factoring the characteristic polynomial

$$|\lambda I - A| = (\lambda - 1)(\lambda - 2 - j)(\lambda - 2 + j)$$

indicates that the eigenvalues of A are $\lambda_1 = 1$, $\lambda_2 = 2 + j$, and $\lambda_3 = 2 - j$. Note that λ_2 and λ_3 are complex but form a conjugate pair.

3. The spectrum of A is simply

$$\sigma(A) = \{\lambda_1, \lambda_2, \lambda_3\} = \{1, 2 + j, 2 - j\}$$

4. For the eigenvalue $\lambda_1 = 1$,

$$(\lambda_1 I - A) = \begin{bmatrix} 0 & 0 & 0 \\ -2 & -2 & 1 \\ -2 & -2 & 0 \end{bmatrix}$$

which, by applying elementary row operations, yields the row reduced echelon form

$$(\lambda_1 I - A)_R = \begin{bmatrix} 1 & 1 & 0 \\ 0 & 0 & 1 \\ 0 & 0 & 0 \end{bmatrix}$$

from which we see that

$$v_1 = \begin{bmatrix} 1 \\ -1 \\ 0 \end{bmatrix}$$

is a nonzero vector lying in $\mathrm{Ker}(\lambda_1 I - A)$ and therefore is a right eigenvector corresponding to $\lambda_1 = 1$. For the eigenvalue $\lambda_2 = 2 + j$,

$$(\lambda_2 I - A) = \begin{bmatrix} 1+j & 0 & 0 \\ -2 & -1+j & 1 \\ -2 & -2 & 1+j \end{bmatrix}$$

which has row reduced echelon form

$$(\lambda_2 I - A)_R = \begin{bmatrix} 1 & 0 & 0 \\ 0 & 1 & -\dfrac{1}{2} - j\dfrac{1}{2} \\ 0 & 0 & 0 \end{bmatrix}$$

and therefore,

$$v_2 = \begin{bmatrix} 0 \\ \dfrac{1}{2} + j\dfrac{1}{2} \\ 1 \end{bmatrix}$$

is a right eigenvector corresponding to $\lambda_2 = 2 + j$. Finally, since $\lambda_3 = \bar{\lambda}_2 = 2 - j$, we can take as a corresponding eigenvector

$$v_3 = \bar{v}_2 = \begin{bmatrix} 0 \\ \dfrac{1}{2} - j\dfrac{1}{2} \\ 1 \end{bmatrix}$$

5. Proceeding in an analogous fashion for A^T and, $\bar{\lambda}_1 = \lambda_1, \bar{\lambda}_2 = \lambda_3$, and $\bar{\lambda}_3 = \lambda_2$, we see that associated left eigenvectors are given by

$$w_1 = \begin{bmatrix} 1 \\ 0 \\ 0 \end{bmatrix} \qquad w_2 = \begin{bmatrix} 1 \\ 1 \\ -\frac{1}{2} - j\frac{1}{2} \end{bmatrix}$$

$$w_3 = \overline{w}_2 = \begin{bmatrix} 1 \\ 1 \\ -\frac{1}{2} + j\frac{1}{2} \end{bmatrix} \qquad\qquad \square$$

Interesting situations arise in the search for eigenvectors when the matrix A has nondistinct or repeated eigenvalues. To explore further, we first require

Definition B.14 *Let $d \leq n$ denote the number of distinct eigenvalues of A, denoted $\{\lambda_1, \lambda_2, \ldots, \lambda_d\}$. Upon factoring the characteristic polynomial of A accordingly,*

$$\det(\lambda_i I - A) = (\lambda - \lambda_1)^{m_1}(\lambda - \lambda_2)^{m_2} \cdots (\lambda - \lambda_d)^{m_d}$$

*m_i denotes the **algebraic multiplicity** of the eigenvalue λ_i for $i = 1, \ldots, d$. The **geometric multiplicity** of the eigenvalue λ_i is defined as*

$$n_i = \text{nullity}(\lambda_i I - A) = \dim \text{Ker}(\lambda_i I - A)$$

for $i = 1, \ldots, d$. $\qquad\qquad \square$

As a consequence, for each distinct eigenvalue λ_i, $i = 1, \ldots, d$, there are n_i linearly independent solutions to the homogeneous equation $(\lambda_i I - A)v = 0$, which we denote by $\{v_1^i, v_2^i, \ldots, v_{n_i}^i\}$. This set of vectors, in turn, forms a so-called *eigenbasis* for the *eigenspace* $\text{Ker}(\lambda_i I - A)$ associated with the eigenvalue λ_i.

The following facts establish some useful relationships.

Proposition B.15 *For $i = 1, \ldots, d$, $1 \leq n_i \leq m_i$* $\qquad\qquad \square$

Proposition B.16 *Eigenbases associated with different eigenvalues are linearly independent.* $\qquad\qquad \square$

We conclude this subsection with an illustrative example.

Example B.11 Consider the following four 3×3 matrices:

$$A_1 = \begin{bmatrix} \mu & 1 & 0 \\ 0 & \mu & 1 \\ 0 & 0 & \mu \end{bmatrix} \quad A_2 = \begin{bmatrix} \mu & 0 & 0 \\ 0 & \mu & 1 \\ 0 & 0 & \mu \end{bmatrix}$$

$$A_3 = \begin{bmatrix} \mu & 1 & 0 \\ 0 & \mu & 0 \\ 0 & 0 & \mu \end{bmatrix} \quad A_4 = \begin{bmatrix} \mu & 0 & 0 \\ 0 & \mu & 0 \\ 0 & 0 & \mu \end{bmatrix}$$

in which $\mu \in \mathbb{C}$ is a parameter. Each matrix has the characteristic polynomial $(\lambda - \mu)^3$, which indicates that each matrix has one distinct eigenvalue $\lambda_1 = \mu$ with associated algebraic multiplicity $m_1 = 3$. Next, with

$$\mu I - A_1 = \begin{bmatrix} 0 & -1 & 0 \\ 0 & 0 & -1 \\ 0 & 0 & 0 \end{bmatrix} \quad \mu I - A_2 = \begin{bmatrix} 0 & 0 & 0 \\ 0 & 0 & -1 \\ 0 & 0 & 0 \end{bmatrix}$$

$$\mu I - A_3 = \begin{bmatrix} 0 & -1 & 0 \\ 0 & 0 & 0 \\ 0 & 0 & 0 \end{bmatrix} \quad \mu I - A_4 = \begin{bmatrix} 0 & 0 & 0 \\ 0 & 0 & 0 \\ 0 & 0 & 0 \end{bmatrix}$$

we see that the eigenvalue $\lambda_1 = \mu$ has geometric multiplicity $n_1 = 1$ for A_1, $n_1 = 2$ for A_2, $n_1 = 2$ for A_3, and $n_1 = 3$ for A_4. Moreover, for each matrix, the eigenvalue $\lambda_1 = \mu$ has an eigenbasis defined in terms of the standard basis for $\mathbb{R}^3$

$$\{e_1\} \qquad \{e_1, e_2\} \qquad \{e_1, e_3\} \qquad \{e_1, e_2, e_3\}$$

for A_1, A_2, A_3, and A_4, respectively. This example indicates that the geometric multiplicity can cover the entire range from 1 through the algebraic multiplicity. In addition, we see by comparing A_2 and A_3 that although the geometric multiplicity is 2 in each case, the structural difference between the matrices leads to a different eigenspace. □

Similarity Transformations and Diagonalization

Definition B.17 *Matrices* $A, B \in \mathbb{C}^{n \times n}$ *are said to be* **similar** *if there is a nonsingular matrix* $T \in \mathbb{C}^{n \times n}$ *for which*

$$B = T^{-1}AT \qquad \textit{equivalently} \qquad A = TBT^{-1} \qquad\qquad \square$$

In this case, T is called a *similarity transformation*. A fundamental relationship between similar matrices is the following.

Proposition B.18   If A and B are similar matrices, then they have the same eigenvalues.   □

This result is proven by using determinant properties to show that similar matrices have identical characteristic polynomials.

A diagonal matrix necessarily displays its eigenvalues on the main diagonal. For a general matrix A, *diagonalization* refers to the process of constructing a similarity transformation yielding a diagonal matrix that, by virtue of the preceding proposition, displays the eigenvalues of A. The ability to diagonalize the matrix A via similarity transformation is closely connected to its underlying *eigenstructure*.

To investigate further, note that, by definition, the distinct eigenvalues of $A \in \mathbb{C}^{n \times n}$ have associated algebraic multiplicities that satisfy

$$m_1 + m_2 + \cdots + m_d = n$$

because the characteristic polynomial of A has degree n. On the other hand, the geometric multiplicities indicate that we can find a total of $n_1 + n_2 + \cdots + n_d$ eigenvectors associated with the distinct eigenvalues $\{\lambda_1, \lambda_2, \ldots, \lambda_d\}$, written

$$\{v_1^1, v_2^1, \ldots, v_{n_1}^1, \quad v_1^2, v_2^2, \ldots, v_{n_2}^2, \quad \cdots, \quad v_1^d, v_2^d, \ldots, v_{n_d}^d\}$$

By Proposition B.16, this is a linearly independent set. The relationships $Av_i^j = \lambda_i v_i^j$, for $i = 1, \ldots, d$ and $j = 1, \ldots, n_i$, can be packaged into

$$AT = T\Lambda$$

in which T is the $n \times (n_1 + n_2 + \cdots + n_d)$ matrix whose columns are the complete set of eigenvectors, and Λ is a diagonal $(n_1 + n_2 + \cdots + n_d) \times (n_1 + n_2 + \cdots + n_d)$ matrix given by

$$\Lambda = \mathrm{diag}\left[\underbrace{\lambda_1, \cdots, \lambda_1}_{n_1 \text{ times}}, \quad \underbrace{\lambda_2, \cdots, \lambda_2}_{n_2 \text{ times}}, \quad \cdots \quad \underbrace{\lambda_d, \cdots, \lambda_d}_{n_d \text{ times}} \right]$$

In the event that

$$n_1 + n_2 + \cdots + n_d = m_1 + m_2 + \cdots + m_d = n$$

which is possible when and only when $n_i = m_i, i = 1, \ldots, d$, the matrices T and Λ are both $n \times n$, and again by Proposition B.16, T is nonsingular because it is square and has linearly independent columns. As an immediate consequence, we have

$$\Lambda = T^{-1}AT$$

so the matrix T we have constructed serves as a similarity transformation that diagonalizes A. We have argued that the existence of a total of n linearly independent eigenvectors is sufficient to diagonalize A, and we have explicitly constructed a suitable similarity transformation. This condition turns out to be necessary as well as sufficient.

Theorem B.19 *A matrix $A \in \mathbb{C}^{n \times n}$ is diagonalizable via similarity transformation if and only if A has a total of n linearly independent eigenvectors, equivalently, the geometric multiplicity equals the algebraic multiplicity for each distinct eigenvalue.* □

Based on this, the following corollary provides an easily verified sufficient condition for diagonalizability:

Corollary B.20 *A matrix $A \in \mathbb{C}^{n \times n}$ is diagonalizable via similarity transformation if it has n distinct eigenvalues.* □

If A has n distinct eigenvalues, then $d = n$ and $m_i = n_i = 1$ for $i = 1, \ldots, n$. This guarantees the existence of n linearly independent eigenvectors from which a diagonalizing similarity transformation is explicitly constructed.

Jordan Canonical Form

Whereas not every square matrix can be diagonalized via a similarity transformation, every square matrix can be transformed to its *Jordan canonical form* defined as follows: We begin by defining a *Jordan block matrix* of size $k \times k$:

$$J_k(\lambda) = \begin{bmatrix} \lambda & 1 & 0 & \cdots & 0 \\ 0 & \lambda & 1 & \cdots & 0 \\ 0 & 0 & \lambda & \ddots & 0 \\ \vdots & \vdots & \vdots & \ddots & 1 \\ 0 & 0 & 0 & \cdots & \lambda \end{bmatrix}$$

displaying the parameter λ on the main diagonal and 1s on the first super-diagonal. An $n \times n$ *Jordan matrix* is a block diagonal matrix constructed from Jordan block matrices:

$$J = \begin{bmatrix} J_{k_1}(\lambda_1) & 0 & \cdots & 0 \\ 0 & J_{k_2}(\lambda_2) & \cdots & 0 \\ \vdots & \vdots & \ddots & \\ 0 & 0 & \cdots & J_{k_r}(\lambda_r) \end{bmatrix} \quad \text{with} \quad k_1 + k_2 + \cdots + k_r = n$$

Note that if each $k_i = 1$ and $r = n$, then J is a diagonal matrix.

Proposition B.21 *Let $A \in \mathbb{C}^{n \times n}$ have d distinct eigenvalues $\lambda_1, \lambda_2, \ldots, \lambda_d$ with associated algebraic and geometric multiplicities $m_1, m_2, \ldots, m_d$ and $n_1, n_2, \ldots, n_d$, respectively. There exists a similarity transformation T yielding a Jordan matrix*

$$J = T^{-1} A T$$

in which there are n_i Jordan blocks associated with the eigenvalue λ_i, and the sum of the sizes of these blocks equals the algebraic multiplicity m_i, for $i = 1, \ldots, d$. The matrix J is unique up to a reordering of the Jordan blocks and defines the Jordan canonical form of A. □

Note that the algebraic and geometric multiplicities alone do not completely determine the Jordan canonical form. For example, the Jordan matrices

$$J_1 = \left[\begin{array}{ccc|cc} \lambda & 1 & 0 & 0 & 0 \\ 0 & \lambda & 1 & 0 & 0 \\ 0 & 0 & \lambda & 0 & 0 \\ \hline 0 & 0 & 0 & \lambda & 1 \\ 0 & 0 & 0 & 0 & \lambda \end{array} \right] \quad J_2 = \left[\begin{array}{cccc|c} \lambda & 1 & 0 & 0 & 0 \\ 0 & \lambda & 1 & 0 & 0 \\ 0 & 0 & \lambda & 1 & 0 \\ 0 & 0 & 0 & \lambda & 0 \\ \hline 0 & 0 & 0 & 0 & \lambda \end{array} \right]$$

each could represent the Jordan canonical form of a matrix having a single distinct eigenvalue λ with algebraic multiplicity 5 and geometric multiplicity 2. An interesting special case occurs when the geometric multiplicity equals the algebraic multiplicity for a particular eigenvalue. In this case, each of the Jordan blocks associated with this eigenvalue is scalar. Conversely, if each of the Jordan blocks associated with a particular eigenvalue is scalar, then the geometric and algebraic multiplicities for this eigenvalue must be equal. If the geometric multiplicity equals the algebraic multiplicity for each distinct eigenvalue, then the Jordan canonical form is diagonal. This is consistent with Theorem B.19.

Cayley-Hamilton Theorem

Theorem B.22   *For any matrix $A \in \mathbb{C}^{n \times n}$ with characteristic polynomial*

$$|\lambda I - A| = \lambda^n + a_{n-1}\lambda^{n-1} + \cdots + a_1\lambda + a_0$$

there holds

$$A^n + a_{n-1}A^{n-1} + \cdots + a_1 A + a_0 I = 0 \qquad (n \times n) \qquad \square$$

By definition, eigenvalues of A are roots of the associated (scalar) characteristic equation. Loosely speaking, the Cayley-Hamilton theorem asserts that the matrix A is itself a root of a matrix version of its characteristic equation. This is not difficult to verify when A is diagonalizable, although the result still holds for nondiagonalizable matrices.

Example B.12   For the matrix A studied in Example B.10, the characteristic polynomial was found to be

$$\lambda^3 - 5\lambda^2 + 9\lambda - 5$$

Then

$$A^3 - 5A^2 + 9A - 5I = \begin{bmatrix} 1 & 0 & 0 \\ 12 & 13 & -11 \\ 22 & 22 & -9 \end{bmatrix} - 5 \cdot \begin{bmatrix} 1 & 0 & 0 \\ 6 & 7 & -4 \\ 8 & 8 & -1 \end{bmatrix}$$

$$+ 9 \cdot \begin{bmatrix} 1 & 0 & 0 \\ 2 & 3 & -1 \\ 2 & 2 & 1 \end{bmatrix} - 5 \cdot \begin{bmatrix} 1 & 0 & 0 \\ 0 & 1 & 0 \\ 0 & 0 & 1 \end{bmatrix}$$

$$= \begin{bmatrix} 0 & 0 & 0 \\ 0 & 0 & 0 \\ 0 & 0 & 0 \end{bmatrix} \qquad \square$$

B.9 NORMS FOR VECTORS AND MATRICES

Definition B.23   A *vector norm* on $\mathbb{C}^n$ is any real-valued function $\| \cdot \|$ that satisfies

Positive Definiteness: $\|x\| \geq 0$ for all $x \in \mathbb{C}^n$ and $\|x\| = 0$ if and
   only if $x = 0 \in \mathbb{C}^n$
Triangle Inequality: $\|x + y\| \leq \|x\| + \|y\|$ for all $x, y \in \mathbb{C}^n$
Homogeneity: $\|\alpha x\| = |\alpha| \|x\|$ for all $x \in \mathbb{C}^n$ and $\alpha \in \mathbb{C}$

$\qquad\qquad\qquad\qquad\qquad\qquad\qquad\qquad\qquad\qquad\qquad\qquad\square$

A special class of norms, called $p-$ norms, is defined by

$$\|x\|_p = (|x_1|^p + |x_2|^p + \cdots + |x_n|^p)^{\frac{1}{p}} \qquad \text{for all } p \in [1, \infty)$$
$$\|x\|_\infty = \max_{1 \leq i \leq n} |x_i|$$

Most common of these are $\| \cdot \|_1, \| \cdot \|_\infty$, and

$$\|x\|_2 = (|x_1|^2 + |x_2|^2 + \cdots + |x_n|^2)^{\frac{1}{2}}$$
$$= (x^*x)^{\frac{1}{2}}$$

which is the familiar Euclidean norm. These $p-$ norms satisfy the so-called Hölder inequality

$$|x^*y| \leq \|x\|_p \|y\|_q \text{ for all } x, y \in \mathbb{C}^n \text{ and for all } p, q$$

$$\text{such that } \frac{1}{p} + \frac{1}{q} = 1$$

The special case $p = q = 2$ yields the familiar Cauchy-Schwarz inequality:

$$|x^*y| \leq \|x\|_2 \|y\|_2 \text{ for all } x, y \in \mathbb{C}^n$$

For a linear transformation $A : \mathbb{C}^n \to \mathbb{C}^m$ represented by a matrix $A \in \mathbb{C}^{m \times n}$, we define:

Definition B.24   A *matrix norm* on $\mathbb{C}^{n \times m}$ is any real-valued function $\| \cdot \|$ that satisfies

Positive Definiteness: $\|A\| \geq 0$ for all $A \in \mathbb{C}^{m \times n}$ and $\|A\| = 0$ if
   and only if $A = 0 \in \mathbb{C}^{m \times n}$

Triangle Inequality: $\|A + B\| \leq \|A\| + \|B\|$ for all $A, B \in \mathbb{C}^{m \times n}$

Homogeneity: $\|\alpha A\| = |\alpha| \|A\|$ for all $A \in \mathbb{C}^{m \times n}$ and $\alpha \in \mathbb{C}$
   $\square$

Corresponding to any pair of vector norms on $\mathbb{C}^n$ and $\mathbb{C}^m$, respectively, we define an associated *induced matrix norm* according to

$$\|A\| = \sup_{x \neq 0} \frac{\|Ax\|}{\|x\|}$$

in which sup stands for *supremum*, or least upper bound, here taken over all nonzero vectors in $\mathbb{C}^n$. It can be shown that this definition is equivalent to

$$\|A\| = \max_{\|x\|=1} \|Ax\|$$

As a direct consequence of the definition, any induced matrix norm satisfies

$$\|Ax\| \leq \|A\| \|x\| \quad \text{for all} \quad x \in \mathbb{C}^n$$

The class of $p-$ norms for vectors define a class of induced $p-$ norms for matrices via the preceding definition. For $p = 1, 2, \infty$, the induced matrix norms can be computed using

$$\|A\|_1 = \max_{1 \leq j \leq n} \sum_{i=1}^{m} |a_{ij}|$$

$$\|A\|_2 = (\lambda_{\max}(A^T A))^{\frac{1}{2}}$$

$$\|A\|_\infty = \max_{1 \leq i \leq m} \sum_{j=1}^{n} |a_{ij}|$$

The induced matrix $2-$ norm is often referred to as the *spectral norm*.

Not every matrix norm is induced by a vector norm. An example is the so-called *Frobenius norm* defined by

$$\|A\|_F = \left(\sum_{i=1}^{m} \sum_{j=1}^{n} |a_{ij}|^2 \right)^{\frac{1}{2}}$$

This can be seen to correspond to the Euclidean vector norm of the $(nm) \times 1$-dimensional vector obtained by stacking the columns of A on top of one another.

CONTINUING MATLAB EXAMPLE M-FILE

Chapter 1 introduced the Continuing MATLAB Example, based on the single-input, single-output rotational mechanical system of Figure 1.9. The input is torque $\tau(t)$, and the output is the angular displacement $\theta(t)$. This example was revisited each chapter, illustrating the important topics in the context of one familiar example for which most computations can be done by hand to check the MATLAB results. MATLAB code segments were presented in each chapter to demonstrate important calculations; in general, each chapter's code did not stand alone but required MATLAB code from previous chapters to execute properly.

This appendix presents the entire m-file for the Continuing MATLAB Example, from Chapter 1 through Chapter 9. No new code is given; rather, the complete code is listed here for the convenience of the student.

```
%- - - - - - - - - - - - - - - - - - - - - - - - - - - - - - - - - - - - -
% Continuing MATLAB Example m-file
% Chapter 1 through Chapter 9
%       Dr. Bob Williams

%- - - - - - - - - - - - - - - - - - - - - - - - - - - - - - - - - - - - -
%   Chapter 1. State-Space Description
%- - - - - - - - - - - - - - - - - - - - - - - - - - - - - - - - - - - - -

J  = 1;
b  = 4;
kR = 40;
```

Linear State-Space Control Systems, by Robert L. Williams II and Douglas A. Lawrence
Copyright © 2007 John Wiley & Sons, Inc.

```
A = [0 1;-kR/J -b/J];   % Define the state-space
                        % realization
B = [0;1/J];
C = [1 0];
D = [0];

JbkR = ss(A,B,C,D);     % Define model from state-space

JbkRtf  = tf(JbkR);     % Convert to transfer function
JbkRzpk = zpk(JbkR);    % Convert to zero-pole
                        % description

[num,den] = tfdata(JbkR,'v');
                        % Extract transfer function
                        % description
[z,p,k] = zpkdata(JbkR,'v');
                        % Extract zero-pole description

JbkRss = ss(JbkRtf)     % Convert to state-space
                        % description

%------------------------------------------------------------
%   Chapter 2. Simulation of State-Space Systems
%------------------------------------------------------------

t  = [0:.01:4];             % Define array of time
                            % values
U  = [zeros(size(t))];      % Zero single input of
                            % proper size to go with
x0 = [0.4; 0.2];            % t Define initial state
                            % vector [x10; x20]

CharPoly = poly(A)          % Find characteristic
                            % polynomial from A
Poles = roots(CharPoly)     % Find the system poles

Eigs0    = eig(A);          % Calculate open-loop
                            % system eigenvalues
damp(A);                    % Calculate eigenvalues,
                            % zeta, and wn from ABCD

[Yo,t,Xo] = lsim(JbkR,U,t,x0);% Open-loop response
```

```
                             % (zero input, given ICs)
Xo(101,:);                   % State vector value at
                             % t=1 sec
X1 = expm(A*1)*XO;           % Compare with state
                             % transition matrix
                             % method


figure;                      % Open-loop State Plots
subplot(211), plot(t,Xo(:,1)); grid;
axis([0 4 -0.2 0.5]);
set(gca,'FontSize',18);
ylabel('{\itx}_1 (\itrad)')
subplot(212), plot(t,Xo(:,2)); grid; axis([0 4 -2 1]);
set(gca,'FontSize',18);
xlabel('\ittime (sec)'); ylabel('{\itx}_2 (\itrad/s)');

%-------------------------------------------------------
%    Chapter 2.  Coordinate Transformations and Diagonal
%                      Canonical Form
%-------------------------------------------------------

[Tdcf,E] = eig(A);           % Transform to DCF via
                             % formula
Adcf = inv(Tdcf)*A*Tdcf;
Bdcf = inv(Tdcf)*B;
Cdcf = C*Tdcf;
Ddcf = D;

[JbkRm,Tm] = canon(JbkR,'modal');
                             % Calculate DCF using
                             % MATLAB canon
Am = JbkRm.a
Bm = JbkRm.b
Cm = JbkRm.c
Dm = JbkRm.d


%-------------------------------------------------------
%    Chapter 3.  Controllability
%-------------------------------------------------------

P = ctrb(JbkR);              % Calculate controllability
                             % matrix P
```

```
if (rank(P) == size(A,1)) % Logic to assess
                          %   controllability
    disp('System is controllable.');
else
    disp('System is NOT controllable.');
end

P1 = [B A*B];              % Check P via the formula

%-------------------------------------------------------------
%    Chapter 3.  Coordinate Transformations and
%                     Controller Canonical Form
%-------------------------------------------------------------

CharPoly = poly(A);      % Determine the system
                         % characteristic polynomial
a1 = CharPoly(2);        % Extract a1

Pccfi = [a1  1;1  0];    % Calculate the inverse of
                         % matrix Pccf
Tccf = P*Pccfi;          % Calculate the CCF
                         % transformation matrix

Accf = inv(Tccf)*A*Tccf;% Transform to CCF via
                         % formula
Bccf = inv(Tccf)*B;
Cccf = C*Tccf;
Dccf = D;

%-------------------------------------------------------------
%    Chapter 4.  Observability
%-------------------------------------------------------------

Q = obsv(JbkR);          % Calculate observability
                         % matrix Q
if (rank(Q) == size(A,1))% Logic to assess
                         %   observability
    disp('System is observable.');
else
    disp('System is NOT observable.');
end
```

```matlab
Q1 = [C; C*A];              % Check Q via the formula


%------------------------------------------------------------
%   Chapter 4.  Coordinate Transformations and Observer
%                  Canonical Form
%------------------------------------------------------------

Qocf = inv(Pccfi);
Tocf = inv(Q)*Qocf;         % Calculate OCF transformation
                            % matrix
Aocf = inv(Tocf)*A*Tocf; % Transform to OCF via formula
Bocf = inv(Tocf)*B;
Cocf = C*Tocf;
Docf = D;


[JbkROCF,TOCF] = canon(JbkR,'companion');
                            % Compute OCF using canon
AOCF = JbkROCF.a
BOCF = JbkROCF.b
COCF = JbkROCF.c
DOCF = JbkROCF.d


%------------------------------------------------------------
%   Chapter 5.  Minimal Realizations
%      The Continuing Matlab Example is already minimal;
%      hence, there is nothing to do here.  The student
%      may verify this with MATLAB function minreal.
%------------------------------------------------------------


%------------------------------------------------------------
%   Chapter 6.  Lyapunov Stability Analysis
%------------------------------------------------------------

if (real(Poles(1))==0 |  real(Poles(2))==0)  % lyap will
   fail
    if (real(Poles(1))<=0 |  real(Poles(2))<=0)
        disp('System is marginally stable.');
    else
        disp('System is unstable.');
    end
else                                         % lyap will
                                             % succeed
```

```
    Q = eye(2);            % Given positive definite
                           % matrix
    P = lyap(A',Q);        % Solve for P
    pm1 = det(P(1,1));     % Sylvester's method to see if
       P is positive definite
    pm2 = det(P(1:2,1:2));
    if (pm1>0 & pm2>0)     % Logic to assess stability
                           % condition
         disp('System is asymptotically stable.');
    else
         disp('System is unstable.');
    end
end

figure;             % Plot phase portraits to enforce
                    % stability analysis
plot(Xo(:,1),Xo(:,2),'k'); grid; axis('square');
   axis([-1.5 1.5 -2 1]);
set(gca,'FontSize',18);
xlabel('{\itx}_1 (rad)'); ylabel('{\itx}_2 (rad/s)');

%-----------------------------------------------------------
%    Chapter 7.  Dynamic Shaping
%-----------------------------------------------------------

PO = 3;  ts = 0.7;                % Specify percent
                                  % overshoot and settling
                                  % time
term = pi2 + log(PO/100)2;
zeta = log(PO/100)/sqrt(term)     % Damping ratio from PO
wn   = 4/(zeta*ts)                % Natural frequency from
                                  % settling time and zeta
num2   = wn2;                     % Generic desired
                                  % second-order system
den2   = [1 2*zeta*wn wn2]
DesEig2 = roots(den2)             % Desired control law
                                  % eigenvalues
Des2 = tf(num2,den2);             % Create desired system
                                  % from num2 and den2

figure;
td = [0:0.01:1.5];
```

```
step(Des2,td);               % Right-click to get
                             % performance measures


%-----------------------------------------------------------
%    Chapter 7.  Design of Linear State Feedback Control
%               Laws
%-----------------------------------------------------------


K    = place(A,B,DesEig2)     % Compute state
                             % feedback gain matrix
                             % K
Kack = acker(A,B, DesEig2);   % Check K via
                             % Ackerman's formula


Ac = A-B*K;   Bc = B;         % Compute closed-loop
                             % state feedback system
Cc = C;       Dc = D;
JbkRc = ss(Ac,Bc,Cc,Dc);      % Create the
                             % closed-loop
                             % state-space system


[Yc,t,Xc] = lsim(JbkRc,U,t,XO); % Compare open-loop and
                             % closed-loop responses

figure;
subplot(211), plot(t,Xo(:,1),'r',t,Xc(:,1),'g'); grid;
axis([0 4 -0.2 0.5]);
set(gca,'FontSize',18);
legend('Open-loop','Closed-loop');
ylabel('{\itx}_1')
subplot(212), plot(t,Xo(:,2),'r',t,Xc(:,2),'g'); grid;
axis([0 4 -2 1]);
set(gca,'FontSize',18);
xlabel('\ittime (sec)'); ylabel('{\itx}_2');

%-----------------------------------------------------------
%    Chapter 8.  Design and Simulation of Linear
%               Observers for State Feedback
%-----------------------------------------------------------


% Select desired observer eigenvalues; ten times
   control law eigenvalues
```

```
ObsEig2 = 10*DesEig2;

L    = place(A',C', ObsEig2)';% Compute observer gain
                             % matrix L
Lack = acker(A',C', ObsEig2)';% Check L via Ackerman's
                             % formula

Ahat = A-L*C;       % Compute the closed-loop observer
                    % estimation error matrix
eig(Ahat);          % Check to ensure desired eigenvalues
                    % are in Ahat

% Compute and simulate closed-loop system with control
% law and observer
Xr0 = [0.4;0.2;0.10;0];      % Define vector of
                             % initial conditions
Ar = [(A-B*K) B*K;zeros(size(A)) (A-L*C)];
Br = [B;zeros(size(B))];
Cr = [C zeros(size(C))];
Dr = D;
JbkRr = ss(Ar,Br,Cr,Dr);     % Create the closed-loop
                             % system with observer
r = [zeros(size(t))];        % Define zero reference
                             % input to go with t
[Yr,t,Xr] = lsim(JbkRr,r,t,Xr0);

% Compare Open, Closed, and Control Law/Observer
% responses
figure;
plot(t,Yo,'r',t,Yc,'g',t,Yr,'b'); grid;
axis([0 4 -0.2 0.5]);
set(gca,'FontSize',18);
legend('Open-loop','Closed-loop','w/ Observer');
xlabel('\ittime (sec)'); ylabel('\ity');

figure;                      % Plot observer errors
plot(t,Xr(:,3),'r',t,Xr(:,4),'g'); grid;
axis([0 0.2 -3.5 0.2]);
set(gca,'FontSize',18);
legend('Obs error 1','Obs error 2');
xlabel('\ittime (sec)'); ylabel('\ite');
```

```
%------------------------------------------------------------
%    Chapter 9.   Linear Quadratic Regulator Design
%------------------------------------------------------------
Q = 20*eye(2);                    % Weighting matrix for
                                  % state error
R = [1];                          % Weighting matrix for
                                  % input effort
BB = B*inv(R)*B';

KLQR = are(A,BB,Q);               % Solve algebraic Ricatti
                                  % equation
ALQR = A-B*inv(R)*B'*KLQR;        % Compute the closed-loop
                                  % state feedback system
JbkRLQR = ss(ALQR,Bc,Cc,Dc);      % Create LQR closed-loop
                                  % state-space system

% Compare open- and closed-loop step responses
[YLQR,t,XLQR] = lsim(JbkRLQR,U,t,XO);

figure;
subplot(211),
plot(t,Xo(:,1),'r',t,Xc(:,1),'g',t,XLQR(:,1),'b');
grid; axis([0 4 -0.2 0.5]);
set(gca,'FontSize',18);
legend('Open-loop','Closed-loop','LQR');
ylabel('{\itx}_1')
subplot(212),
plot(t,Xo(:,2),'r',t,Xc(:,2),'g',t,XLQR(:,2),'b');
grid; axis([0 4 -2 1]);
set(gca,'FontSize',18);
xlabel('\ittime (sec)'); ylabel('{\itx}_2');

% Calculate and plot to compare closed-loop and LQR
% input efforts required
Uc = -K*Xc';                      % Chapter 7 input effort
ULQR = -inv(R)*B'*KLQR*XLQR';     % LQR input effort

figure;
plot(t,Uc,'g',t,ULQR,'b'); grid; axis([0 4 -10 6]);
set(gca,'FontSize',18);
legend('Closed-loop','LQR');
xlabel('\ittime (sec)'); ylabel('\itU');
```

REFERENCES

Anderson, B. D. O., and J. B. Moore, 1971, *Linear Optimal Control*, Prentice-Hall, Englewood Cliffs, NJ.

Bupp, R. T., D. S. Bernstein, and V. T. Coppola, 1998, "A Benchmark Problem for Nonlinear Control Design", *International Journal of Robust and Nonlinear Control*, 8: 307–310.

Bernstein, D. S., editor, 1998, Special Issue: A Nonlinear Benchmark Problem, *International Journal of Robust and Nonlinear Control* 8(4–5): 305–461.

Bryson, A. E., Jr., and Y.-C. Ho, 1975, *Applied Optimal Control, Hemisphere: Optimization, Estimation, and Control*, Hemisphere, Washington, DC.

Chen, C. T., 1984, *Linear System Theory and Design*, Holt, Rinehart, and Winston, New York.

Dorato, P., C. Abdallah, and V. Cerone, 1995, *Linear Quadratic Control: An Introduction*, Prentice-Hall, Englewood Cliffs, NJ.

Dorf, R. C., and R. H. Bishop, 2005, *Modern Control Systems*, 10th ed., Prentice-Hall, Upper Saddle River, NJ.

Friedland, B., 1986, *Control System Design: An Introduction to State-Space Methods*, McGraw-Hill, New York.

Graham, D., and R. C. Lathrop, 1953, "The Synthesis of Optimum Response: Criteria and Standard Forms, Part 2," *Transactions of the AIEE* 72: 273–288.

Kailath, T., 1980, *Linear Systems*, Prentice Hall, Upper Saddle River, NJ.

Khalil, H. K., 2002, *Nonlinear Systems*, 3rd ed., Prentice-Hall, Upper Saddle River, NJ.

Kwakernaak, H., and R. Sivan, 1972, *Linear Optimal Control Systems*, John Wiley & Sons, Inc.

Lewis, F. L., 1992, *Applied Optimal Control and Estimation*, Prentice-Hall, Englewood-Cliffs, NJ.

Ogata, K., 2002, *Modern Control Engineering*, 4th ed., Prentice Hall, Upper Saddle River, NJ.

Reid, J. G., 1983, *Linear System Fundamentals: Continuous and Discrete, Classic and Modern*, McGraw-Hill.

Rugh, W. J., 1996, *Linear System Theory*, 2nd ed., Prentice-Hall, Upper Saddle River, NJ.

Zhou, K., 1995, *Robust and Optimal Control*, Prentice-Hall, Upper Saddle River, NJ.

INDEX

Ackermann's formula, 258, 262, 274, 277, 279, 282, 289, 294, 310–311, 321, 394–395, 352

Adjoint, 64, 187, 414

Adjugate, 414

Algebraic output equation, 5, 7, 9–10, 13–14, 16, 29, 35–36, 39, 49, 104, 109

Algebraic Riccati equation, 390–392, 395–399, 401, 403–404

Asymptotic Stability, 200, 203–204, 211–212, 216, 220, 222–224, 227, 236, 278, 338–339

Asymptotic stabilization, 263, 312

Asymptotically stable
equilibrium, 199–204, 207–208, 211, 216, 276
error dynamics, 302, 308, 313
state equation, 216, 221–222, 231, 264, 269, 274, 295, 392, 395, 397
(sub)system, 203–204, 225–231, 254, 265, 278, 293, 314

Ball and beam system, 20–24, 46, 105, 148, 183, 233, 298, 355, 405

Basis, 130, 394, 419–430, 433–434
change of, see Change of basis

orthonormal, see Orthonormal basis
standard, see Standard basis

Bass-Gura formula, 255–257, 267, 274, 277, 294, 305–310, 315, 352

Bounded-input, bounded-output stability, 198, 218–224, 228–231

Canonical form
controller, see Controller canonical form
diagonal, see Diagonal canonical form
observer, see Observer canonical form
phase-variable, see Phase-variable canonical form

Cauchy-Schwarz inequality, 445

Cayley-Hamilton theorem, 57, 123, 131–132, 260, 443

Change of basis, 72, 422, 430

Characteristic
equation, 436, 443
polynomial, 57, 65–66, 73–74, 77, 79–81, 85–88, 90–91, 93, 118, 127–128, 131, 140, 142, 144–145, 161, 170, 216, 221, 237–238, 241, 243, 249, 251–258, 260, 262, 276, 279, 285, 289, 293, 304–307, 309, 311, 321, 327–328, 348, 351, 435–437, 439–441, 444

Linear State-Space Control Systems, by Robert L. Williams II and Douglas A. Lawrence
Copyright © 2007 John Wiley & Sons, Inc.

Closed-loop
  eigenvalue(s), 236, 243, 250, 252–253,
    255–257, 262–263, 267–268, 271,
    274–275, 289, 292–293, 298–299,
    309, 327, 334, 336, 339, 345, 347,
    349, 353, 399, 404
  state equation, 234–235, 254, 256, 264,
    268–270, 273–277, 295, 325–326,
    328, 330, 332, 334–335, 337–339,
    342, 353, 384, 386, 391–392, 395,
    397, 404
  system, 236, 242–244, 254, 263, 273,
    278, 282, 286–288, 292–293,
    347–351
  stability, 236, 272–273, 278, 326,
    339
  transfer function, 243, 269, 271, 295,
    337
Cofactor, 412–414
Conjugate transpose, 393, 408, 425, 436
Control law, 234, 287
Controllability, 108–109, 119–120, 133,
    138, 141, 149–150, 163, 166, 168,
    185, 187–193, 234, 250, 263–264,
    274, 301, 312–314, 323, 392
  Gramian, 110–111, 114, 120, 145, 154,
    375–376, 403
  matrix, 110, 112, 114–121, 125, 130,
    132–134, 138, 141–143, 161, 211,
    256, 259
  rank condition, 110, 113
Controllable
  pair, 110, 123, 130, 133, 136, 146–147,
    193, 232, 250–252, 255–256, 259,
    261, 264–265, 267, 272, 274, 295,
    300, 392, 397, 403
  state, 109
  realization, 122, 127, 147, 188–192
  state equation, 109, 115, 123, 125, 263,
    268, 375–376
Controller canonical form, 17, 124–125,
    127–128, 138–144, 147–148, 168,
    170–171, 175, 179, 181, 183–184,
    187–191, 195, 230, 252–255, 259,
    262–263, 270–271, 276–277, 306,
    309, 327
Coordinate transformation, 72–75, 77–78,
    80, 82, 88, 91, 109, 119–124, 127,
    129, 134, 137–140, 142–143, 150,
    165–172, 174, 176, 178–180, 185,
    188, 255, 264–265, 267, 313, 315,
    319–320, 326–327, 332, 337,
    353
Cost function, 358

dc gain, 269–271, 275, 337
Damping ratio, 79, 81, 87, 238–242, 245,
    254, 284, 297
Detectability, 301, 312–316
Determinant, 41, 60, 64, 73, 110,
    117–118, 138, 152, 160–161, 178,
    188, 212, 227, 274, 407, 412–414,
    416, 440
Diagonalization, 440
Diagonal canonical form, 49, 72, 74–75,
    77, 80, 82–83, 88, 91–92, 99, 101,
    103, 105, 107, 121–122, 127, 129,
    141, 143, 170–171, 178–179,
    183–184, 257–258
Differential Riccati equation, 383–385,
    387, 390–391, 397
Dimension, 3, 17, 25, 34, 38, 74, 76, 110,
    113, 149, 151, 163, 185, 187, 192,
    247, 269–270, 301, 316, 318, 331,
    353, 361, 407–409, 411, 419, 431
Direct transmission matrix, 49
Dual state equation, 163–165, 167
Duality, 150, 163, 165–169, 171–175,
    183–184, 188–189, 190, 303, 312,
    314, 348–349, 351

Eigenbasis, 439–440
Eigenstructure, 127, 190, 387, 440
Eigenvalue, 59–61, 65, 73–74, 77–82,
    84–87, 89–92, 121–122, 127–129,
    134–136, 141, 143, 145, 147,
    170–171, 174, 178–179, 197, 198,
    202–205, 208–209, 211–213, 216,
    221, 223–224, 227–229, 231–232,
    236, 243, 250, 252–257, 262–263,
    267–268, 271, 274–275, 289,
    292–293, 298–299, 309, 327, 334,
    336, 339, 345, 347, 349, 353, 399,
    404, 435–443
Eigenvalue placement, 234, 250–252,
    256, 263–264, 303, 313, 323, 326,
    399, 401
Eigenvector, 73–74, 77–78, 84, 88, 92,
    145, 213, 392, 441–442
  left, 133–136, 173, 265–266, 268, 315,
    436, 438
  right, 173–174, 213, 314–316,
    394–395, 436–438
Elementary row operations, 160,
    431–434, 437
Energy, 5–6, 8–9, 12, 28, 34, 198,
    204–211, 359, 392
Equilibrium condition, 1, 20, 24, 44, 270,
    275, 339

Equilibrium state, 18, 199–201, 236
  asymptotically stable, *see*
    Asymptotically stable equilibrium
  exponentially stable, 200–201, 216
  stable, 199–201
  unstable, *see* Unstable equilibrium
Error dynamics, 301–302, 306, 308–309,
  312–313, 315, 318, 323, 337, 339,
  343–344, 348, 351–352
Estimation error, 301, 304, 317
Euclidean norm, 111–112, 156, 425,
  445–446
Euclidean space, 360–361, 363, 367–368,
  418
Euler-Lagrange equation, 366–367, 369,
  374, 381
Exponential stability, 200–201, 216, 222

Feedback,
  state, *see* State feedback
  output, *see* Output feedback
Frobenius norm, 446

Gâteaux variation, 364–367, 373, 380
Gramian
  controllability, *see* Controllability
    gramian
  observability, *see* Observability gramian
  reachability, *see* Reachability gramian

Hamiltonian
  function, 380
  matrix, 385–387, 390, 392
Hölder inequality, 445
Homogeneous state equation, 52, 59, 61,
  150–151, 153, 159, 161–162,
  171–173, 182, 198–199, 201, 203,
  205, 231, 301, 318, 328, 373–374,
  382, 384

Identity matrix, 25, 40–41, 52, 140, 270,
  408, 416, 421, 434
Image, 42, 114, 430, 433
Impulse response, 48, 51–52, 63, 65, 70,
  74, 76, 79, 98–99, 124, 146, 192,
  218–222, 298–299
Inner product, 211, 424
Input
  gain, 137, 243, 268–271, 288, 292, 337
  matrix, 4, 36, 49, 397
  vector, 4, 7, 49, 52, 163
Input-output behavior, 17, 48, 70, 76, 186,
  198, 222, 295, 337

Internal stability, 198–199, 218
Inverted pendulum, 44–45, 101–103, 148,
  183, 233, 298, 355, 404–405
ITAE criteria, 249–250, 276, 279, 283,
  285–286, 289, 294, 296–298

Jordan canonical form, 202, 388,
  390–391, 442–443
Jordan block matrix, 41, 202–203, 388,
  442–443

Kalman, R. E., 108
Kernel, 430

Lagrange multiplier, 361–362, 368–372,
  379, 382
Laplace domain, 17, 51–52, 63, 66,
  69–70
Laplace transform, 15, 37, 48, 50–52, 61,
  63–67, 273
Linear dependence, 112, 134, 136–137,
  152, 154, 266, 319, 395
Linear independence, 74, 77, 110,
  118–119, 130, 133, 146, 151–152,
  162–163, 224, 419, 425, 431,
  433–434, 439, 441–442
Linear quadratic regulator, 357, 359–360,
  377, 382–383, 385, 403
  steady-state, 390–392, 404–406
  MATLAB, 397–402
Linear time-invariant
  state equation, 20, 24, 42–43, 61–62,
    72–73, 75, 77, 119, 137, 150, 165,
    270, 198, 201, 235, 300, 359
  system, 1, 3, 5, 26–27, 32, 48–49, 51,
    70, 78, 105, 107, 149, 186
Linear transformation, 41, 48, 72, 426,
  428–431, 434–435, 445
Linearization, 1, 17–20, 44, 46–47, 102,
  105–107
Luenberger, D. G., 300
Lyapunov, A. M., 198, 210–211
Lyapunov
  function, 211
  matrix equation, 213–216, 227, 229,
    231–232, 393, 395

Marginal stability, 204, 229, 277, 293, 327
MATLAB,
  controllability, 138
  controller canonical form, 138
  diagonal canonical form, 80
  introduction, 24

MATLAB, (*continued*)
  m-files for Continuing MATLAB
    Example, 447
  minimal realizations, 194
  observability, 174
  observer canonical form, 174
  observer design, 343
  optimal control, 397
  shaping the dynamic response, 278
  stability analysis, 225
  simulation, 79
  state feedback control law design, 279
  state-space description, 26
Matrix arithmetic, 409–412
Matrix exponential, 48, 53–55, 57–61,
  64, 93, 95–96, 100, 113, 146, 182,
  201–203, 216, 220–222, 376, 386,
  390, 396
Matrix inverse, 13, 41, 54, 64, 68, 74, 84,
  127, 142–143, 259, 275, 332, 387,
  414–416
Matrix norm, 42, 219, 223, 231, 445–446
Minor, 412–413
Minimal realization, 185–187, 189,
  191–197, 222, 230–231, 299
Minimum energy control, 357, 371,
  374–377, 379, 403

Natural frequency
  damped, 239, 241, 243, 245–246
  undamped, 79, 81, 238–239, 241, 243,
    245, 249, 254, 276, 283–286,
    297–298
Negative (semi)definite, 211–212
Nominal trajectory, 1, 17–24, 46–47, 102,
  105
Nonlinear state equation, 1, 18–21, 44,
  46–47, 102
Null space, 113, 153–154, 430, 436
Nullity, 153–154, 158, 160, 193, 431,
  433, 439

Objective function, 249, 358, 371–372
Observability, 149–151, 154, 159, 161,
  163–166, 171, 173, 175, 177,
  179–180, 183–184, 185, 187–189,
  191–193, 300–301, 307, 309,
  312–314, 320, 323, 338–339, 353,
  356, 392
  Gramian, 154–157, 166, 181
  matrix, 151–152, 158–163, 166, 169,
    174, 177–179, 188, 307, 321
  rank condition, 151–152

Observable
  pair, 151, 171, 173–174, 182, 193,
    303–305, 310, 312–313, 318,
    320–321, 323, 392, 397
  realization, 167, 169, 188–190, 354
  state equation, 151, 157, 169, 301, 312
Observer, 300–302, 306–309, 312–313,
  324–325, 327–328, 338,
  341–343,345–346, 348, 350
  reduced-order, 316, 318, 320–323, 331,
    334
Observer-based
  compensator, 301, 323, 325, 327–328,
    330–331, 335, 337–338, 341, 349,
    353–356
  servomechanism, 338, 340–341,
    354–355
Observer canonical form, 168–171,
  174–181, 183–184, 187–190, 231,
  304
Observer error, 310, 326–327, 330, 339,
  346, 354–356
  dynamics, 301–302, 306, 308–309,
    312–313, 315, 318, 323, 337, 339,
    343–344, 348, 351–352
Observer gain, 301–304, 306–307, 310,
  312–314, 316, 318, 322, 330, 334,
  339, 341, 343, 345, 348–351, 352
Optimal control, 357–360, 371, 397, 404
Orthogonal
  matrix, 95
  complement, 425, 435
  vectors, 133–136, 173, 265–266, 268,
    314–316, 425, 435
Orthonormal basis, 425
Output
  feedback, 197, 325
  matrix, 49
  vector, 4, 163

Peak time, 236, 240–241, 244–247, 279,
  284, 297
Percent overshoot, 236, 240–242, 244,
  247, 254, 271, 279–280, 283–286,
  288, 294–297, 309
Performance index, 2, 358–359, 379–381,
  384–385, 390–392, 396, 401, 404
Phase portrait, 207–209, 226, 229, 337
Phase-variable canonical form, 17, 125
Pole, 2, 26, 29, 76, 181, 185–186, 190,
  194, 196, 222, 234, 272–273, 276,
  338
Popov-Belevitch-Hautus
  eigenvector test for controllability, 133,
    136, 145, 147, 173, 265

eigenvector test for detectability, 314
eigenvector test for observability, 150, 173, 181, 190, 395
eigenvector test for stabilizability, 265, 314
rank test for controllability, 136, 137–138, 145, 174, 224, 265, 393
rank test for detectability, 314
rank test for observability, 150, 174, 181, 224, 318, 393
rank test for stabilizability, 265, 314
Positive definite, 210–216, 225, 227–228, 231–232, 295, 359, 363, 381–382, 392, 396–397, 403–404
Positive semidefinite, 110, 154, 359, 391, 396
Proof-mass actuator system, 47, 105, 148, 184, 197, 233, 299, 356

Quadratic form, 111, 154, 211–213, 359, 363, 379, 381, 384, 393

Range space, 114, 430
Rank, 431, 433
Reachability Gramian, 146, 375–377
Reachable
  state, 115
  state equation, 115
Reduced-order observer, 316, 318, 320–323, 331, 334
Reference input, 137, 235–236, 255, 268–270, 272, 274, 276, 287–289, 292, 328, 336, 338–339, 343
Riccati, J. F., 383
Riccati matrix equation, 295
Rise time, 236, 239–245, 248, 279, 285, 297
Robot joint/link, 46, 103, 148, 183, 233, 298, 356, 405
Rotational mechanical system, 27, 80, 139, 175, 225, 279, 345, 398
Rotational electromechanical system, 36, 89, 143, 179, 228, 290, 350
Row-reduced echelon form, 160, 432, 437–438

Separation property, 327, 331
Servomechanism,
  observer-based, 338, 340–341, 354–355
  state feedback, 271, 276, 296–297
Settling time, 236, 241–247, 254, 271, 279–280, 282–286, 288, 294–295, 297, 309

Similarity transformation, 73, 255, 332, 388–391, 440–442
Span, 130–131, 418–419, 426, 431
Spectral norm, 231, 446
Spectral radius, 42
Spectrum, 436–437
Stability,
  asymptotic, *see* Asymptotic stability
  bounded-input, bounded-output, *see* Bounded-input, bounded-output stability
  closed-loop, *see* Closed-loop stability
  exponential, *see* Exponential stability
  internal, *see* Internal stability
  marginal, *see* Marginal stability
Stabilizability, 234, 263–265, 295, 301, 312–315
Standard basis, 63, 77, 133, 201, 421, 423, 428, 430, 440
State
  differential equation, 5, 7, 9–10, 13–14, 16, 28, 35–36, 39, 49, 65, 104, 109, 273
  equation solution, 48–49, 52, 61–62, 109, 149, 201
  estimate, 300–302, 317, 320, 323–324, 331, 338, 34
  feedback (control law), 137, 234–236, 242, 250–252, 254, 263–265, 268, 270, 272–273, 279, 281, 292, 295–299, 300–301, 309, 323–325, 327, 330–331, 337–338, 341, 347–351, 353–356, 383–384, 391, 397–399, 401
  feedback gain matrix/vector, 234–236, 250–259, 262–267, 269, 271, 274, 276–277, 279, 281, 287, 289, 292, 294–295, 303, 314, 327–329, 334, 339–341, 383, 386, 391, 396–397, 399
  feedback-based servomechanism, 271, 276, 296–297, 338, 341–342
  variables, 4–6, 8–10, 12, 14, 16, 20, 28, 34, 38–40, 44–46, 61, 66–67, 76, 90, 101–102, 104, 116, 122, 149, 159, 171, 205, 210, 273, 295, 300, 317
  vector, 3–4, 6–7, 48–49, 65, 72, 127, 149, 163, 273, 300, 317, 324, 326
  transition matrix, 61, 79, 87
State-space realization, 17, 26, 32, 39–40, 44–46, 48, 70–71, 74, 76–78, 80, 91, 96–98, 117–118, 122–124, 127, 147, 161, 164, 167, 169, 186, 354

State-space realization, (*continued*)
  minimal, *see* Minimal realization
Stationary function, 366–367
Steady-state tracking, 268–269, 272, 276,
  337, 339–340
Subspace, 41–42, 394, 419–421, 425,
  430–431, 435
Supremum, 218–219, 223, 445
Symmetric matrix, 110, 127, 154, 169,
  211–213, 215–216, 231–232, 295,
  306, 359, 363, 383, 392–395,
  403–404, 420
System dynamics matrix, 34, 49, 73,
  78–79, 91, 93, 95, 100, 102, 121,
  198, 214, 227, 229–230, 236, 252,
  255, 273–274, 277, 279, 293–294,
  327, 339, 397, 401

Taylor series, 19
Time constant, 92, 237, 244, 290–291,
  294, 296, 298
Transfer function, 2, 5, 14–15, 17, 26–27,
  29, 37–39, 42–43, 45, 48, 52, 65, 70,
  74, 76, 86, 88, 91, 92, 96–98, 103,
  117, 122–123, 125, 127, 142, 144,
  147, 161, 163–164, 167–169,
  179–181, 185–190, 192, 196,
  221–222, 230–232, 238, 249,
  284–285, 291, 352, 354
  closed-loop, 243, 269, 271, 295, 337
  open-loop, 271, 273, 276, 295
Transpose, 112–113, 165, 168, 172, 212,
  225, 305, 314, 408, 411, 414, 416
  conjugate (Hermitian), 393, 408, 425,
  436

Two-mass translational mechanical system,
  32, 84, 141, 177, 227, 283, 348, 399
Two-point boundary value problem, 382
Three-mass translational mechanical
  system, 44, 99, 147, 183, 233, 297,
  355, 405

Uncontrollable state equation, 129, 132,
  135, 172, 263, 267, 193
Unobservable state, 151, 153–154, 159,
  160, 165
Unobservable state equation, 171–172,
  194, 312, 315
Unstable
  equilibrium, 44, 199–201, 209, 211
  system, 204, 239, 257

Vandermonde matrix, 59–60, 121, 128
Variational calculus, 357, 359–360, 368
Vector norm, 219, 444–445
  bound, 206, 216
  Euclidean, 111–112, 156, 425,
  445–446
Vector space, 364, 417–422, 424

Zero-input response, 48, 50–51, 62,
  64–65, 69, 87, 95, 150–151, 157,
  159, 198, 222, 328, 399
Zero-state response, 48, 50–51, 62–65,
  69–70, 76, 95, 150, 198, 218–220,
  222, 232, 327
Zero, 26, 29, 76, 181, 185–186, 190, 194,
  196, 222, 250, 272–273, 276, 338